# Advanced Sciences and Technologies for Security Applications

The series Advanced Sciences and Technologies for Security Applications comprises interdisciplinary research covering the theory, foundations and domain-specific topics pertaining to security. Publications within the series are peer-reviewed monographs and edited works in the areas of:

- biological and chemical threat recognition and detection (e.g., biosensors, aerosols, forensics)
- crisis and disaster management
- terrorism
- cyber security and secure information systems (e.g., encryption, optical and photonic systems)
- traditional and non-traditional security
- energy, food and resource security
- economic security and securitization (including associated infrastructures)
- transnational crime
- human security and health security
- social, political and psychological aspects of security
- recognition and identification (e.g., optical imaging, biometrics, authentication and verification)
- smart surveillance systems
- applications of theoretical frameworks and methodologies (e.g., grounded theory, complexity, network sciences, modelling and simulation)

Together, the high-quality contributions to this series provide a cross-disciplinary overview of forefront research endeavours aiming to make the world a safer place.

More information about this series at http://www.springer.com/series/5540

Margaret E. Kosal
Editor

# Technology and the Intelligence Community

## Challenges and Advances for the 21st Century

Springer

*Editor*
Margaret E. Kosal
Georgia Institute of Technology
Atlanta, Georgia, USA

ISSN 1613-5113 ISSN 2363-9466 (electronic)
Advanced Sciences and Technologies for Security Applications
ISBN 978-3-030-09177-4 ISBN 978-3-319-75232-7 (eBook)
https://doi.org/10.1007/978-3-319-75232-7



Printed on acid-free paper

This Springer imprint is published by the registered company Springer International Publishing AG part of Springer Nature.
The registered company address is: Gewerbestrasse 11, 6330 Cham, Switzerland

# Acknowledgments

This work was initiated while the editor was serving as Director of the Sam Nunn Security Program (SNSP) at Georgia Institute of Technology (aka Georgia Tech), where she is also an Associate Professor of International Affairs. The mission of the SNSP is to strengthen and foster scholarship and policymaking at the intersection of science, technology, and security that address grand challenges of today and for tomorrow through the creation of educational opportunities for pre-doctoral to mid-career scientists and engineers, structured research projects, and promotion of opportunities to inform and advise the policy community.

In addition to the contributing authors, the following persons are gratefully acknowledged for making this work a success:

- Mr. Joe Bankoff, Chair of Sam Nunn School of International Affairs, for his generous efforts to see the SNSP continues into its next decade and thrives.
- Dr. Robert Kennedy, Emeritus Professor of International Affairs and currently President of the Atlanta Council on International Relations, for initial guidance on understanding the intelligence community and its challenges.
- Ms. Alaina Totten and Ms. Anmol Soni, who served as successive editorial assistants on this project. While each is an independent scholar in her own right, I am ever grateful for their assistance and amazing organization.

# Contents

# List of Contributors

**Shai Bernstein** Georgia Institute of Technology, Atlanta, Georgia, USA

**Meghan A. Check** Georgia Institute of Technology, Atlanta, Georgia, USA

**Karan P. Jani** Georgia Institute of Technology, Atlanta, Georgia, USA

**Abdalla Abou-Jaoude** Georgia Institute of Technology, Atlanta, Georgia, USA

**Ben Johnson** Georgia Institute of Technology, Atlanta, Georgia, USA

**Nabil Kleinhenz** Georgia Institute of Technology, Atlanta, Georgia, USA

**Margaret E. Kosal** Georgia Institute of Technology, Atlanta, Georgia, USA

**Margaret L. Loper** Georgia Tech Research Institute, Atlanta, Georgia, USA

**Allison J. Mahvi** Georgia Institute of Technology, Atlanta, Georgia, USA

**Jenna K. McGrath** Georgia Institute of Technology, Atlanta, Georgia, USA

**Leah J. Ruckle** Georgia Institute of Technology, Atlanta, Georgia, USA

**Jon Schmid** Georgia Institute of Technology, Atlanta, Georgia, USA

**Lindsey R. Sheppard** Georgia Institute of Technology, Atlanta, Georgia, USA

**Janille Smith-Colin** Southern Methodist University, Dallas, Texas, USA

**Anmol Soni** Georgia Institute of Technology, Atlanta, Georgia, USA

**Supraja Sudharsan** Georgia Institute of Technology, Atlanta, Georgia, USA

**Alaina Totten** Georgia Institute of Technology, Atlanta, Georgia, USA

# Introduction

**Margaret E. Kosal and Alaina Totten**

**Abstract** Technology has become a cornerstone for gathering effective intelligence to combat threats to national security. Rapid technological development in the twentieth century enabled the intelligence community to gather, analyze, and disseminate information at a pace and magnitude that was never before possible. Examples of this include the role of radio technology in revolutionizing communication intelligence as well as aircraft and satellite reconnaissance that provided much-needed intelligence on Soviet weapons capabilities and enabled an increase in intelligence gathering from denied areas across the world. More recent examples that demonstrate the prominence of technology in intelligence gathering include the operation of Unmanned Aerial Vehicles for surveillance of Iran's weapons capabilities and of Osama bin Laden's complex in Pakistan. The advent of the information age has further revolutionized these processes. According to the testimony of then-Director of National Intelligence, James Clapper to the Senate Armed Services Committee on worldwide threat assessment, "The consequences of innovation and increased reliance on information technology in the next few years on both our society's way of life in general and how we in the Intelligence Community specifically perform our mission will probably be far greater in scope and impact than ever" (Clapper 2016).

Technology has become a cornerstone for gathering effective intelligence to combat threats to national security. Rapid technological development in the twentieth century enabled the intelligence community to gather, analyze, and disseminate information at a pace and magnitude that was never before possible. Examples of this include the role of radio technology in revolutionizing communication intelligence as well as aircraft and satellite reconnaissance that provided much-needed intelligence on Soviet weapons capabilities and enabled an increase in intelligence

M. E. Kosal (✉)
Sam Nunn School of International Affairs, Georgia Institute of Technology, Atlanta, Georgia, USA
e-mail: margaret.kosal@inta.gatech.edu

A. Totten
School of Economics, Georgia Institute of Technology, Atlanta, Georgia, USA

 
M. E. Kosal (ed.), *Technology and the Intelligence Community*, Advanced Sciences and Technologies for Security Applications,
https://doi.org/10.1007/978-3-319-75232-7_1

gathering from denied areas across the world. More recent examples that demonstrate the prominence of technology in intelligence gathering include the operation of Unmanned Aerial Vehicles for surveillance of Iran's weapons capabilities and of Osama bin Laden's complex in Pakistan. The advent of the information age has further revolutionized these processes. According to the testimony of then-Director of National Intelligence, James Clapper to the Senate Armed Services Committee on worldwide threat assessment, "The consequences of innovation and increased reliance on information technology in the next few years on both our society's way of life in general and how we in the Intelligence Community specifically perform our mission will probably be far greater in scope and impact than ever" (Clapper 2016).

This volume explores three dimensions of the technology-intelligence community relationships: expansions in the intelligence capabilities due to technology in the past and future and the implications thereof; the impacts of the institutional and organizational infrastructure within which the intelligence community operates; and ethical implications of the growing presence of security agencies in day-to-day functions of modern society.

This book examines the impact of technology on the intelligence process. Through qualitative case studies and process tracing methods, it reviews the role of technology in the effective gathering, analysis, and dissemination of new intelligence information. Beginning as early as the late 1800s and early 1900s and including the period from World War II to the present, this analysis focuses on several technologies that marked a breakthrough in the intelligence process and for the intelligence community. The lessons learned can then be used to understand implications for the future of the intelligence community and derive effective ways to utilize emerging, disruptive technologies in the intelligence process.

The analysis in these chapters shows how technology greatly expands the scope of intelligence capabilities. Although technology is necessary for these many new areas of intelligence, lessons from the past and present indicate that it is insufficient in and of itself. In conjunction with innovative technology, adequate human capital is required for analysis and interpretation. Necessary skills include the adaptability and the ability to understand new information in terms of the overall situation. To create a fully functioning intelligence effort, the correct technology needs to be in place to obtain the necessary information within the right organizational structure, which then allows the analysis to flow to the human decision makers who use the intelligence. A failure in technology, organizational structure, or the human decision making process can lead to intelligence failures. New technologies can disrupt these processes in unexpected ways. Several key new technologies are considered with respect to the lessons learned from historical examples in order to forecast potential pitfalls in the intelligence process.

This compilation fits in with the current literature including texts such as Richelson's *The US Intelligence Community* (2015) which looks at organization and procedure of the US intelligence agencies; Kleinman et al.'s *Controversies in Science & Technology: From sustainability to surveillance* (2014) which presents current challenges for the community; and Johnson's *The Oxford Handbook of National Security Intelligence* (2010) which analyzes case studies of challenges presented by technological development in the realm of security. This book contributes to the literature by looking at past, present, and future innovations which have disrupted the intelligence community's status quo. By pursuing in-depth analysis such as Richelson's *The Wizards of Langley* (2002) this text seeks to explain components of the intelligence community's development and adaptation to a diverse collection of technological advancements.

In order to facilitate a comprehensive study of technology's role within the United States intelligence community, it is beneficial to begin with a historical review of the intelligence community's evolution. This book begins with a discussion of the intelligence cycle in terms of intelligence gathering, analysis, and decision making processes. Next, the organizational structure within the intelligence community is reviewed including the legal, policy, and ethical aspects related to intelligence gathering. Case studies and examples are used throughout the text to assist in demonstrating the role intelligence has played during a period of advancing technology and analyzing the implications. We summarize these issues here.

## 1 Past

### *1.1 The Origin of Centralized Intelligence Gathering in the United States*

World War II and the early years of the Cold War were critical to the development of the United States' intelligence community. The US did not have any peacetime civilian agencies prior to the war and the creation of the Central Intelligence Agency in 1947. The role of the United States and the newly evolved intelligence community, both military and civilian, began to experience a powerful transformation for safeguarding national security. New considerations for best practices included: interception had to be tightly coupled with processing and analysis, dissemination was critical, and irrespective of technology, the role of all intelligence gathering was to provide useful and timely information to a decision maker. Later chapters look at specific case studies, such as radio technology or the post WWII period, to illustrate how the role of technology influenced the organization of the intelligence agencies during this time period.

### *1.2 Growth in the Use of Overhead Surveillance Technology in Intelligence Gathering*

At the start of the Cold War, the role of technology in intelligence gathering had not yet reached the level of prominence that it would come to see in later years. Director of Central Intelligence (DCI) Allen Dulles, along with many members of the intelligence community, held the belief that the role of agencies such as the CIA is to collect human-based intelligence (HUMINT). This slowed the use of technology.

As the Iron Curtain closed and the USSR became increasingly aggressive and secretive, traditional espionage tactics were not sufficient to produce the needed information. Additionally, technologies adapted to intelligence gathering tasks from other purposes were becoming unusable. For example, upgrades to telephone lines made wiretapping more difficult and increased aggression in defending Soviet airspace made the use of modified bomber aircraft very dangerous for photo-reconnaissance.

Starving for information on Soviet military and nuclear programs and subsisting primarily on information brought back from released German scientists and captured Luftwaffe photo-reconnaissance, the intelligence community made a dedicated shift to include more technology in order to enhance its intelligence gathering efforts. Opinion on the role and use of technology for intelligence gathering shifted to the pro-technology stance of current day. The US Air Force (USAF) made some of the first forays into technology acquisition specifically for intelligence gathering. However, the CIA under the leadership of DCI Dulles was more reluctant but followed suit after prodding from President Eisenhower.

A number of technology-based projects sprang up after the intelligence community culture shift and the increasing difficulty in obtaining intelligence on the highly secretive USSR. Some notable aerial reconnaissance programs were the Project GENETRIX balloon program (first launch 1955), the U-2 Dragon Lady reconnaissance aircraft (first operational flight 1956), and the Corona satellite program (first successful use 1960).

Implementation of these new technologies was paired with a need for human analysis to make sense of new data collection. The employment of reconnaissance aircraft and satellites for intelligence gathering did not end in the Cold War. These are technologies that are still used today. The secretive RQ-170 Unmanned Aerial Vehicle (UAV) has been the reconnaissance aircraft of choice during the "War on Terror," taking on notable missions such as surveillance of Osama bin Laden's complex in Pakistan during the 2011 Navy SEAL raid.

Current use of technology in intelligence gathering is certainly not exclusive to overhead reconnaissance efforts, but these reconnaissance technologies, particularly since they can trace their ancestry back to the effects of the pro-technology culture change are valuable case studies. They show that a technology can aid in the gathering, analysis, or dissemination of intelligence. Technology cannot function without the people guiding it, so it is still subject to the faults of the people and policies that govern it and its use. Moreover, it has also become that the people guiding the technology cannot function without the technology. Though each component has flaws, humans and technology are both necessary elements in the intelligence community.

## 1.3 *Importance of Integrating Technology with Traditional Intelligence Gathering*

An assessment of the evolving relationship between technology and the intelligence community requires consideration on how the intelligence agencies in the United States have incorporated technological advancements into their intelligence gathering techniques in the past. As in other fields and areas of work, there is a risk of becoming over-reliant on technology and overlooking basic, though often just as critical, forms of information. Do non-technical methods, such as human, cultural, and ethnic intelligence get overlooked? Examining past historical events, such as so-called "intelligence failures," provides opportunities to assess how the intelligence community integrates new technology into their methodologies. Furthermore, a historical assessment provides insights on the continued importance on non-technology based intelligence.

The collapse of the Soviet Union in 1991 is a useful example of the importance of human and cultural intelligence. The US intelligence agency's technology-based intelligence gathering methods had been very successful at evaluating the Soviet military's threat, including nuclear capabilities (Diamond 2008). This was the case until the Soviet Union began to show signs of deterioration. They exhibited ethnic conflicts across the empire's regions and an increase in military spending despite a struggling economy. The demographic data analysis indicated civil and political unrest. Despite these signals, the US intelligence community maintained that the Soviet Union was a long-term threat (Diamond 2008; Westad 2000). Even as Mikhail Gorbachev declared troop reduction and withdrawal of Eastern Europe in 1988 towards the aim of improving international relations, which was followed by the breakdown of the Berlin Wall in 1989, the US intelligence community issued a memo warning of a possible Warsaw Pact attack (Diamond 2008). The December 25, 1991 declaration that the Soviet Union was dissolving and that the transition of power to new regions went relatively smoothly and therefore served as a great surprise to the US intelligence community.

The misplaced focus of the CIA during the final years of the Cold War demonstrates that there was a possible over reliance on technology informing the decision making process for intelligence gathering and important indicators on a human- and cultural-intelligence level were overlooked by the intelligence community. The reason why the non-technology based intelligence indicators were overlooked can be traced further than just the CIA misjudging the significance of the weakening Soviet economy or the increasing frequency of ethnic conflicts. This example instead provides insight into the dysfunction and misguided efforts from the intelligence community as a whole. The US intelligence community had a predetermined understanding of how the Soviet Union operated and was unwilling or unable to adapt that mindset even as the Soviet Union's government, economy, and citizens began to evolve (Betts 1978; Westad 2000; Diamond 2008). The prevalence of pre-conceived notions in the US intelligence community results from biases during the intelligence analysis process and pre-determined forms of action which can negatively impact the decision making process.

## 1.4 Intelligence Analysis

Obtaining data is only one side of intelligence assessments. Intelligence analysis plays an equally important role in the decision making process of a state. However, as Montgomery and Mount discuss in their article on nuclear intelligence failures, intelligence can be distorted by a range of political, cultural, bureaucratic, or organizational factors (Montgomery and Mount 2014). These challenges are important to understand as they can often lead to fallacious responses from states.

Biases are often the main cause of incorrect intelligence estimates. Intelligence agencies often have preconceived notions of programs. For instance, the CIA underestimated the North Korean nuclear program in the late 1980s due to a preconceived notion that an underdeveloped country could never develop nuclear capacity (Central Intelligence Agency 2014). Another failure can occur when an intelligence agency falls victim to a single person dictating its position on an issue (e.g. Joe T with Iraq intel in the 2000s) and ignoring the advice of experts that are knowledgeable on relevant subjects (e.g., Department of Energy Laboratory scientists on the Iraqi aluminum tubes) (US Congress 2004). These findings show how the 'human in the loop' aspect plays a crucial role in bias. Intelligence on secretive programs is often sparse and the analysis highly speculative. Ensuring sensible opinions are formed about the information is essential.

While technologies hold a very important role in intelligence gathering, analysis is arguably as or even more important than the technologies used to capture that data. Institutions should strive to be less biased in their opinions, more skeptical of the intelligence they gather, and heed advice of technical experts more carefully. This is crucial in ensuring the resulting responses are well grounded and effective.

## 1.5 Decision Making Based on Intelligence Information

The value of intelligence depends on how the decision makers choose to use the information provided (Betts 1978). Even though the gathering of intelligence and its analysis can be completely accurate, and the findings effectively communicated, the final decision still rests on those with authority. Decision makers have a variety of responses they can opt for, ranging from the military (sabotage, invasions, airstrikes, etc.) to the diplomatic (economic sanctions, negotiations, etc.) options. Unfortunately, there can be a disconnect between the results of intelligence and the decisions being made.

## 1.6 Legality and Policy Decisions

Through history, changing technology has often outpaced policy, leaving many new technologies unregulated until effective policies can be implemented. As an example, wiretapping policy, technology, and public opinion have been complex over time. Technology gives law enforcement and intelligence officials the enhanced ability to monitor individuals. While at the same time, public opinion pushes for privacy rights and limited authority of law enforcement. Shortly after the invention of the telephone, it was discovered you could easily listen in on phone conversations by attaching a listening piece to the wire carrying the phone conversation. This gave law enforcement the ability to collect evidence without being physically in the room themselves, giving birth to the beginning of executive freedoms exercised through wiretapping.

As technology progressed, pro-wiretapping or pro-surveillance policies became significantly more powerful. The regulating policies became stricter as observed in the Foreign Intelligence Surveillance Act (FISA). However, despite the stricter policies, new technological developments allowed the government to be much more efficient in their surveillance operations. Developments in computing technology and data analytics no longer meant the government needed the staff to manually listen in on phone calls, but rather, computers could be used to sift through large amount of data in the form of phone call, text messages, and other types of electronic data with the goal of identifying patterns related to terrorism. One such program developing these technologies was created by the Total Information Awareness Office developed under DARPA (Wyden 2016; TIA 2016). The program was developed and used after the passage of the Patriot Act, which was used as a law to circumvent some of the restrictions written into FISA ("Why the FISA court is not what it used to be" 2013).

Changes in policy through the years have intended to both tighten and loosen the authority of the executive branch for surveillance at different points in history. While the balance between executive authority and privacy continue to be debated, one item which is not debatable is the impact technology has had on these policies. The advent of new technologies and surveillance capabilities has significantly expanded the importance and scope of these regulating policies by vastly expanding the government's ability to conduct surveillance.

## 1.7 Ethical Implications

A democratic free society requires open and ongoing debate around government undertakings including the actions of the intelligence community. Intelligence decisions are informed by theories of ethics including public attitude towards the intelligence community itself. Other factors that contribute to the ethical view of intelligence activities include the following: view of the enemy, political system of the target nation, sense and severity of the threat, intended use for intelligence information, and the prevailing climate and conditions with respect to war (Johnson 2006). Intelligence activities, because of their inherently secret nature, are often spared intense public scrutiny. In the absence of the public system of checks and balances, intelligence officers' philosophies and the prevailing views on intelligence ethics and morality, often guide national intelligence decision-making.

In 2013, National Security Agency (NSA) contractor Edward Snowden leaked thousands of documents revealing the global surveillance activities of NSA which include harvesting millions of emails, instant messaging contact lists, tracking and mapping cell phone locations, accessing phone conversations including those of thirty-five world leaders, and PRISM, a program which allows court-approved access to personal communications through online accounts including Google and Yahoo accounts. In the information age of the twenty-first century, where leaks such as those perpetrated by Snowden have become more common, the public now has greater opportunity to weigh in on actions taken by the intelligence community.

The ethical standards used by intelligence officials differ from those applied by the general public. While the rhetoric among public officials is heated as ascertained through official statements and global opinions strongly view NSA surveillance activities of foreign citizens or heads of state as "unacceptable"(unless specifically targeting a terrorist), there are currently few discernible changes in intelligence activities or strategic relationships since the Snowden leaks began.

## 2 Future Trends in Technology and Intelligence

Technological developments of the modern era have brought the intelligence community to a transitional time. The community must adapt and evolve to innovations in intelligence gathering and analysis, while staying abreast of emerging requirements that will drive future technology needs. The increased pace of technological development has brought about this transition, but factors outside of the technical realm impact how the community responds and evolves. As detailed previously, the intelligence community in the US has embraced the technological advances while maintaining their human-in-the-loop drivers where needed.

Over the last thirty years, information technology has become more prominent than ever in the lives of people around the globe. With the invention of mobile computing platforms (e.g., wireless networking, inexpensive cameras, cellphones) and social media (e.g., Facebook, Twitter, Instagram), openly available data and global connections has become ubiquitous. The abundance of data to analyze and view, and the security of the networks communicating the data have become critical collection opportunities for the intelligence community. In addition to the information revolution, economic, political and environmental factors are driving people around the world to migrate to urban environments. The concentration of people into highly dense areas presents new challenges for validating, analyzing, securing, and viewing the data generated by these new populations. This represents emerging requirements for the intelligence community.

Transformative intelligence gathering technologies that will continue to be prominent in the twenty first century and will require strategies on how to integrate these technologies into the intelligence community. Past intelligence successes and failures should guide future innovation implementation. These emerging trends will drive the need for new intelligence gathering and analysis techniques over the next ten years.

## 2.1 Technological Innovations in Intelligence Gathering

Technologies such as sophisticated malware and augmented reality are changing both the methods and content of intelligence gathering. These technologies have enormous potential benefits but are not without limits. This section will discuss new intelligence gathering methods and will extrapolate from past cases to understand how they can be successfully integrated into the existing intelligence organizations.

### 2.1.1 Strategic Cyber Operations

One game-changing technology for the intelligence community is the use of malicious code or malware for espionage or sabotage missions. Over the past decade, the capabilities of malware have improved dramatically. Today, sophisticated semi-autonomous or autonomous codes can infect computer networks and record spoken language, take screen shots of important programs, export files stored on the device, or even cause physical damage to industrial equipment. This has opened up opportunities for the intelligence community to collect critical information or conduct covert missions without sending people into potentially dangerous situations. Although cyber operations will affect intelligence gathering and sabotage efforts in the future, the development of these codes is not a straightforward task and there are some practical limitations.

These types of malware cannot be developed in a vacuum and need a 'human in the loop' to ensure their success. The leaders of cyber operation missions must be able to accurately identify the targets using information from other sectors of the intelligence community and find effective ways to analyze and disseminate the information that is gathered. Looking back at past intelligence failures, states must be wary of becoming over-reliant on cyber tools as they become easier to develop and implement. Sophisticated malware will have its place in intelligence gathering and sabotage efforts in the twenty first century, but it will only be successful in conjunction with traditional methods.

### 2.1.2 Augmented Reality

If cyber operations require a 'human in the loop' for facilitation, augmented reality is downright dependent upon individuals as operators. Augmented reality is a variation of virtual environments that "enhances a user's perception of and interaction with the real world" (Azuma 1997). For intelligence purposes, it is a situation awareness *tool* that has the potential to significantly enhance intelligence collection efforts and operations in the field.

Of the intelligence collection disciplines, augmented reality has the greatest potential to influence geospatial intelligence; this is due in large part to the technology's reliance on GPS and its ability to function best outdoors. Since augmented reality requires a human wearer and operator, the technology also has the added benefit of helping an individual contextualize their observations in the field in real time.

As a disruptive technology, augmented reality presents additional challenges to just those posed by conditions in the field. In a world that is becoming more and more dependent on technology for previously analog tasks, what is the risk of intelligence workers becoming dependent on it? Furthermore, and perhaps most compelling, what does augmented reality mean in the hands of the enemy? The US government's use of augmented reality has thus far steered clear of major ethical controversies. Dual use technologies beyond augmented reality, such as human performance modification, have considerable ethical implications for which other nations have demonstrated a lack of concern, possibly setting the stage for an 'arms race' of disruptive technology as we edge further into this millennium (Brimley et al. 2013).

## 2.2 *Emerging Intelligence Analysis Techniques*

Intelligence analysis is changing as the types and magnitude of the information gathered transitions with the use of new technologies. An unprecedented amount of information is now available to the intelligence community through advances in gathering technologies such as the use of malicious code or augmented reality to mine detailed information, and through changes in open source intelligence. This section will discuss big data analysis, and how big data can be leveraged to understand cities of geopolitical importance.

### 2.2.1 Open Source Intelligence and Big Data

The evolution of intelligence technology has historically been driven by military or clandestine government needs where case officers and handlers had the advantage of time in determining how best to integrate developments into the field. The scientists, engineers, and technologists of the CIA in the Cold War responded to the needs of the operatives to drive technology development and deployment strategy. The findings and developments of the intelligence community stayed securely behind closed doors, hidden from the view of the domestic and international general public (Wallace and Melton 2009). The advent of the Information Age, however, presents opportunities and challenges for the intelligence community with many of the same technologies used by the intelligence community undergoing development, available to, and used by the general public.

Not even two decades ago, the information that was relevant to both the intelligence community and the general public was limited to well-known or easily monitored sources and mediums. Open Source Information, or "publicly available information that anyone can lawfully obtain by request, purchase, or observation" (Benes 2013), is used to produce Open Source Intelligence (OSINT) products and enhance or supplement traditional intelligence products.

The future of intelligence gathering and processing is rapidly being shaped by the emergence of new technologies and the vast quantities of information that follows. The amount of open source information now available to an analyst far exceeds that of their Cold War counterparts. In addition, much of this information now at the disposal of intelligence analysts is also available to the general public, at home and abroad. Further, the computational power and prowess necessary to leverage the advantage of the information itself is more widely available. While the 'human in the loop' plays an important part of the intelligence community's processing, exploitation, disseminating cycle, the capabilities of technology are more closely intertwined with the analyst.

There are number of challenges that intelligence community faces in terms of the long-term reliability of big-data sciences for anticipating surprise. The analytics of such massive data can fill the information gap, but to connect the dots and bridge the knowledge gap still fundamentally relies on experts in the subject. For example, open source data, particularly through social media, has led to more understanding about the military acquisitions in otherwise secretive regime such as China, North Korea, and Iran. However, such information simply based on analytics does not shed much light to intelligence community on the strategies behind such acquisition, and can often lead resources to a false alarm. The biggest problem of big data is the inability to predict whether the current analytic tools and technology investment will be functional more than half a decade from now. With the Internet of Things, smart cities, and virtual/augmented reality tools gearing up, the complexities of open source data will scale much beyond that for which the intelligence community or anyone out there is prepared.

### 2.2.2 Situational Awareness in Megacities

In the future, the "big data problem" will expand beyond data collected from open sources and digital spying missions and into the analysis of cities and regions of geopolitical importance. A historic transition is underway; each day an estimated 180,000 people across the globe migrate to cities (United States Agency for

International Development 2013). By 2030 cities will account for 60% of the world's population and 70% of the world's GDP (Heilig 2012). Cities with populations of ten million or more are called megacities. There are currently over twenty megacities in the world, and by 2025 there will be close to forty (Harris et al. 2014). In the next century, the problems found in megacities: explosive growth rates, vast and growing income disparity, and a security environment that is increasingly attractive to the politically dispossessed, are likely to present great challenges to national and international security. Therefore, large scale situational awareness will be crucial to provide decision makers with effective predictors of looming instability with global impact.

Traditional military intelligence approaches often focus on the concept of Areas of Operation, which emphasizes discrete problem sets and well-defined regions. The problem is that megacities are neither discrete nor well-defined. They are multidimensional, interconnected through globalization and information, and uncontrollable due to issues such as connectivity and informal economies. Therefore, new ideas for collecting intelligence are needed that capture the dynamic nature of the modern urban environment. Both the US Army and the World Bank are studying megacities, in an effort to define indicators through which to manage or describe these urban environments. There is some overlap, but each group defines their indicators from a single perspective.

The future of megacities indicates that instability could come from a variety of areas, including rapid population growth, income disparity, environmental vulnerabilities, racial or ethnic separation, infrastructure capacity, or hostile actors. In other words, megacities will face the triggers for instability across many different operational interests – military, emergency response, health and human affairs, and city management. Based on the rise of the Internet of Things and Smart City technologies, future intelligence techniques for megacities will be data focused. Therefore, there is a need for a common intelligence framework that captures military operations, urban operations, emergency response operations, and city behavior. By creating such a framework, the sharing of information and common understanding of potential instabilities can be recognized and communicated across interested parties. The rise of smart cities has spurred new work in defining the anatomy of a city (City Protocol Society 2015). A later chapter looks at the city behavior from an urban design and city operations perspective, which puts forward a new framework, referred to as an ontology, to capture the dynamic nature of the modern urban environment inherent to megacities with a variety of potential technologies that could support intelligence gathering.

## 3 Summary

Technology has become a cornerstone for gathering effective intelligence to combat threats to national security. Rapid technological development in the twentieth century enabled the intelligence community to gather, analyze, and disseminate information at a pace and magnitude that was never before possible. Better understanding of how technology and intelligence activities intersect is needed and is important for future security.

Using multiple methods such as narrative context setting, process tracing and qualitative case studies, this anthology aims to explore the dominant paths in military intelligence in the course of history, the key shifts in approach, and cross-cutting themes across technology, and putting together the past and the future. Beginning in the early 1900s and including the period from World War II to the present, this book focuses on several technologies that marked a breakthrough in the intelligence process and for the intelligence community. The lessons learned can be used to understand implications for the future of the intelligence community and derive effective ways to utilize emerging, disruptive technologies in the intelligence process.

## References

Azuma, R. T. (1997, August). A survey of augmented reality. *Presence: Teleoperators & Virtual Environments, 6*(4), 355–385.

Benes, L. (2013). OSINT, New technologies, education: Expanding opportunities and threats. A new paradigm. *Journal of Strategic Security, 6*(3.) Supplement.

Betts, R. K. (1978). Analysis, War, and Decision: Why intelligence failures are inevitable. *World Politics, 31*(1), 61–89. Web.

Brimley, S., FitzGerald, B., & Sayler, K. (2013). Game changes: Disruptive technology and US Defense Strategy (Disruptive Defense Papers). Center for a New American Security. Retrieved from http://www.cnas.org/sites/default/files/publications-pdf/CNAS_Gamechangers_BrimleyFitzGeraldSayler.pdf

Central Intelligence Agency. (2014, September). North Korea: Potential for Nuclear Weapons Development.

City Protocol Society. (2015). City anatomy: A framework to support city governance, Evaluation and transformation. City Protocol Agreement (CPA-I_001-v2).

Clapper, J. (2016). *Worldwide Threat Assessment of the US Intelligence Community*. Director of National Intelligence. Retrieved from http://www.armed-services.senate.gov/imo/media/doc/Clapper_02-09-16.pdf

Diamond, J. (2008). *The CIA and the culture of failure: US intelligence from the end of the Cold War to the Invasion of Iraq*. Stanford: Stanford Security Series. Print.

Harris, M., Dixon, R., Melin, N., Hendrex, D., Russo, R., & Bailey, M. (2014). *Megacities and the United States Army: Preparing for a Complex and Uncertain Future*. Arlington: Chief of Staff of The Army Strategic Studies Group.

Heilig, G. K. (2012). World urbanization prospects: The 2011 revision. United Nations, Department of Economic and Social Affairs (DESA), Population Division, Population Estimates and Projections Section, New York.

Johnson, L. K. (2006). Ethics of covert operations. In J. Goldman (Ed.), *Ethics of spying a reader for intelligence professionals*. Lanham: Scarecrow Press, Inc..

Montgomery, A. H., & Mount, A. (2014). Misestimation: Explaining US failures to predict nuclear weapons programs. *Intelligence and National Security, 29*(3).

TIA. (2016). (aka Topsail) unveiled: the real scope of the NSA's domestic spying program. ars technica website http://arstechnica.com/uncategorized/2006/05/6813-2/ Updated May 11, 2006. Accessed April 25 2016.

United States Agency for International Development. (2013). Sustainable service delivery in an increasingly urbanized world, USAID Policy.

US Congress. (2004). Senate Select Committee on Intelligence, Report on the US Intelligence Community's Prewar Intelligence Assessments in Iraq.

Wallace, R., & Melton, H. K. (2009). Spycraft: The secret history of the CIA's Spytechs, from communism to Al-Qaeda. Plume.

Westad, O. A. (2000). The New International History of the Cold War: Three (Possible) Paradigms. *Diplomatic History, 24*(4), 551–565.

"Why The FISA Court Is Not What It Used To Be." *NPR.org*. (2013). Web. 6 Nov. 2017.

Wyden Calls for Congressional Oversight, Accountability of Total Information Awareness Office. (2016) Ron Wyden Senate website https://www.wyden.senate.gov/news/press-releases/wyden-calls-for-congressional-oversight-accountability-of-total-information-awareness-office. Updated January 15, 2003. Accessed April 25 2016

# Interaction of Technology and Organization: Case Study of US Military COMINT in World War II

**Shai Bernstein**

**Abstract** This chapter is motivated by the question of whether an intelligence agency must adapt itself - its organization, its hierarchy, its operation - at a fundamental level in order to best make use of a new technology. This question is investigated by looking at two past examples as a case study comparing how the intelligence apparatuses of two different countries adapted to a technology. Specifically, the chapter compares how the UK's Government Code and Cypher School and the collective civilian and military intelligence apparatuses of the US both adapted to the rise of radio-based intelligence gathering (more generally known as Signals Intelligence or Communications Intelligence) during World War II. This comparison is worthwhile because whereas the UK's radio-intelligence gathering capability is quite well regarded, the US's was fraught with challenges (though with its share of successes). The chapter traces the evolution, throughout the war, of the organizational hierarchies of the various US intelligence groups and notes how information was shared, which group had authority over which other, and how these changed over time. A comparison to the UK's organization reveals that the new technology did not fundamentally alter how intelligence ought to be gathered - it only served to exacerbate extant organizational challenges. The continued inefficiency stemming from these challenges was what led to the eventual restructuring and newfound efficiency - not the technology itself. This conclusion will hopefully serve as a reminder that whenever a new technology comes along, the intelligence community must remember the basics first.

## 1 Introduction

From James Bond's gadgets to the National Security Agency (NSA), intelligence gathering has been associated with technology and high technology, in particular, for several decades now. However, although an intelligence agency utilizes technology in order to produce its product, it is not unreasonable to expect that the

S. Bernstein (✉)
Georgia Institute of Technology, Atlanta, Georgia, USA

 
M. E. Kosal (ed.), *Technology and the Intelligence Community*, Advanced Sciences and Technologies for Security Applications,
https://doi.org/10.1007/978-3-319-75232-7_2

technology can also affect the agency using it – at an operational level, a cultural level, or an organizational level. The investigation the interaction of technology and the organization of intelligence agencies or an intelligence community is the primary goal of this paper.

To do this, it is necessary to determine where to start. Although human intelligence (HUMINT), forensic and biometric intelligence (FABINT), measurements and signatures intelligence (MASINT), and geographic intelligence (GEOINT) can all involve technological aspects, signals intelligence (SIGINT), which includes electronic and communication intelligence (ELINT and COMINT), has been in use since the early 1900s with the advent of radio communications and, naturally, the interception thereof. The early inception of radio-based intelligence gathering, nearly-simultaneous with the rise of intelligence agencies as we know them today, allows for an analysis of how fledgling intelligence agencies organized themselves to best take advantage of this new wealth of information traveling unseen through the air, and whether the technology influenced how the agencies were organized.

## 2 Motivation

Although an intelligence agency utilizes technology in order to generate its product, this very use could have an effect on the organization using it. If there is some sort of meaningful interaction between the intelligence agency and the technology it employs to gather intelligence, it is possible that different types of technology could have a different interaction effect or more specifically that an intelligence agency must operate, behave, or be structured differently in order to best employ the technology. Put another way: how must intelligence agencies change in order to best utilize a new technology?

This question must necessarily become more relevant as the pace of technological advance accelerates (Anderson and Tushman 1990). As more and more technologies such as cyber, big data, biotechnology, nanotechnology, augmented reality, smart cities, etc., mature and potentially become tools or subjects of intelligence gathering, it may become even more necessary to understand any such potential interactions because these may significantly affect the effectiveness of agencies utilizing these new technologies.

A useful starting point for studying this problem is a case study of effects that a single technology had on an organization of limited scope within the intelligence community. This case study may locate multiple avenues of further investigation or may perhaps determine that the question does not merit further investigation.

As will be seen in the chapter, its age gives SIGINT the distinction of arising at the same time as many of our common conceptions of what an intelligence agency is (Finnegan 2009). SIGINT accompanied the increasingly widespread use of cryptography, translation, and analysis in the intelligence process. As a physics-based application of intelligence gathering, it is proposed that early SIGINT challenges are a prime candidate for analysis of how exactly a new technology can interact with an intelligence gathering agency.

Government organizations are well known for the all-too-ubiquitous "org charts," or organizational charts, detailing the hierarchy of organizations that lead to the commander in chief or the chain of command within the organization itself. Although organizational charts are occasionally mocked in informal circles, they have an inherently useful purpose since they detail official control and management structures and authority. They can aid one in understanding priorities within an organization, conceptions of the ideal operation of the organization, and areas where inefficiencies may arise within the organization. The same can also be said the "meta-organization," or the structure of organizations within an area of the government.

Thus, by analyzing how the structure of the intelligence community changed over the course of the adoption of new technology, light will be shed on exactly what effects the technology had on the organization.

This combination of the development of SIGINT capabilities and change in organization of an agency naturally leads to the military intelligence community of the United States in World War II, specifically those that dealt with SIGINT and COMINT. As many scholars of that time period have noted, this was a time of great disorganization, rivalry and infighting, and inefficiency within the intelligence community (Benson 1997; Finnegan 1993). Many of today's commonly-thought-of organizations weren't in existence yet (e.g., CIA and NSA) while others existed but no longer do (OP-20-G, MID chief amongst them). Examining the causes for such dynamic changes within the military intelligence community specifically could lead to insights into the nature of these changes and whether some of them could be directly attributed to radio communication interception.

## 3 Background and History

### *3.1 Summary of Historical Context, Properties of Early Radio Communication, and the Intelligence Cycle*

As this chapter deals with the interaction of technology and intelligence organizations, it is important to first address what sorts of technological aspects exist which could interact with intelligence organizations. Secondly, it is important to address what intelligence is and how information is manipulated by an organization in order to produce finished intelligence.

### 3.1.1 Important Factors in Transmission and Interception of Radio Communications

Although a thorough discussion of the physics of radio transmission and interception could take many chapters, it is important to make a few important notes. Radio communications is, of course, a highly technical field requiring plenty of training and skill. Relevant factors include the frequency (or wavelength) of the transmission, the size of the receiving antenna aperture, the curvature of the earth, atmospheric effects, etc. These variables affect, in practice, which transmissions one can intercept given a certain antenna length, at a certain distance and power. If a signal is sent at one frequency, an antenna must be able to receive at that frequency and be close enough (US Congress Office of Technology Assessment 1995). In situations with limited man power, the operator must also be listening to that frequency to receive the message, unless there is a means of storing the signal. Finally, pinpointing the direction of the signal, known as direction finding (sometimes referred to as DF), required specialized equipment and training and necessitated an operator be listening for the signal, hear for the signal, and have at least a few seconds or up to 60 to find the direction (Benson 1997; Finnegan 1993).

These facts had very important consequences in World War II. The tactical transmissions of most armies could not be intercepted at long distances – mobile antennas were smaller by necessity and therefore could only intercept high frequency communications, which did not travel well over long distances. Long-range communications, on the other hand, required large antennas for interception and therefore could not be very mobile. This created a geographic component to interception that necessitated interception capabilities at both short and long distances (Finnegan 1993).

### 3.1.2 The Nature of Intelligence Gathering and the Intelligence Cycle

The concept of the intelligence cycle is in use today to help guide the processes of intelligence agencies. While there is some disagreement, the major parts of the intelligence cycle consist of: (Intelligence Branch, FBI n.d.)

1. Planning and Direction
2. Collection
3. Processing
4. Analysis and Production
5. Dissemination

These elements are all critical. Planning and direction is required to reduce duplication of effort and focus the elements of the organization in order to best satisfy requirements placed on it. Collection is self-explanatory and likely gets most of the media attention, as it is perceived as the flashiest and most clandestine part of intelligence. After information is collected, it must be processed, which means one or more of the following: decryption, translation, data reduction, and generally making the data available for analysis. Analysis is also highly critical to the process – without analysis, there is no finished intelligence, because analysis is what converts the information into an actionable, timely, relevant product. As the FBI (Intelligence Branch, FBI n.d.) notes:

> The information is logically integrated, put in context, and used to produce intelligence. This includes both 'raw' and finished intelligence. Raw intelligence is often referred to as 'the dots' – individual pieces of information disseminated individually. Finished intelligence reports 'connect the dots' by putting information in context and drawing conclusions about its implications.

Finally, intelligence that is held too 'close to the chest' is not useful: it can only be used by decision makers whose actions it informs. However, there is a challenge in determining who the potential customers of a piece of intelligence are – often, they themselves may not know that they could benefit from intelligence until they are given access to it. Customers might also not need finished intelligence but rather directives that originated from finished intelligence. For example, if a blindfolded person is being guided through a room of obstacles by another person, they can either be told "there is a chair right in front of you, go around it" or they can be told "turn left, walk a step, turn right, walk two steps, turn right again, take another step, and back to the left." Both sets of information involved walking around an obstacle, but only one involved sharing intelligence.

## *3.2 Summary of History of US Naval and Military Intelligence Prior to and During World War II*

The history of the US military COMINT community in WWII could, and has, filled volumes. Since an in-depth explanation of the history is outside the scope of this chapter, only a relatively brief summary, sufficient for the purpose of explaining any relevant effects, will be undertaken. Much of the content in this summary comes from Benson's *History of US Communications Intelligence During World War II: Policy and Administration* (Benson 1997) and Finnegan's *US Army Signals Intelligence in World War II: A Documentary History*, (Finnegan 1993) which describe the relevant history in great detail.

### 3.2.1 Brief Summary Up to the End of World War I

Guglielmo Marconi, the man chiefly credited with the development of radio transmission, began his work in the mid-1890s, although his work was based on numerous other advances carried out by others in the field. Less than a decade later, during the Boer wars, radio communications were already in use on Royal Navy vessels in South Africa (Lee 2002).

Use of radio transmission for military matters accelerated quickly after that. The famed case of the Russo-Japanese war, where Japan gained the edge on Russia partly through monitoring unencrypted transmissions comes to mind, though in the later aughts and teens there were more cases with Austria, France, and the US (Lee 2002).

The British developed some of the most advanced SIGINT methods during World War I. They cut undersea cables at the start of the war, meaning Germans had to send messages via telegraph cable systems and so-called "wireless telegraphy." This enabled Britain to spy on these systems, providing British spies with an invaluable source of intelligence leading to, among other things, the famous Zimmerman Telegram (Lee 2002; Wheeler 2012).

British intercept stations were the first to employ long-range interception (in their "Y" stations) as well as intercepting higher-frequency transmissions. As these stations intercepted more and more transmissions, they forwarded them to Room 40, which decrypted and analyzed this intelligence. The "Y" station-Room 40 combination led to some vital naval victories, such as Dogger Bank. This is in part because, since the Germans were prone to sending daily position reports, British spies could map out German movements. Analysis of these movements in the context of other information, especially which areas of water the Germans avoided and how frequently they sent out information, could give the British context clues about where a defensive minefield was located or when an attack was likely to take place (Beesley 1982).

US, French, and German capabilities were also advancing. French interception of German messages was effective for the same reasons as it was for the British. On the other hand, the Germans employed many similar tactics, including deception, against the Russians (Wheeler 2012).

### 3.2.2 The Interbellum Period

Prior to the Second World War, the United States had a noteworthy history with COMINT. Having seen the benefits of successful exploitation of COMINT during World War I with their victory over the Central Powers, both the UK and the US created peacetime agencies responsible for handling code-breaking. While the UK created the Government Code and Cypher School (GC&CS, also known as Bletchley Park or BP), which would go on to become highly successful during World War II and which will be discussed towards the end of the paper, the US created the US Cipher Bureau, jointly-funded by the State Department and the Army, which had demobilized its own COMINT and cryptanalytic services after the war. This bureau was successful, reportedly braking the systems of two dozen countries, according to its head. However, in 1929 the US Cipher Bureau was cancelled by newly-appointed Secretary of State Henry Stimson, who famously wrote in a memoir about the incident that "gentlemen do not read other gentlemen's mail." This no doubt affected the military's abilities at the start of the war (Lee 2002).

Another advance before the war was the development of the Federal Communication Commission's Radio Intelligence Division (FCC RID) – a purely civilian branch (as evidenced by its parent organization) that with the aid of a high budget, proceeded to develop significant expertise in the area of cryptanalysis and COMINT. The RID was so successful that its products and services were in use by many other organizations, both civilian (State Department, the Office of Strategic Services, and the Federal Bureau of Investigation) and military (Navy and Army) (Benson 1997).

## Army COMINT at the End of 1941

By WWII, however, the Army had developed something akin to a COMINT gathering capability. Army efforts were complicated by a few things – firstly, tactical information of relevance to Army units was often transmitted via frequencies that could not be intercepted over long distances. Secondly, there was significant expertise with foreign diplomatic transmissions, but this was clearly not solely an Army area of interest, so the primary Army COMINT organization often fought for relevance against other branches' agencies. This created a confusing organizational structure whereby Army COMINT was under two distinct hierarchies – there were the Signal Radio Intelligence (SRI) companies, which were organic to numbered armies and under the command of the field commanders of those armies, and the primary Army COMINT group, Signals Intelligence Service (SIS), which was totally distinct from the SRIs and controlled the 2nd Signal Service Battalion, which operated the various long-range monitoring stations the Army controlled around the world. This did in fact result in administrative confusion when on occasion SRI companies would intercept diplomatic communications and send these to SIS.

A further point of contention was who would command the SIS. The Army's primary intelligence agency was the Military Intelligence Division (MID), and as such it was the one that primarily consumed SIS intelligence. However, after the dissolution of the Cipher Bureau, Signal Corps took control of cryptanalysis: as it already dealt with radio communications, this may have seemed a logical decision at the time. Thus, SIS was actually under the command of the Army Signal Corps, specifically the Office of the Chief Signal Officer (OCSigO) (Benson 1997; Finnegan 1993).

## Navy COMINT in the Pacific at the End of 1941

The Navy's situation mirrored the Army's in a few key ways. The equivalent to the 2nd Signal Service Battalion's monitoring stations were the Navy's processing centers, of which we will focus on CAST, HYPO, and NEGAT (located at Corregidor Island in the Philippines, Pearl Harbor in Hawai'i, and Washington D.C. respectively). These stations were under the operational control of the Director of Naval Communications (DNC, or OP-20) instead of the Office of Naval Intelligence (ONI, or OP-16), mirroring the Army's conception that COMINT was primarily a communications technology task instead of an intelligence task. However, they were under administrative control of the 16th, 14th, and 5th Naval Districts respectively, which presumably handled staffing and budgets.

Unlike Army monitoring stations, which only handled interception, the processing centers were far more self-sufficient: they handled interception, translation, traffic analysis, and cryptanalysis. Furthermore, unlike the schizophrenic split between SRI companies and monitoring stations, the processing centers directly supported the fleets, with CAST supporting the Asiatic fleet (stationed in the Philippines) and HYPO supporting the Pacific fleet (in Hawai'i). This was due to the beneficial physical fact that Japanese communications with Japanese fleets necessarily had to utilize longer-range frequencies, and thus were susceptible to interception from a longer range, too.

The processing centers had their own informal structure as well. NEGAT (also designated OP-20-G) would often provide guidance and directives to the other processing centers, implying some degree of superiority of that center compared to the others.

However, the Navy had a second, smaller intelligence organization within the office of the Commander in Chief, US Fleet (COMINCH). This office had its own intelligence staff, designated F-11 (F was the designation for COMINCH staff). This staff had operational control of the Operational Information Section (F-35), but as evidenced by the numbering structure, F-35 was actually placed within F-3, the Operations Division, and supported it with partially processed intelligence. F-3 in turn turned this intelligence over to F-11. Both F-35 and F-11 were supported by NEGAT (Benson 1997).

## Civilian COMINT Agencies at the End of 1941

The FCC's RID was serving many civilian organizations. At the same time, it was serving the Army and other military organizations with partially processed intelligence (Benson 1997) (Fig. 1).

## Army COMINT at the End of 1942

By the start of the war, there were already changes taking place, though not necessarily for the better. The Army created the Special Branch in order to directly interface with SIS, which that summer underwent three name changes (to SSD, SSS, and finally SSA). SIS (now the Signal Security Agency, or SSA) provided cryptanalyzed, translated intelligence to the Special Branch, which was controlled by the newly formed Military Intelligence Service (MIS). The MIS had the same functional role as the MID had had previously; a newly-formed MID now served in a purely administrative capacity for Army intelligence (now MIS). Thus, on the face of it,

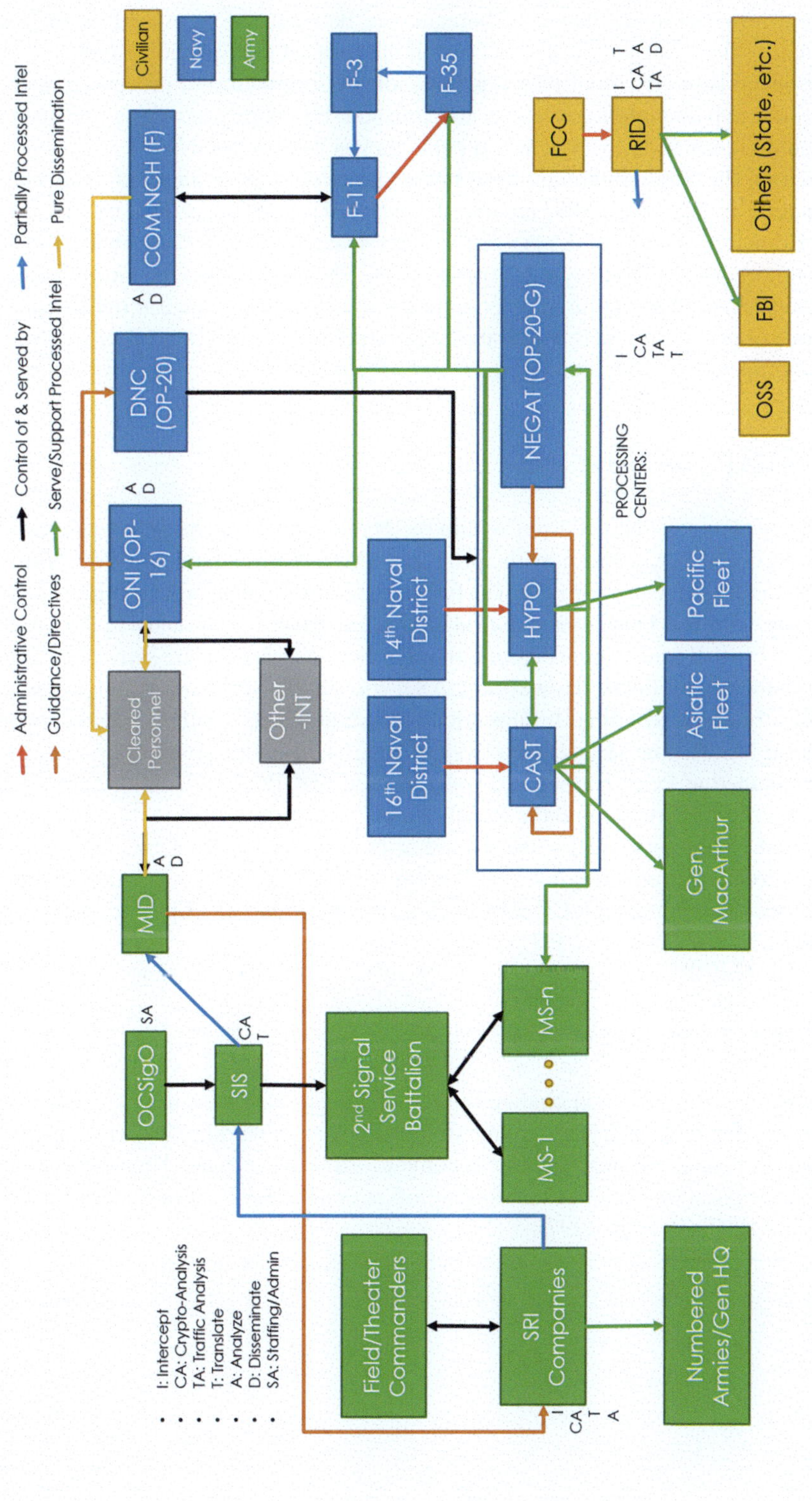


**Fig. 1** Structure of the US intelligence community prior to the start of World War II. Compiled from information found in (Benson 1997) and (Finnegan 1993)

there were now two extra layers of bureaucracy between the SSA and the MID, though really there was only one extra one in the form of the Special Branch.

The Special Branch's other role was to handle dissemination of processed intelligence (although this intelligence was not 'finished' by today's standards). As such, it handled dissemination to any cleared personnel as well as to ONI.

As before, the SRI companies were under direct control of the field or theater commanders, though they now received occasional guidance or directives from Special Branch instead of MID. As before, they would occasionally forward intercepted diplomatic and strategic intelligence to the SSA, which had to start referring to any intelligence thus acquired as coming from a monitoring station (technically under control of the 2nd SSB which the SSA directly controlled) in order to avoid undue administrative confusion (Benson 1997; Finnegan 1993).

## Navy COMINT at the End of 1942

The primary changes that occurred within the Navy concerned the processing centers. OP-20-G (NEGAT) was now the primary COMINT organization for the Navy, mirroring the SSA for the Army. It had full control of the other processing centers, doing away with the unnecessary Naval District control. CAST and HYPO were now called Fleet Radio Units (FRU), although they operated in the same way and supported the same fleets. However, CAST moved to Melbourne and became FRUMEL (though it still directly supported the Asiatic Fleet, which was also known as "General MacArthur's Fleet" in a twist of accidental inter-service cooperation). HYPO remained at Pearl Harbor and was renamed FRUPAC, indicating its role in supporting Admiral Nimitz and the Pacific Fleet.

A second change was ONI's loss of authority in providing the DNC with guidance and directives, reflecting the deteriorating role of the Navy's primary intelligence service in matters of COMINT. This theme would continue to play out throughout the war (Benson 1997) (Fig. 2).

## Army COMINT at the End of 1943

By the end of 1943, little had changed in the organizational structure of the Army, although not for lack of trying. Many studies and recommendations were undertaken and proposed, but they would not come to fruition this year. There were also efforts on the parts of many leaders to develop structures for inter-service cooperation on COMINT (Benson 1997).

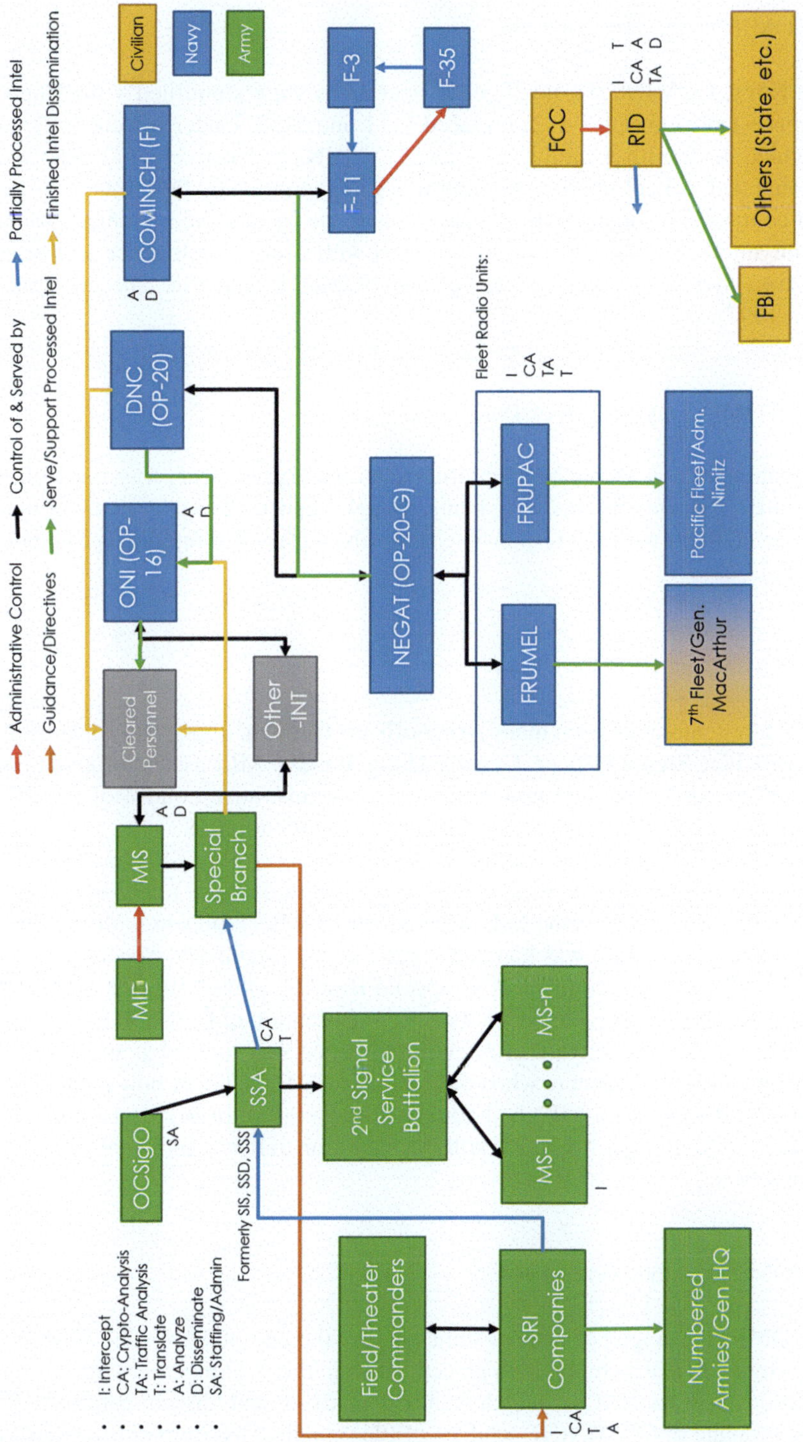


**Fig. 2** Structure of the US intelligence community by the end of 1942. Compiled from information found in (Benson 1997) and (Finnegan 1993)

## Navy COMINT at the End of 1943

For the Navy, COMINCH's intelligence operations were simplified – the Central Intelligence Division (CID, F-2) replaced F-11 and F-35. Under it were F-21 and F-22, which dealt with Atlantic and Pacific COMINT respectively. F-2 dealt with operational and tactical intelligence, and was still supported by OP-20-G.

ONI continued its decline: the decision was made by DNC to no longer provide ONI with any COMINT. ONI would now deal with counterintelligence and strategic studies, though it continued to receive COMINT from the Special Branch (Benson 1997).

## Civilian COMINT at the End of 1943

The only thing of note is that Navy-civilian relations began to truly fray, particularly with the FBI. Reasons for this were many, though importantly many in the military did not see civilians as being responsible stewards of classified intelligence (Benson 1997) (Fig. 3).

## Army COMINT at the End of 1944

By 1944, after entreaties from many Army officers (though, importantly, against the wishes of Chief Signal Officer, General Harry Ingles), MIS assumed operational control of the SSA. OCSigO now had purely administrative control of the SSA, although this was to the satisfaction of neither OCSigO nor MIS.

Importantly, MIS absorbed the Special Branch, so control of SSA lost a layer of complexity. However, a new Special Branch was created whose role was to handle COMINT dissemination to the field armies and COMINT liaison with the Navy. Dissemination to the field was handled via the newly established Special Security Officer (SSO) program, controlled by Special Branch, whose purpose was to furnish SSOs to key personnel to enable better dissemination of COMINT.

Thus, the resulting structure of Army intelligence was somewhat simplified by 1944, primarily with respect to who controlled the SSA and under how many layers of command it was. Now, MID had administrative control of MIS, which directly controlled SSA, though OCSigO still had administrative control of the latter (Benson 1997).

## Navy COMINT at the End of 1944

Navy COMINT changed structures somewhat by the end of 1944. The Navy now began to move to a structure which placed FRUs in the field together with Combat Information Centers (CICs), which handled analysis and dissemination to F-2 (under command of COMINCH) while the FRUs handled dissemination upwards to

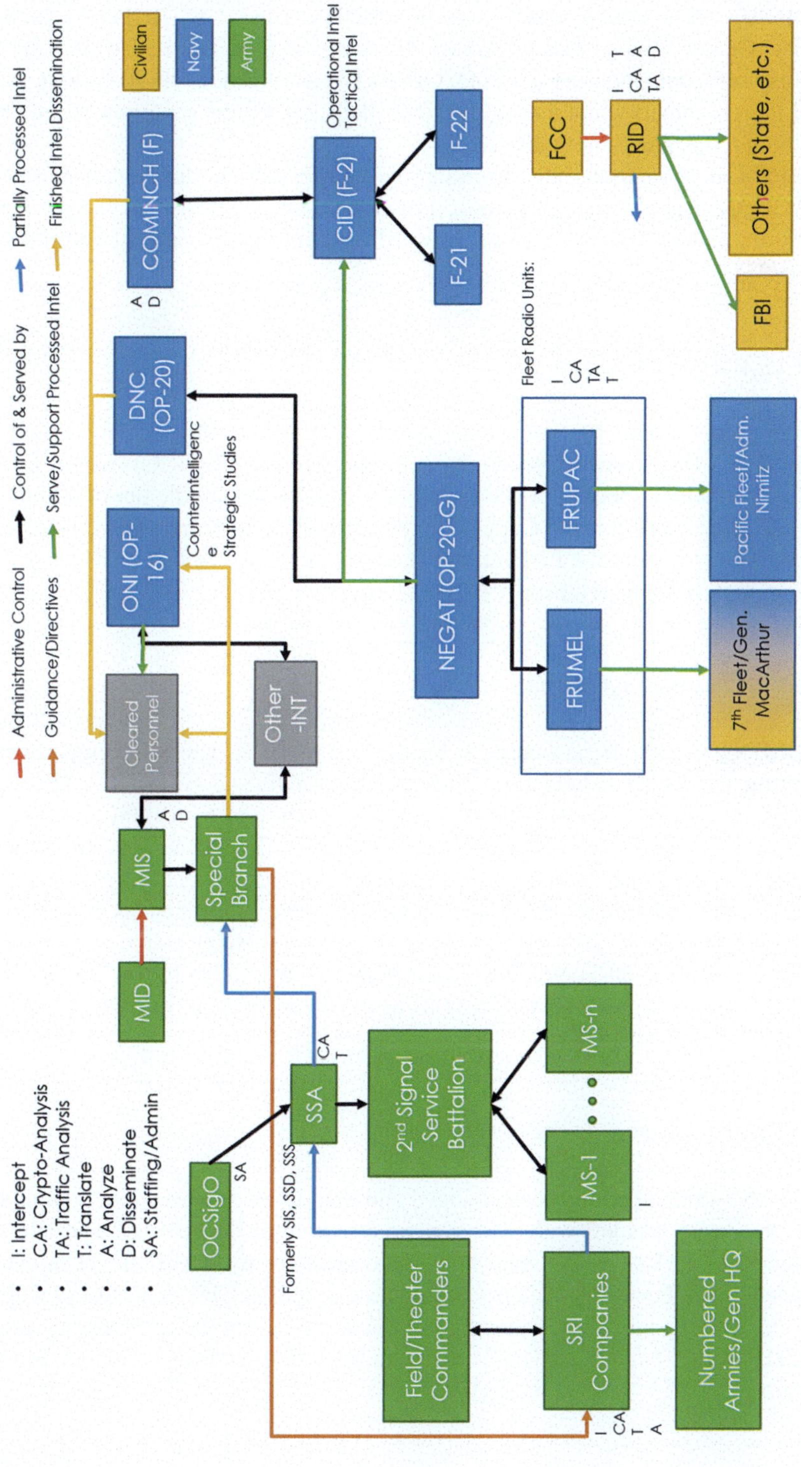


**Fig. 3** Structure of the US intelligence community at the end of 1943. Although the Army showed relatively little structural change, the Navy's organization was undergoing some streamlining. Compiled from information found in (Benson 1997) and (Finnegan 1993)

DNC (OP-20). This structure resulted in FRUPAC and CIC-JICPOA (Joint Intelligence Center Pacific Ocean Areas) operating out of Pearl Harbor. CIC-JICPOA also served an inter-service role. OP-20-G was renamed OP-20-3 (with its personnel largely intact) while a new OP-20-G likely served the same role as other FRUs, although this part is not quite clear.

FRUMEL was downgraded due to changing requirements in the field, but in its place RAGFOR (Radio Analysis Group, Forward) was established in Guam. As implied by its name, RAGFOR was a forward element, specifically of FRUPAC (Benson 1997).

## Civilian COMINT at the End of 1944

A decision was made by the joint Army-Navy committee (Army Navy Communications Intelligence Coordinating Committee, or ANCICC), which had been formed to foster cooperation between the services at the middle tier of authority (Colonels, Commanders, and Captains from MIS, SSA, F-22, and OP-20-G), to keep COMINT away from the FBI and to continue to prevent OSS COMINT capabilities from materializing (Benson 1997) (Fig. 4).

## Army COMINT at the End of 1945

Army COMINT saw the bigger changes by the end of 1945. The SRI companies, long under the command of field commanders, were finally subordinated to MIS in the form of the Army Security Agency (ASA), which was to handle all Army COMINT and COMSEC (communications security). The ASA was, in some respects, another layer of bureaucratic complexity, and yet by totally centralizing the chain of command under the MIS, operations were simplified as stated by Benson: "The creation of ASA was no more than a name change because the MIS had intended to exercise its centralized control through an enlarged SSA." It should also be noted that this removed OCSigO from the chain of command completely (Benson 1997; Finnegan 1993).

## Navy COMINT at the End of 1945

Navy COMINT remained largely unchanged and thereby somewhat more complex than what the Army was left with. The general structure of FRUs/RAGs and CICs was maintained – RAGFOR was now accompanied by CICFOR (Combat Information Center, Forward), which served the same role as CIC-JICPOA did with FRUPAC: analyzing and disseminating both Army and Navy intercepts to the COMINCH side of Navy COMINT, while sharing this analysis with the SSO representatives.

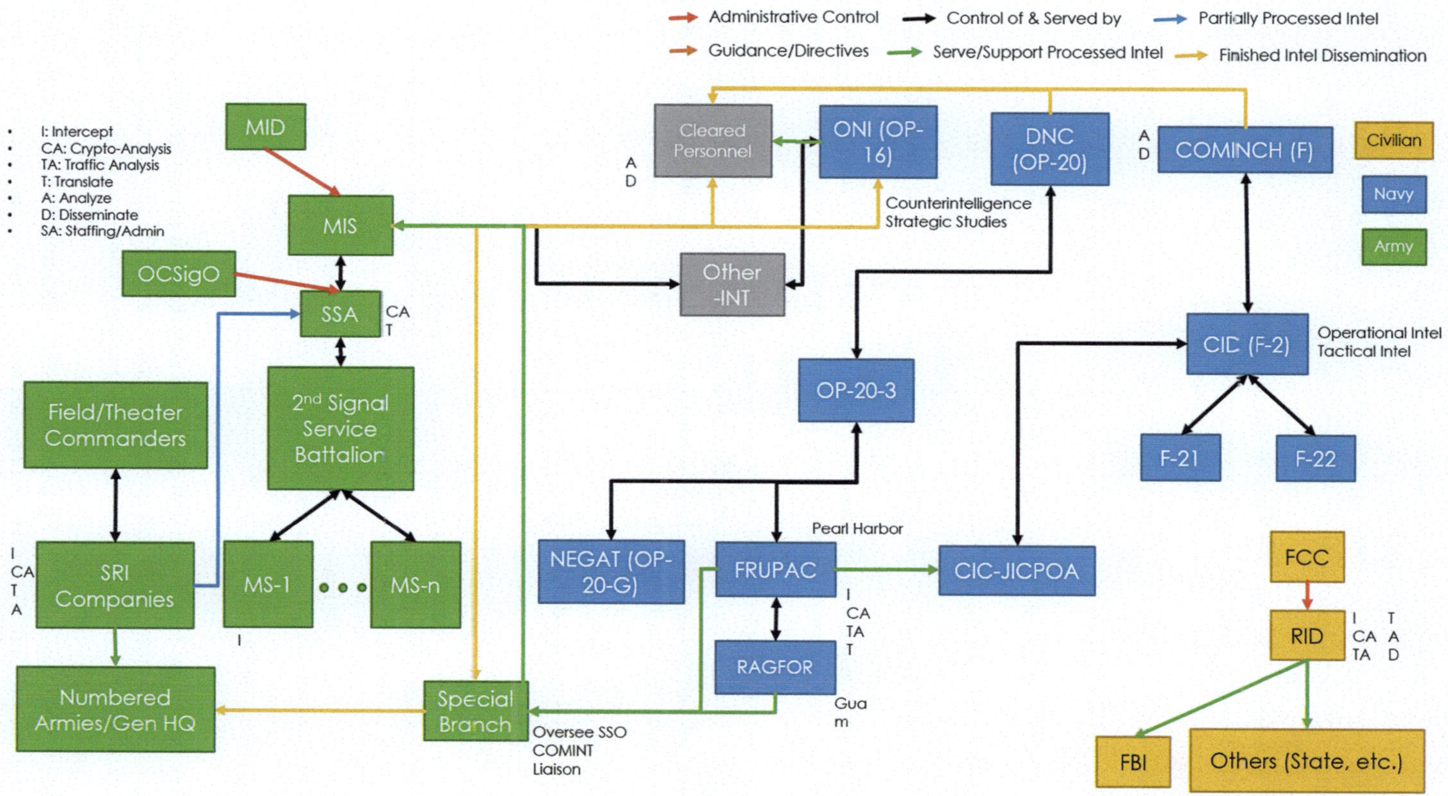


**Fig. 4** US intelligence agency structures by the end of 1944, showing increasing simplification both in Army and Navy intelligence. Compiled from information found in (Benson 1997) and (Finnegan 1993)

In terms of inter-service cooperation, the Army Navy Communications Intelligence Board (ANCIB) had been approved at the highest tier (Admiral King and General Marshal). Its membership consisted of Maj. Gen. Bissel, ACS, G-2 (the group responsible for administrating MIS, etc., as opposed to OCSigO), Rear. Adm. Redman, DNC, and Col. Corderman, Commander, SSA. However, a "new" ANCICC was formed as the ANCIB's working committee – this "new" committee consisted of the same mid-tier members as before, which showed an acknowledgement of the value of this committee and its members (Benson 1997) (Fig. 5).

## Overall Summary

The Army was by far "most improved" and perhaps even outdid the Navy in the resultant simplification and centralization of its COMINT organization. The Army started the war with multiple SRI companies, which answered only to field or theater commanders but occasionally received guidance or directives from the prime COMINT organization, and sporadically intercepted and sent strategic COMINT back. At the same time the prime COMINT organization (whatever its name at the time) had its own monitoring stations, its own battalion which manned these stations, and its own superior officers in the form of MID or MIS or OCSigO. These were combined under the control of the primary COMINT organization, SSA, which was under the control of Military Intelligence Service (the Army intelligence organization). Dissemination was handled by a system of SSO representatives with direct communication capability to Special Branch, stationed with various cleared personnel (in Army command, in the Executive, as well as in the field armies). OCSigO was completely removed from the equation, in recognition of the fact that COMINT had to be viewed as a primarily intelligence-related task in order to be most effectively leveraged by decision makers. As Benson (Benson 1997) writes:

> It was again becoming apparent that the subordination of the (newly renamed) Signal Security Agency (SSA) to the chief signal officer and the Army Service Forces was a problem. This had been recognized by General Strong in 1942 when he attempted to have the SIS placed under MIS. The issue was now raised by Colonel Otto Nelson, assistant to the deputy chief of staff. On 18 October 1943 he wrote General Strong citing the personnel allotment problems of the SSA. He noted that the SSA obtained its personnel through the OCSigO and the Army Services Forces, while it existed mainly to serve not these organizations, but rather the Special Branch, MIS. "Recommendations were sought. General Strong replied on 23 October. He suggested that the SSA be removed from the Signal Corps and made an independent agency." As the SSA was "our most important source of secret intelligence," it ought not to be "under the command of those who have no concern with the intelligence produced."

Possibly due to the fact that it started in a far better position than the Army, with far more cross-communication and more local processing capabilities, and therefore there was less of an obvious need to improve, the Navy's improvement was slower than the Army's. The Navy effectively evolved into having a chain of intercept cites extending out into the Pacific, with each successive one nominally under the command of the one previous. These intercept cites would process and disseminate up to

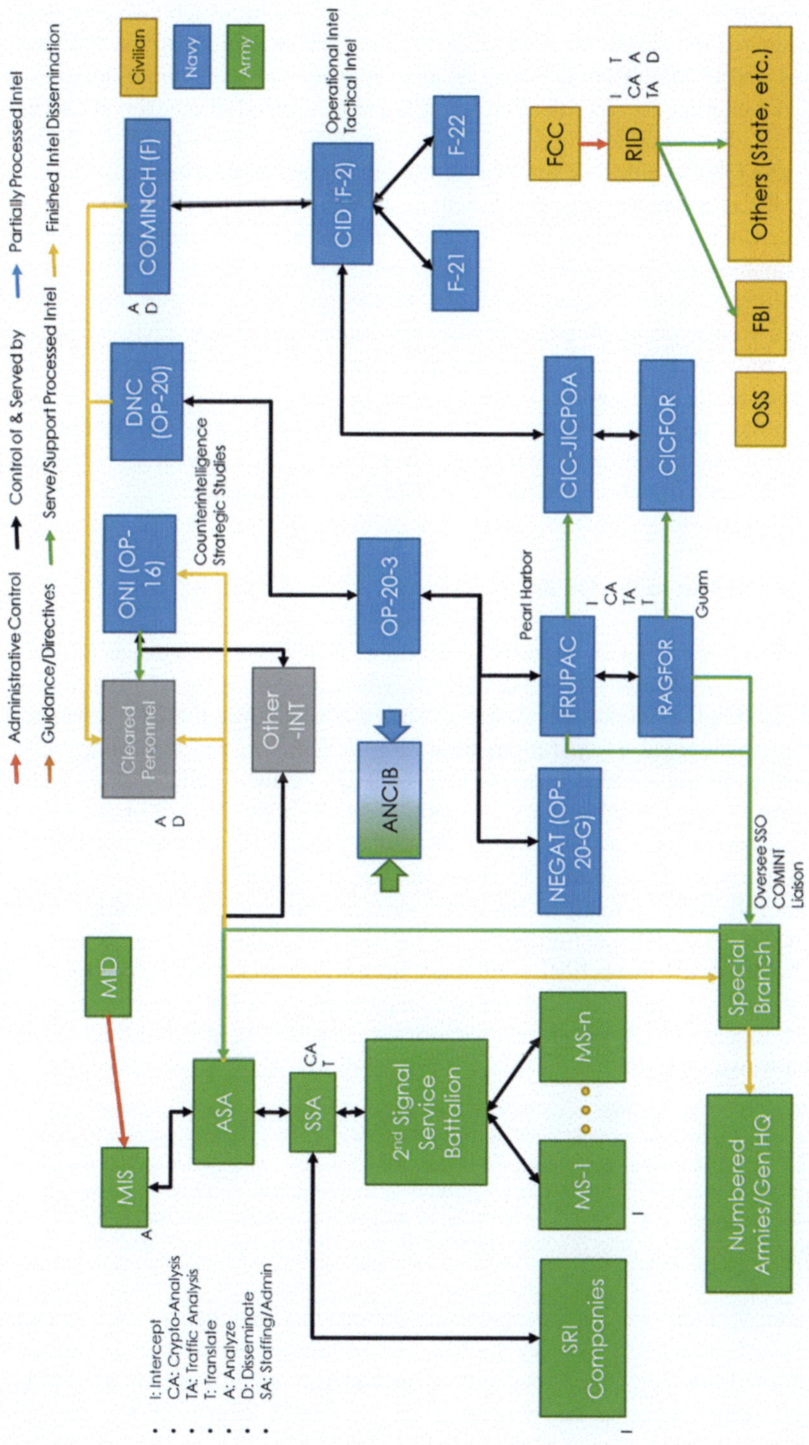


**Fig. 5** Final structure by the end of 1945. Compared to 1941 and to the Navy's eventual structure, the Army's intelligence community was greatly simplified. Compiled from information found in (Benson 1997) and (Finnegan 1993)

DNC. The sites also had organic Combat Information Centers which also processed and disseminated the intelligence, albeit in an inter-service manner, to the SSO representatives and COMINCH. The biggest achievements were the simplification of COMINCH's intelligence organization, as well as the removal of any Naval District command of the processing centers. However, DNC command of COMINT remained, and ONI's COMINT role was thoroughly weakened during the war.

Finally, the relationship with civilian agencies like the RID, FBI, and OSS was quite strained, with the war seeing a breakdown in sharing of intelligence. There were legitimate concerns about the civilian desire to prosecute in court based on secret intelligence, since this could require revealing sources or that the US had broken enemy encryption, though there was no doubt a sense that intelligence was primarily a military task – the concept of peacetime civilian intelligence agencies was nascent (albeit such agencies were becoming quite successful). This tension between military and civilian matters also existed where civilians were employed by the military. Such tensions were a notable factor in terms of how an agency might be organized, and with whom it could or should share intelligence.

### 3.2.3 Comparison to the GC&CS

A comparison to the UK Government Code and Cypher School, known as GC&CS or alternatively as Bletchley Park (BP) due to where it was located, is also useful. Unlike the US Cipher School, which was closed prior to the war, the UK's cryptanalytic service maintained its operations until after WWII, when it was renamed the Government Communications Headquarters (GCHQ). Furthermore, BP was created by fusing Army and Navy post-WWI capabilities (MI1b and NID25 or Room 40) into one civilian organization employing the military personnel of those agencies.

BP's primary aim, as indicated by its name, was cryptanalysis. However, in support of this role BP also handled interception. Continuing the British tradition of primacy in SIGINT, BP was highly successful in codebreaking, famously responsible for breaking the theoretically-unbreakable Enigma codes due to lax German security practices. It can be argued, rather compellingly, that BP at least among the few highly successful COMINT organizations during WWII, if not the most successful. However, despite this success, its organization was by no means perfect.

> "Ratcliff has put much emphasis on the centralization of Sigint at BP, the harmony of inter-service relations, and the extent to which staff were aware of the significance of their work [8, pp.72-105]. But this assessment appears to be made relative to what obtained in German, and to a lesser extent US SIGINT organizations at the time, and no doubt it is correct. However, compared to a 'normal' commercial or military organization it certainly appears anomalous… Indeed, in the official internal history, its organization was described as 'freakish.'" (Sturdy 2008)

Furthermore, there were a few organizational peculiarities that set it aside from the others such as the US military's. One such peculiarity to BP was the mix of civilian and military personnel. As can be seen from the organizational chart of BP, reproduced from a paper on the reorganization of GC&CS (Sturdy 2008), the mili-

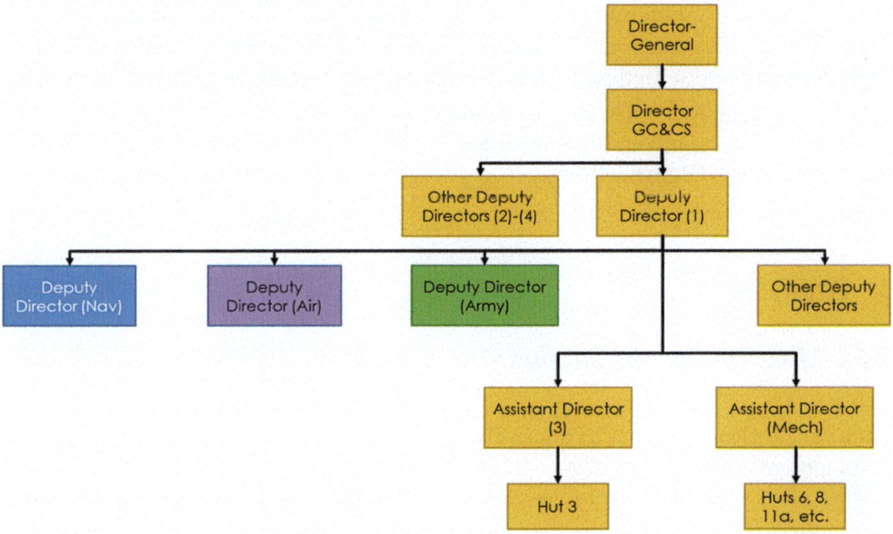


**Fig. 6** Organizational chart showing hierarchy and chain of relationships among different offices and people in the UK Government Code and Cypher School, known as GC&CS or alternatively as Bletchley Park (BP) (Sturdy 2008)

tary services' COMINT capabilities were incorporated into a civilian structure. This meant that there were military personnel serving under civilians or vice versa. There were even senior officers serving under junior officers at some points. Many note that it was primarily because BP was a civilian organization that so little attention was paid to rank as opposed to completing the job (Sturdy 2008) (Fig. 6).

Another peculiarity was that despite the successes of BP, its true structure was highly informal and based on personal arrangements between section chiefs, so much so that this was beginning to be recognized as a serious problem, with memos to that effect being circulated by leadership. As Sturdy (Sturdy 2008) puts it:

> In any case, much of how BP was organised could not be captured by organization charts. The ambiguity of BP's formal structure is illustrated in an incident recalled by the American, Joseph Eachus, who, tasked in his capacity as the US Navy's liaison officer to obtain an organisational chart of BP, was told: "'I don't believe we have one'. I didn't pursue this with him, but I was never quite sure whether he meant we don't have a chart or we don't have an organization."

This informality and casual relationships at times aided the function of the organization but at other times undermined it by creating confusion.

Other challenges faced by BP were how to facilitate the type of creative work necessary for highly intellectual tasks, such as code breaking, versus the more procedural and regimented work of the rest of the organization. This challenge mirrors more closely the issues faced by the US agencies, and yet in and of itself belies how centralized BP already was – in the US, the problems faced by the community weren't how to enable different kinds of work styles in the same organization, but how to effectively communicate, prioritize tasks and focus the organizations in order to better divide labor, who was in command of who, and who needed to have intelligence shared with them versus who didn't. "Although attempts to draw exact parallels are probably useless, it is worth saying that ANCIB brought the US COMINT services into an organizational and policy position reached by the British ten years before" (Benson 1997).

# 4 Analysis

## *4.1 Most Important Factors*

What can be gleaned from this relatively brief summary of Army and Navy COMINT developments during World War II? The question at hand is whether there was a change in the operation or organization of these branches' intelligence gathering during the span of the war, and what role the technical aspects of radio interception played in this change.

It can be seen that there most certainly were drastic organizational changes in Army and Navy COMINT as a result of wartime experiences. Many forces were at play, among them:

1. Budget, which supplied manpower and equipment to carry out the organizations' functions.
2. Availability and proximity of expertise for cryptanalysis, translation, and analysis, which determined the efficiency of the intelligence cycle.
3. Organization, including the military or civilian nature, the technical or non-technical nature, the creativity or standardization of the work, the morale, and all of the personalities involved.
4. Communication, including how compartmentalized the organization was and whether it had a mandate to inform superiors and decision-makers of incremental progress.
5. The meta-organization, or how all the branches, divisions, battalions, and companies were organized, including which organization was in control of which other.
6. And finally the physics, which determined the location of monitoring stations and placed training requirements on staff.

Most of these factors are not technical in nature. Even the physics of the problem served merely to determine location of offices and the required training of collections personnel. Generally speaking, of course, personnel must always be trained – even in non-technical -INTs such as HUMINT.

Instead, of the factors listed above perhaps the most poignant were the organizational and 'meta-organizational' aspects:

1. Whether the control of COMINT should rest in the hands of 'Communications' or 'Intelligence' people.
2. The number of parallel, unique chains of command.

These factors incorporate a few elements within them.

Firstly, the number of parallel, unique chains of command indicate the degree of centralization in the service's organizations. Decentralization introduced problems of over-compartmentalization, especially when there was no clear order or mandate to inform higher-ups or other stakeholders of cryptographic progress or intelligence. Decentralization also led to a misalignment of roles and priorities which created duplication of effort or gaps in coverage. Upon centralization, not only was duplication eliminated, but data could be pooled (resulting in larger samples for cryptanalysis, for example). Centralization also allowed higher-ups to more effectively manage the efforts of lower-level organizations whereas previously they might have been ignored by those organizations, as in the case of CAST and HYPO staff sometimes feeling an understandable desire to serve field commanders over intelligence higher-ups.

Next, the subordination of COMINT under communications-technology personnel (and organizations) as opposed to intelligence ones further created problems. While communications personnel insisted that the task was primarily a technical one and this was quite true, the analysis and dissemination of such information was less a technical task than an intelligence one, since such information must necessarily must be collated with information received from other processing centers, background cultural, political, economic, military knowledge of the enemy, and indeed other –INTs as well.

## 4.2 *The Role of the Technical in COMINT*

It might be best to view the shifting COMINT understanding through the lens of the Army's hierarchy. There is an argument to be made that the initial organization of the Army was, in fact, the best from a technical point of view. The SRI companies, organic to the field armies, intercepted all tactical or local signals, as these could not be intercepted at long range. The SSA and the 2nd Signal Service Battalion intercepted long-range diplomatic communications, which could be intercepted from further-away fixed stations. From a technical point of view, this organization was quite sufficient, as it guaranteed the interception of enemy signals at all ranges. However, as became quite clear to Army leadership over the course of the war, this was grossly insufficient for addressing the rest of the 'intelligence cycle' (not a widespread term at the time), as the multiple hierarchies, duplication of effort, and lack of inter-organizational communication severely hindered processing and, more importantly, analysis and dissemination.

The final organization reflected the growing understanding that interception at long ranges and interception at short ranges did not matter from an organizational point of view – it was all a part of the 'gathering' task, and it all provided information required for processing, analysis, and dissemination. This organization also began to mirror in some ways Bletchley Park's organization, although BP also included Navy and Air Force intelligence and was a civilian organization first and foremost.

Pessimistically, one might say that technical aspects muddied the waters and complicated efforts at centralization and efficient channeling of overall intelligence gathering. The highly technical aspect obfuscated the fact that COMINT was still an –INT, and that not only did it benefit the usual consumers of intelligence (who were not Comms personnel) but that its own quality was improved when combined with COMINT from other sources and other types of intelligence and military intuition of commanders. COMINT also still required analysis, which had never been a communications technology task.

## 5 Conclusion

This goal of this chapter was to answer the question of what the effect of technology on the organizational aspects of intelligence is. After reviewing the history of the US military's COMINT agencies in WWII, the most logical conclusion to draw is that, if it had any effect at all, technology merely served to exacerbate already-existing organizational issues. Most of the military's issues with gathering COMINT were not technical ones, nor was intelligence gathering affected much by radio technology except by adding another source of intelligence. Instead, the military's problems were those of organizations where many of the personnel are not aware of who the customer is, and what the product of the organization is. COMINT agencies exist to provide intelligence to decision makers and commanders, and any issues they faced were a direct result of confusion over how best organize themselves to do this.

If instead, agencies had focused on the 'intelligence cycle,' although there was not such a term at the time, perhaps they would have centralized more quickly, with their numerous intercept (and perhaps processing) stations at the bottom of the hierarchy all serving centralized analysis and dissemination centers at the top. There would have been no parallel and disparate chains of command for short-range vs. long-range COMINT, and the communications directorates of both services would not have had any influence on the intelligence collection process except, perhaps, by means of training and furnishing of technical personnel and expertise. Perhaps, the services would have even united their communications intelligence capabilities with the considerable expertise that existed on the civilian side in the form of the FCC's RID. In fact, this was more or less the organization of the Government Code and Cypher School, Britain's COMINT organization, famed for its numerous successes in the decryption and exploitation of communications intelligence. Instead, there were numerous instances of bickering, turf-wars, and refusals to disseminate. The communications directorates contributed to the turf-wars by continually attempting to assert that "intelligence personnel just don't speak [the] language of communications" because they were not technical, thus maintaining control of COMINT for themselves despite the fact that they were not in charge of dissemination and had no role in the other –INTs.

As the rise of social media, cyber warfare, and other new trends continues, it is important to remember the lessons gleaned from this analysis. New intelligence-gathering technology should not be allowed to muddy organizational waters. Although new types of technology may change how intelligence is gathered and may demand different low-level organizational structures to meet these ends, it must be remembered that processing, analysis, and timely dissemination to any relevant parties are still critical functions of intelligence agencies. Although the technology used can change, the job remains the same.

## References


Anderson, P., & Tushman, M. L. (1990). Technological discontinuities and dominant designs: A cyclical model of technological change. *Administrative Science Quarterly*., December, *35*(4), 604–633.

Beesley, P. (1982). *Room 40: British naval intelligence, 1914–1918*. London: Hamish Hamilton Ltd.

Benson, R. L. (1997). *A History of US Communications Intelligence during World War II: Policy and Administration*. s.l. Center for Cryptologic History, National Security Agency.

Finnegan, J. P. (1993). *US Army Signals Intelligence in World War II: A Documentary History*. Washington (D.C.): Center of Military History, United States Army.

Finnegan, T. J. (2009). Origins of modern military intelligence. *Studies in Intelligence, December, 53*(4), 25–40.

Intelligence Branch, FBI (n.d.) *Intelligence Primer: Intelligence Cycle*. [Online] Available at:. https://www.fbi.gov/about-us/intelligence/intelligence-cycle. Accessed Apr 2016.

Lee, B. (2002). Radio Spies – Episodes in the Ether Wars. *Antique Wireless Association Review*. National Security Agency, n.d. *Pearl Harbor Review – The Black Chamber*. [Online] Available at: https://www.nsa.gov/about/cryptologic-heritage/center-cryptologic-history/pearl-harbor-review/black-chamber.shtml. Accessed April 2016.

Sturdy, A. G. C. (2008). The 1942 reorganization of the government code and cypher school. *Cryptologia, 32*(4), 311–333.

US Congress Office of Technology Assessment. (1995). *Wireless Technologies and the National Information Infrastructure*. Washington, DC: US Government Printing Office.

Wheeler, D. L. (2012). A guide to the History of Intelligence 1800–1918. *The Intelligencer: Journal of US Intelligence Studies, 19*(1), 47–50.

# Intelligence Innovation: Sputnik, the Soviet Threat, and Innovation in the US Intelligence Community

**Jon Schmid**

**Abstract** It is well-documented that the 1957 launch of Sputnik I initiated a flurry of US government activity aimed at reducing a perceived shortfall in US scientific, technological, and military capacity vis-à-vis the Soviet Union. Less well known, however, is that Sputnik's launch immediately preceded a period of rapid organizational and technological innovation within the US intelligence community. This article investigates the contribution of the Sputnik scare to this innovation. In particular, this article applies Barry Posen's model of innovation to the historical case of post-Sputnik innovation in the US intelligence community. I find the historiographic and documentary record to indicate that Posen's theory of innovation has substantial explanatory power in the empirical context examined here. In particular, the US intelligence services' improved capacity to collect and analyze information regarding Soviet rocket and missile programs appears to have been initiated by a process of external auditing motivated by an increase in the perceived level of threat posed by the USSR.

## 1 Introduction

The catalytic role of the 1957 launch of Sputnik I in initiating modern US science and technology (S&T) policy is well documented (Dickson 2001; Neal et al. 2008; Yanek 2013). An abbreviated version of this historical narrative might proceed as follows. Prior to 1957, the majority of the intellectual and institutional scaffolding for a national project of scientific and technological advancement were in place. The intellectual rationale is typically sourced to Vannevar Bush's "Science the Endless Frontier," which in the words of Neal and colleagues outlines the, "foundation for modern American science policy" (Neal et al. 2008: 4). Bush's document provides the justification for the funding of basic science based on its role in driving

J. Schmid (✉)
Georgia Institute of Technology, Atlanta, Georgia, USA

 
M. E. Kosal (ed.), *Technology and the Intelligence Community*, Advanced Sciences and Technologies for Security Applications,
https://doi.org/10.1007/978-3-319-75232-7_3

innovation, which is described in the document as being critical to national economic welfare and post-World War II security.[1]

The early institutional framework for modern US S&T policy emerged during the late 1940s and early 1950s. In 1946, the Atomic Energy Commission (later the Department of Energy) and the Office of Naval Research were established. In 1951, the US Army established a research unit; a year later the US Air Force (USAF) did the same. In 1950, the National Science Foundation (NSF) was founded based on Bush's prioritization of basic science and his principal that scientists, rather than elected officials, should determine the direction of federally funded scientific research. However, according to prevailing account, prior to 1957, the pieces of what would become US S&T policy lay dormant. Large scale funding, political mobilization, and the adoption of Bush's vision would require an exogenous shock: the launching of Sputnik I.

Changes in levels of government funding for R&D supports this narrative. In 1935, US gross domestic expenditure on R&D (GERD) was only 0.05% and the government accounted for just 13% of R&D spending (Brooks 1996). By 1952, GERD has increased to over 1% and the federal government accounted for 60% of national R&D expenditure (Godin 2003). During the early 1960s (post-Sputnik), GERD reached 2.9% per year, a rate of R&D spending that remains the highest in the country's history. Finally, the NSF's budget was increased dramatically following Sputnik's launch; increasing from $40 million in 1958 to $134 million in 1959.

Besides increasing funding, Sputnik's launch is also argued to have initiated a period of institution and policy genesis. In 1958, the Advanced Research Projects Agency (later DARPA) and the National Aeronautics and Space Agency (NASA) were established. In the same year, the first special assistant to the president for S&T was appointed and the National Defense Education Act was passed. Indeed, DARPA is explicit in citing the contribution of Sputnik to its creation, "DARPA's original mission, established in 1958, was to prevent technological surprise like the launch of Sputnik" (DARPA: Bridging The Gap 2005: 1).

In short, Sputnik's launch is argued to have initiated a flurry of government activity to address a perceived shortfall in US scientific, technological, and military capacity vis-à-vis the Soviet Union. Less well known however, is that the launch of Sputnik resulted in significant innovation within the US intelligence community. Prior to 1957, the United States' capacity to collect and analyze intelligence on Soviet rocket and missile development was low. Examining the declassified intelligence products produced during this period reveals that the CIA possessed little information regarding the Soviet rocket and missile programs that would, within a four-month span beginning in August 1957, launch an inter-continental ballistic missile (ICBM), launch a 183-pound satellite (Sputnik I), and launch a half-ton satellite (Sputnik II). However, the US agencies charged with the gathering and analysis of Soviet capabilities in these areas responded relatively quickly following

[1] Bush is explicit in linking basic science to innovation, stating, "New products and new processes … are founded on new principles and new conceptions, which in turn are painstakingly developed by research in the purest realms of science" (Bush 1945).

Sputnik's launch. Within ten years, the CIA had developed significant novel capacity to gather and analyze imagery, electronic, and communications intelligence and had issued multiple intelligence documents that demonstrate an improvement in the agency's understanding of Soviet capabilities. Given that the "US intelligence infrastructure was in disarray" immediately following World War II (Gordin 2009: 80) and the newly formed CIA (established in 1947) had little capacity to gather intelligence regarding Soviet rocket and missile programs prior to Sputnik's launch, what explains this rapid improvement?

This article attempts to answer this question. Specifically, the sections that follow apply Barry Posen's model of innovation in military doctrine to the historical case of post-Sputnik innovation in US intelligence.[2] Towards this end, I utilize two sources of evidence: the historical research undertaken by other scholars over the years and primary source documents. To preview the results, the historiographic and documentary record indicate that Posen's theory of doctrinal innovation has substantial explanatory merit in the present empirical context. Namely, the US intelligence services' improved capacity to collect and analyze information regarding Soviet rocket and missile programs appears to have been initiated by a process of external auditing motivated by an increase in the level of threat posed by the USSR. That is, a change in the perceived balance of power caused by Sputnik's launch, spurred political scrutiny of the activities of the US intelligence community, which, in turn, led to innovation.

## 2 Posen's Model of Doctrinal Innovation

In *Sources of Military Doctrine* (1986), Barry Posen proposes a model of innovation in military doctrine based on civilian-military relations and the balance of power.[3] Posen defines military doctrine as the component of a country's national security strategy that determines *what* and *how* military means are employed towards the realization of the security priorities contained in a country's national security strategy (Posen 1986:13).[4] Innovation is defined as "large change" and is contrasted with

[2] Posen's model seeks to explain innovation in *military* doctrine. Here his model is applied to military and non-military organizations. However, these organizations are similar in that they possess certain traits that should make them (in the absence of external intervention) resistance to change. Specifically, Posen explains that militaries are "parochial, closed, large, endowed with all sorts of resources, and masters of a particularly arcane technology" (Posen 1986: 39). With exception of resources, these traits characterize the pre-Sputnik intelligence community.

[3] While the term doctrine is typically not used to refer to the activities of the intelligence services, the changes outlined here largely correspond to changes that would in a military context constitute doctrine. Specifically, the employment of novel means (e.g. imagery collection and analysis) to realize a stated end (understanding Soviet missile capacity) represents the kind of innovation that Posen aims to describe.

[4] Posen's definition of military doctrine innovation refers to a departure from the status quo and not necessarily an improvement. At any given time, a state's appropriate course of action might be

incremental change (Posen 1986:47). The relevant portion of Posen's causal logic can be summarized as follows.

Organizations such as militaries or intelligence agencies have, internal, equilibrating characteristics that promote organizational and doctrinal stasis. In the author's words, "organizations place a premium on predictability, stability, and certainty" (Posen 1986: 46). Innovation in this context is rare and is unlikely to originate from within the organization. When innovation does occur it is initiated by a source that is external to the organization. Posen cites the political leadership as the most common source of such auditing. However, external intervention does not occur randomly. Rather, civilian scrutiny of the military is initiated by a deterioration in a country's international security environment. That is, civilians audit the military during times of increased threat or, according to Posen, "anything that increases the perceived threat to state security is a cause of civilian intervention in military matters and hence a possible cause of integration of innovation" (Posen 1986: 79).

Posen's description of the process by which the British Royal Air Force (RAF) reoriented its doctrine immediately prior to World War II is illustrative of his model of innovation. During this period Britain's grand strategy was to dissuade aggression through the threat of a long war. Towards this end, Britain sought to maintain a large and protected industrial base and to defend the sea-lanes to her colonies. In sum, the British grand strategy was fundamentally defensive.

However, during the 1920s and 1930s the doctrine of the RAF was *offensive*, centered on executing a first strike. Implicit in RAF doctrine was that an enemy bombing campaign would meet little defensive resistance due to the impracticability of defending against such attacks. Thus achieving coherence between Brittan's grand strategy and the doctrine of the RAF required innovation.

In response to "increases in advisory capabilities and evidence of malign intent," in 1934 civilian officials, along with a handful of champions within the military, began to take positive steps towards bringing RAF doctrine and, the associated technologies, in-line with British grand strategy (Posen 1986: 166). In 1934, the Secretary of State for Air established the Committee for the Scientific Study of Air Defense (CSSAD), to investigate methods for detecting incoming enemy aircraft. Within three months of the CSSAD's formation, Robert Watson-Watt had demonstrated that radio waves could be used to detect aircraft. A few months later, a well-funded research institute was established. By 1939, the Chain Home radar system had been erected on the Sothern and Eastern coasts of Brittan.

The sections that follow apply Posen's model of doctrinal innovation to the case of the post-Sputnik changes in the US intelligence community. If Posen's model has

either stagnation (although Posen advocates a sort of intentional stagnation that is the result of careful deliberation) or innovation depending on the external and internal conditions facing the state. Because whether or not a given change will increase military effectiveness can only be determined following the change (i.e. once the change has been tested in the setting for which it is intended), I also use doctrinal innovation to refer to intentional and significant departures from the status quo.

explanatory merit, the following three propositional claims should hold. First, some demonstrable innovation in the doctrine of the intelligence community should be manifest. Second, the actors primarily responsible for change in the intelligence community should be located outside of that community. Third, the intervention of external actors should trace to a change in threat perception. The sections that follow indicate that each of these claims hold in the case of post-Sputnik innovation in US intelligence gathering on Soviet rocket and missile programs.

## 3 Pre-Sputnik Intelligence on Soviet Rocket and Missile Programs

Assessing what the US intelligence community knew of Soviet rocket and missile programs prior to Sputnik's launch requires some historical background regarding the state of these programs during this period of concern. Following World War II, the Soviets took up the task of rocket development in earnest. Immediately following World War II, the USSR established a Scientific-Technical Council for Rocket Development and the Zentralweke rocket development agency both of which sought to replicate German achievements in rocketry, especially the V-2 (Bulkeley 1991:60). Bulkeley explains that during the immediate postwar period, "High priority was given to the establishment of a nationwide complex of rocket research facilities" (Bulkeley 1991:60). In March 1947, Stalin personally took steps to establish a state commission for rocketry. Progress was rapid. By 1947, V-2 test flights had begun. Soviet progress in this scientific and technical realm did not heavily rely on German knowhow.

Awareness regarding these programs within the US intelligence community, however, was relatively low. Both primary sources and the work of other scholars attest to the paucity of reliable intelligence in the pre-Sputnik period. This shortcoming owes largely to a dearth of technical intelligence collecting systems during the 1940s and 1950s. During this period, the US lacked the radio stations and other technological resources required to conduct signals intelligence. For example, the AN/FPS-17 fixed antenna listening station at the Pirinçlik Air Base in Turkey that was eventually used to monitor Soviet missile test flights was not operational until 1955 (Bulkeley 1991:61–62). The National Security Agency (NSA) was not established until 1952 thus leaving the CIA bereft of a reliable means of gathering signals intelligence. The National Photographic Interpretation Center was not established until 1953. Besides the lack of technical means of intelligence gathering, the Soviet rocket and missile programs were not an intelligence priority until the latter part of the 1950s. In 1950, the CIA had only three analysts concentrated on Soviet missile intelligence (Prados 1982: 58).

The United States' lack of intelligence gathering resources during the late 1940s and early 1950s led US intelligence services to miscalculate the USSR's indigenous capacity to develop long-range missiles. A 1955 National Intelligence Estimate

(NIE) predicted that the Soviets would launch their first ICBM by 1960 (NIE 11–12-55*)*. In fact, the USSR achieved this feat in 1957, a few months prior to the launch of Sputnik I.

In other cases, errors were made in the opposite direction. In 1949, a Department of Defense technical evaluation group predicted that the Soviets would have long-range guided missiles by 1951–1952 (Gainor 2014). In a similar instance of overestimating Soviet capacity, in a 1955 US Air Force intelligence estimate, analysts incorrectly predicted that the Soviet Long Range Air Force would surpass US Strategic Command by 1960. In fact, the US held a substantial advantage in this area at the time. Such exaggerated estimates contributed to the myth of a US "bomber gap" and "missile gap" relative to the USSR and were used by supporters such as General Curtis Lemay of a pre-emptive strike against the USSR (Andrew 1998).

The lack of sound intelligence during this period is also evident in the commentary of those tasked with the external evaluation of Soviet programs. For example, the failure of the early intelligence community to gain access to useful information related to Soviet rocketry is evident in the records of the Guided Missiles Committee (GMC). In 1947, the GMC described the status of US intelligence on the Soviet missile program as follows, "it is evident that little or no direct knowledge of work being done at Russian guided missile test ranges can be obtained" (quoted in Gainor 2014: 43). Gainor summarizes the overall state of US intelligence during this time, stating, "Before embarking on their own ICBM program in 1954, decision makers in the US government had very little solid information on the state of Soviet missile programs" (Gainor 2014: 42). By 1950, Fred Darwin, executive director of the GMC was still concerned about the lack of US intelligence about Soviet missile technology and expressed a concern that the absence of intelligence made comparison of the US program with the Soviet program impossible (Gainor 2014). In 1952, the GMC also expressed concern about a lack of intelligence on Soviet surface-to-surface missiles.

In October 1953, the US Air Force (USAF) assembled a committee to evaluate three nascent missile projects: the Navaho (a rocket ramjet cruise missile), Snark (a cruise missile), and Atlas (an ICBM). The "Teapot Committee" (so named because of its code name) evaluated available intelligence and decided to prioritize the Atlas program. The committee's recommendations are useful in discerning the extent of knowledge in regards to Soviet missile technology in the pre-Sputnik period. As with the GMC, members of Teapot Committee complained about the lack of credible direct evidence related to Soviet space and rocket technology. Specifically, the committee writes, "The available intelligence data are insufficient to make possible a positive estimate of the progress being made by the Soviets in the development of intercontinental ballistic missiles. Evidence exist of an appreciation of this field on the part of the Soviet and of activity in some important phases of guided missiles which could have as an end objective the development by the Soviet of intercontinental missiles. While the evidence does not justify a conclusion that the Russians are ahead of us, it is also felt by the Committee that this possibility cannot be ruled out" (quoted from Rosen 1991: 214).

In the absence of technical means of monitoring Soviet activities, the American intelligence community sought to leverage human sources.[5] Specifically, it was hoped that defected German scientists that had worked in the USSR could fill in intelligence gaps. However, the information possessed by these individuals proved not to be current. Gainor explains that such scientists were unable to provide relevant information because they "had been separated from Soviet rocket work since the Soviets had succeeded in launching recovered Germany V-2s in 1947" (Gainor 2014: 43).

Besides a lack of high quality human intelligence sources, Gordin (2009) explains that during the immediate post-war period US intelligence services suffered from a lack of appropriate aircraft. Attempts during the early 1940s to capture images of the USSR by launching high altitude balloons form Europe (with the hope of collecting the descended balloons once they had drifted to Japan) failed. While the US and UK flew planes "crammed with electronic and photographic equipment" along the border of the Soviet Union, President Truman did not authorize a program of shallow flights over Soviet territory until 1950 (Bulkeley 1991:62). It was not until 1953 (seven years after its opening) that Western surveillance planes passed directly over the Kapustin Yar missile development site in Znamensk. Indeed, during the early 1950s the best images of the USSR that the US possessed were those taken during World War II by German reconnaissance planes. Gordin summarizes this early period, stating, "Much of the infrastructure now used to gather intelligence of any kind simply did not exist" (Gordin 2009: 82).

While the evidence provided above supports the contention that prior to Sputnik's 1957 launch, US intelligence on Soviet rocket and missile programs was relatively space, one intelligence document issued seven months prior to the launch of Sputnik appears, at first blush, to indicate otherwise. Specifically, a NIE titled "Soviet Capabilities and Probable Program in the Guided Missiles Field" informed the US government on March 12, 1957 that the Soviets would likely launch a satellite within a year. This document states: "The USSR will probably make a major effort to be the first country to orbit an earth satellite. We believe that the USSR has the capability of orbiting, in 1957, a satellite vehicle which could acquire scientific information and data of limited military value. A satellite vehicle possessing substantial reconnaissance capabilities of military value could probably be orbited in the period 1963-1965." However, a closer examination of the document raises additional question regarding the quality of the evidence on which this conclusion was made.

[5] On occasion, human intelligence was successfully used to overcome the shortage of technical means of assessing Soviet capabilities. The International Geophysical Year (IGY) was an international scientific research initiative (lasting from July 1, 1957 to December 31, 1958) in which Soviet and American scientists (as well as researchers from other countries) collaborated on a variety of scientific research projects. During this project, Hugh Odishaw, the head of the US National Committee for the IGY, required that all American participants send him any Soviet scientific documents that they may have obtained during the course of the collaboration (Bulkeley 1991:151).

The document – recently released under the CIA's historical review program – is explicit in acknowledging an intelligence shortfall vis-à-vis the Soviet programs: "Although some new intelligence has strengthened our previous estimate that the USSR has an extensive guided missile program, intelligence on specific guided missile systems continues to be deficit" (11-5-57: 1). Later in the document, analysts acknowledge a shortage of evidence regarding the Soviet ICBM program, stating, "We have no direct evidence that the USSR is developing an ICBM, but we believe its development has probably been a goal of the Soviet missile program" (11-5-57: 3) The report then provides an estimate for the timing of the technology's completion, "We estimate that the USSR could probably have a 5,500 n.m. ICBM ready for operational use in 1960-1961" (11-5-57: 3–4). This estimate of the timing of the completion of the first Soviet ICBM turned out to be incorrect. Indeed, only five months after the NIE was issued, the Soviets launched an ICBM that travelled 3000 miles. Thus while the March 12, 1957 NIE accurately predicts Sputnik's launch, it had little direct evidence on which to base this conclusion and made other predictions regarding related Soviet programs that were inaccurate.

In summary, while the March 12, 1957 NIE accurately predicted the launch of Sputnik, the overall quality of US intelligence about the Soviet rocket and missile programs during this period was low. Evidence for this claim comes from erroneous intelligence estimates, the frustrated testimonies of those tasked with evaluating the Soviet programs, and from secondary historical scholarship. However, beginning with Sputnik's 1957 launch, external scrutiny of the status of US intelligence on Soviet programs would increase. Increased scrutiny, in turn, would lead to improved intelligence. The next section traces the process of auditing that followed the satellite's launch.

## 4 The Launch of Sputnik I, the Heightened Soviet Threat, and Increased Scrutiny

At 11:56 pm on October 4, 1957, from the newly completed Baikonur rocket testing facility in Soviet *Kazakhstan,* Soviet engineers and military officers launched a three-stage R-7 rocket. The rocket, which generated one million pounds of thrust, delivered its payload, a 22 inch 183 pound aluminum sphere into low earth orbit. Tellingly, US was not monitoring for satellites when Sputnik I was launched (Dickson 2001: 11). Indeed, the satellite passed over the US twice before, the Moscow bureau of New York Times broke the news in the US (Bulkeley 1991: 3).

Sputnik's launch was followed in quick succession by the launch of the 1120-pound Sputnik II on November 3, 1957. Indeed, due to its weight, Sputnik II may have had a greater effect on US leaders' perception of the USSR as a potential threat to reaching the continental US with an ICBM (United States 1958: 8). This section will demonstrate that these events had two effects. First, the launch of these satellites increased the perceived military threat posed by the USSR amongst US political leaders. Second, the demonstration of Soviet rocketry capacity elicited an immediate political reaction. Specifically, the launch of Sputnik I and II resulted in high-level political scrutiny into the manner in which the intelligence services gathered and analyzed information pertaining to the Soviet rocket and missile programs.

The threat posed by the Sputniks' launch had little to do with the satellites and much to do with the rockets that sent them into orbit. Launching a heavy object into orbit required sophisticated, or at least powerful, rocketry technology, which suggested that the newly nuclear USSR was making progress at constructing rockets capable of reaching the US. In a 1958 article in *International Affairs*, Denis Healey, a member of the British parliament who would later become Defense Secretary of the UK, describes the security ramifications of Sputnik is stark terms, stating, "From the military point of view, the sputnik means that Russia has the capacity to produce a missile which is capable of carrying a thermonuclear warhead a distance of some five thousand miles in something like twenty minutes, and of guiding that missile with sufficient accuracy to destroy the Capitol building in Washington" (Healey 1958: 145).

The military consequences of Soviet rocketry capacity were also evident to members of the US Congress. According to Galloway, "the news struck Capitol Hill like a thunderbolt because thrusting the 184-pound satellite into outer space was evidence of the capability of launching intercontinental ballistic missiles, and therefore instantly perceived as a crisis for US national defense" (Galloway 2000: 209). Then-Senator Lyndon Johnson expressed his concern in particularly colorful terms, fearing that that the Soviets would, "Soon … be dropping bombs on us from space like kids dropping rocks onto cars from freeway overpasses" (quoted in Kuhn 2007: 12).

The threat to national security was also perceived in the executive office. Eleven days following the launch of Sputnik I, Vice-President Nixon delivered a speech to the International Industrial Development Conference in San Francisco in which he described Sputnik's launch as a "a grim and timely reminder of a truth that we must never overlook -- that the Soviet Union had developed a scientific and industrial capacity of great magnitude" (Nixon 1957: 2). A few months later this sentiment was echoed by President Eisenhower who on January 20, 1958 characterized the USSR space dominance as a direct military threat (Peoples 2008: 60).

Besides increasing the perceived threat posed by the USSR, the launch of the Sputnik satellites was immediately followed by increased high-level scrutiny into the activities of the intelligence community. Dickson characterizes the political reaction to Sputnik as "instantaneous" (Dickson 2001: 11). On October 11, 1957 the Senate Armed Services Committee requested that the Department of Defense provide them with a report "on the Soviet satellite and missile program furnishing all information available" (quoted Prados 1982: 64). Additional scrutiny came in the form of a December 1957 request for the CIA to compile a Special National Intelligence Estimate (SNIE).

The activities of the National Security Council (NSC) within the two years following Sputnik were dominated by the topic of Soviet rocketry. In January 1958,

Director of the CIA Allen Dulles personally gave the NSC a briefing on the Soviet programs. The topic of the USSR's missile programs was raised at least nine times at NSC meetings in 1958 (Prados 1982: 82). In the following year, there were ten NSC meetings dedicated to the topic of the Soviet programs and four "detailed exchanges" on this topic between the CIA and the President's Science Advisory Committee (Prados 1982: 85).

Congressional oversight of the CIA's post-Sputnik activities extended beyond hearings. Senator Symington, President Truman's Secretary of the Air Force, personally visited with the Board of National Estimates at the CIA's headquarters in order to learn the process by which NIEs were produced. This was the first time any senator had met with the Board of National Estimates on official business (Prados 1982: 83).

Scrutiny of intelligence estimates extended to the office of the President. Following Sputnik, President Eisenhower became increasingly interested in the process by which intelligence estimates were produced. In fact, in March 1958, Eisenhower described one CIA estimate as comparable to the work of high school students (Prados 1982: 78).

Post-Sputnik scrutiny extended beyond the work of the intelligence community. Of particular attention was the state of the US missile and space programs. Less than two months following the launch of Sputnik I, the Senate Armed Services Committee convened a hearing to assess the state of US aerospace capacity relative to the USSR. Specifically, on November 25, 1957, then-Senator Lyndon B. Johnson (Chairman of the Preparedness Investigating Subcommittee) began the "Inquiry into Satellite and Missile Programs." These meetings continued into January 1958 and the transcription of the various testimonies filled 2476 pages. Scrutiny into the status of the US programs also came from the US House of Representatives. On March 5, 1958 the Committee on Astronautics and Space Exploration was formed to evaluate the future of the US space program.

## 5 The Post-Sputnik Improvement in US Intelligence

In the period following the 1957 launch of Sputnik, the US intelligence services made significant improvements in their capacity to monitor and analyze the activities of the USSR. Specifically, during the immediate post-Sputnik period three novel intelligence resources came online: novel NSA Soviet monitoring facilities, the advent of a US spy satellite program, and an improved institutional capacity to interpret imagery intelligence. These novel resources were supplemented by an increased propensity for the individuals involved in the task of Soviet intelligence gathering to communicate across institutions.

One major reason for the post-Sputnik intelligence improvement is the actions taken by the NSA to gather additional electronic and communications intelligence on the USSR. During the late 1950s, the NSA established facilities in the UK, West Germany, Turkey, Japan, Italy, Greece, and Ethiopia. In 1958, a radar facility to monitor Soviet rocket launches was established on the Aleutian Islands (Prados 1982: 103). In 1959 a listening post with a similar aim was established in Peshawar, Pakistan. In 1960, the NSA established an additional facility in Iran.

Additional information regarding Soviet activities came from the United States' own reconnaissance satellite program. After various failed attempts to put a reconnaissance satellite into orbit, the Discoverer 13 and 14 satellites were successfully launched in August 1960. Their film capsules – the first manmade objects ever retrieved from space – were recovered later that month. The timing of the program's onset was fortuitous as the US stopped flying U-2 s over the USSR after a crashed plane, its camera, and pilot were recovered in Soviet territory in 1960. In fact, the satellite program produced higher resolution images than the U-2 flights and, at that point, there was no threat that a satellite would face Soviet counter-measures. The satellite program cost an estimated $1.3 billion from 1957–1960.

The post-Sputnik intelligence community also took actions to increase their capacity to analyze images. In January 1961, President Eisenhower established the National Photographic Intelligence Center within the CIA to provide expertise in the interpretation of incoming satellite images. Prados contends that these efforts to increase the capacity to interpret photographic images were largely successful, stating, "[b]y 1960 the essential of a massive intelligence-collection and-interpretation capability were in place" (Prados 1982: 110).

Finally, the post-Sputnik improvement owes, in part, to an increased willingness by individuals involved in the task of monitoring Soviet missile development to work across institutional boundaries. For example, during this period the chief of Soviet intelligence within the CIA's Deputy Director for Plans began to attend meetings of the Board of National Intelligence Estimates (Prados 1982). White House staff members with Soviet intelligence responsibilities attended these meetings as well.

Evidence of the post-Sputnik intelligence improvement is found in the increased sophistication of the CIA intelligence estimates produced during this period. For example, a declassified series titled the "Soviet Space Program," demonstrates the efficacy of the novel intelligence resources described above. Beginning in 1962, reports for this series were issued every two to three years until 1985. Each release of the series demonstrates an enhanced technical understanding of the Soviet programs.

Improved intelligence following Sputnik, however, was not immediate. In a December 1957 Scientific Intelligence Memorandum (SIM) the CIA estimates were imprecise. The SIM concludes that Sputnik III would deliver one of four payloads: a 160–300 lb. scientific earth satellite, a large satellite containing a live mammal, a 1000–5000 lb. reconnaissance satellite, or a lunar impacting payload (Future Soviet Earth Satellite Capabilities 1957: 1–3). While the satellite weighed 3000 pounds, the breadth of options provided by the SIM suggests that the immediate post-Sputnik CIA still lacked significant intelligence capacity. Similarly, the first post-Sputnik Soviet Space Program NIE, issued in 1962, still uses a strongly worded caveat regarding the conclusions drawn, "Our evidence as to the future course of the Soviet space program is very limited. Our estimates are therefore based largely on extrapolation from past Soviet space activities and on judgments as to likely advances in Soviet technology (11-1-62: 1)." Indeed, the content of this document primarily comes from publically available resources such as the official statements of the USSR.

By 1965 such caveats had been completely removed from the NIEs, and the technical sophistication of the estimates had been significantly improved. The twelve technical appendices in a 1965 NIE illustrate such an improvement (NIE 11–1-65). These appendices cover, in detail, topics such as Soviet scientific and technical capabilities for space flight, new propulsion and guidance systems, and tracking and communications systems.

In a 1967 Special Report titled "The Soviet Space Program Ten Years After Sputnik I," that was declassified in November 2006, the CIA demonstrates a post-Sputnik improvement in understanding of Soviet programs. In particular, the report notes that the Soviets had launched over 250 satellites in the ensuing ten years and correctly observes that the Soviet's had been able to exploit eight of the nine ideal Mars and Venus launch windows since 1960. The accuracy of these detailed observations demonstrates the extent to which the CIA expanded its intelligence gathering capabilities in the ten years following Sputnik.

Besides describing the scale of the Soviet space program, the Special Report shows a sophisticated understanding of the technical difficulty associated with certain Soviet accomplishments. Specifically, the report describes the Soviet's ability to capture images of the hidden side of the moon as "a brilliant achievement" (The Soviet Space Program 1967: 2). The report also describes the Soviet's frustrations and failures. In particular, the report observes that during this period nine Soviet attempts at interplanetary exploration failed to exit the earth's orbit and describes the Soviet frustration with "the fact that every probe put into an interplanetary trajectory suffered a communications failure prior to reaching its objective (The Soviet Space Program 1967: 2).

The 1967 report also proved prescient. In particular, the report predicted that the Soviets would attempt to launch probes to Venus during the January 1969 launch window. In fact, during this window the Soviets would deploy the Venera 5 (January 5, 1969) and Venera 6 (January 10, 1969) atmospheric probes which transmitted data on Venus' atmosphere (Harvey and Zakutnyaya 2011). The report also predicts the USSR's attempt at a manned lunar landing within the next five years (1968–1973). Indeed, the period in question witnessed a failed Soviet attempt to send manned spacecraft to the lunar surface (Hardigree 2010).

## 6 Conclusions

In their book on US science and technology (S&T) policy, *Beyond Sputnik: US Science Policy in the Twenty-First Century*, Neal et al. (2008) articulate the prevailing account of Sputnik as impetus for change, stating, "More than any other event in US history, the Sputnik crisis focused the attention of the American people and policymakers on the importance of creating government policies in support of science and of education, with the aim of maintaining US scientific, technological, and military superiority over the rest of the world" (Neal et al. 2008: 3). The preceding analysis has attempted to demonstrate that the result of this focused political attention extends beyond the domain of S&T policy to the intelligence community. In particular, it has attempted to determine whether Barry Posen's model of doctrinal innovation holds explanatory merit in the empirical case of the post-Sputnik improvements in the capacity of the US intelligence services. Towards this end, it is argued that Sputnik increased the perceived threat posed by the USSR. This increase threat led US policy makers to direct their attention to the United States' capacity to collect and analyze intelligence on Soviet missile and rocket programs. This external auditing resulted in improved intelligence estimates on the topic of Soviet rocket and missile capacity.

However, the importance of the improvement in US intelligence about the activities of the USSR extends beyond the theoretical. Gaining a more accurate picture of Soviet capabilities may have played a critical role in ensuring the Cold War remained cold. As described above, in the absence of sound empirical evidence, distorted understandings of the capabilities of a given adversary may prevail. Such distorted estimates – in either direction – can increase the possibility of conflict (Renshon 2009). For example, the myth of the Soviet "missile gap" from 1957–1961 is argued to have motivated both the Eisenhower and Kennedy administrations to propose larger defense expenditures than they would have in the absence of exaggerated perceptions of Soviet capabilities (Wenger 1997). Andrew concurs with the contention that improved intelligence may have decreased Cold War tension. In particular, Andrew suggests that if the immediate post-war dearth of intelligence had continued, the Cold War may have reached a heightened state, asserting, "If all presidents had possessed as little intelligence on the Soviet Union as Truman, there would have been many more missile-gap controversies and much greater tension between the superpowers" (Andrew 1998: 328).

## References


Andrew, C. (1998). Intelligence and international relations in the early cold war. *Review of International Studies, 24*, 321–330.

Brooks, H. (1996). The evolution of US science policy. In B. Smith & C. Barfield (Eds.), *Technology, R&D, and the economy* (pp. 15–48). Washington DC: The Brookings Institution.

Bulkeley, R. (1991). *The sputniks crisis and early United States space policy: A critique of the historiography of space*. Indiana University Press.

Bush, V. (1945). *Science, the endless frontier: A report to the president*. Washington, D.C.: Govt. Print. Off.

DARPA: Bridging The Gap Powered By Ideas. (2005). Defense technical information center. Accessed on 3 Jan 2017 https://archive.org/stream/DARPA--Bridging-The-Gap-Powered-By-Ideas--2005/ADA433949_djvu.txt.

Dickson, P. (2001). *Sputnik: The shock of the century* (First ed.). New York: Walker & Co.

Director of Central Intelligence, National Intelligence Estimate NIE 11–12-55. Soviet Guided Missile Capabilities and Probable Programs, 20 December 1955.
Director of Central Intelligence, National Intelligence Estimate 11–5-57. Soviet Capabilities and Probable Programs in the Guided Missile Field. March 12, 1957.
Director of Central Intelligence, National Intelligence Estimate 11-1-62, *The Soviet Space Program*, December 5, 1962.
Director of Central Intelligence, National Intelligence Estimate 11-9-63, Soviet Capabilities and Intentions to Orbit Nuclear Weapons, July 15, 1963.
Director of Central Intelligence, National Intelligence Estimate 11–1-65, *The Soviet Space Program*, January 27, 1965.
Director of Central Intelligence, National Intelligence Estimate 11–1-67, *The Soviet Space Program*, March 2, 1967.
Director of Central Intelligence, Weekly Summary Special Report, The Soviet Space Program Ten Years After Sputnik, October 6, 1967.
Director of Central Intelligence, National Intelligence Estimate 11-1-69,*The Soviet Space Program*, June 19, 1969.
Director of Central Intelligence, National Intelligence Estimate 11-1-71, *The Soviet Space Program*, July 1, 1971.
Director of Central Intelligence, National Intelligence Estimate 11-1-73, *Soviet Space Programs*, December 20, 1973.
Director of Central Intelligence, National Intelligence Estimate 11-1-83, *The Soviet Space Program - Key Judgments and Summary*, July 19, 1983.
Director of Central Intelligence, National Intelligence Estimate 11-1-83, *The Soviet Space Program*, July 19, 1983.
Director of Central Intelligence, National Intelligence Estimate 11-1-85J, *Soviet Space Programs, Volume I - Key Judgments and Executive Summary*, December 1983.
Director of Central Intelligence, Scientific Intelligence Memorandum (SIM), Future Soviet Earth Satellite Capabilities, December 1957.
Gainor, C. (2014). American intelligence on soviet missile programs, 1945–1954. *Quest: The history of spaceflight quarterly, 21*(3), 37–46.
Galloway, E. (2000). Organizing the United States government for outer space: 1957-1958. In R. D. Lanius, J. M. Logsdon, & R. W. Smith (Eds.), *Reconsidering sputnik: Forty years since the soviet satellite*. Routledge.
Godin, B. (2003). The most cherished indicator: Gross domestic expenditures on R&D (GERD). Project on the history and sociology of S&T Statistics Working Paper No. 22.
Gordin, M. D. (2009). *Red cloud at Dawn: Truman, Stalin, and the end of the atomic monopoly*. Macmillan.
Hardigree, Matt. (2010). "Inside the Soviets' Secret Failed Moon Program." *WIRED*. October 15. http://www.wired.com/2010/10/russian-moon-mission/
Harvey, Brian, and Olga Zakutnyaya. (2011). Russian space probes: Scientific discoveries and future missions. Springer science & business media.
Healey, D. (1958). The sputnik and western Defence. International affairs (Royal Institute of International Affairs) 34, 145–156. https://doi.org/10.2307/2606689.
Kuhn, Betsy. ( 2007). The race for space: The United States and the soviet union compete for the new frontier. Twenty-First Century Books.
National Photographic Interpretation Center. (1963). Central intelligence agency, NPIC/R-1567/63, Analysis of Soviet Manned Space Flight Launch Facilities, December 1963.
Neal, H. A., Smith, T. L., & McCormick, J. B. (2008). *Beyond sputnik: US science policy in the twenty-first century*. University of Michigan Press.
Nixon, Richard. (1957). Text of the address of the vice-president of the United States before the international industrial development conference, October 15, 1957. Accessed 6 Jan 2017 https://www.cia.gov/library/readingroom/docs/CIA-RDP80B01676R004200150002-0.pdf.

Peoples, C. (2008). Sputnik and "skill thinking" revisited: Technological determinism in American responses to the soviet missile threat. *Cold War History, 8*, 55–75. https://doi.org/10.1080/14682740701791334

Posen, B. R. (1986). *The sources of military doctrine: France, Britain, and Germany between the world wars, Cornell studies in security affairs*. Ithaca: Cornell University Press.

Prados, J. (1982). *The soviet estimate: US Intelligence Analysis & Russian Military Strength*. Dial Press.

Renshon, J. (2009). Assessing capabilities in international politics: Biased overestimation and the case of the imaginary "missile gap". *Journal of Strategic Studies, 32*, 115–147. https://doi.org/10.1080/01402390802407475

Rosen, S. P. (1991). *Winning the next war: Innovation and the modern military. Cornell studies in security affairs*. Ithaca: Cornell University Press.

United States. ( 1958). Inquiry into satellite and missile programs: hearings before the preparedness investigating subcommittee of the committee on armed services, United States Senate, Eighty-fifth Congress, first and second sessions ... U. S. Govt. Print. Off., Washington.

Wenger, A. (1997). Eisenhower, Kennedy, and the missile gap: Determinants of US military expenditure in the wake of the sputnik shock. *Def. Peace Econ., 8*, 77–100. https://doi.org/10.1080/10430719708404870

Yanek, M. (2013). *Eisenhower's sputnik moment the race for space and world prestige*. Ithaca: Cornell University Press.

# Organizational Process, Leadership, and Technology for Intelligence Gathering: Development of Photo-Reconnaissance Satellites in the United States

**Supraja Sudharsan**

**Abstract** This chapter explains the role of organizational leadership in the development of covert scientific programs. The chapter traces the development of Corona and Future Imagery Architecture satellite programs. It finds that, under similar threat conditions, the two programs, which were similar in many respects, led to different outcomes. This chapter reveals that existing explanations based on organizational process do not completely explain the outcomes of the two programs. Through process tracing, the chapter articulates the role that those in leadership positions played in the development of the Corona and FIA programs for intelligence gathering, thereby providing evidence for an explanation based on bureaucratic politics.

## 1 Introduction

Why did two scientific programs, both conceived in response to prevailing international threat environment, by similar organizations, lead to different outcomes? Historical narrative of the development of satellite-based photo-reconnaissance programs lead to explanations based on organizational process.[1] However, questions

---

[1] Allison (1971) uses the term "Organizational Process" in his seminal work titled, "Essence of Decision", in which he explores decision-making by then-White House Administration during the Cuban Missile Crisis. The organizational process framework views government action as the result of decisions made by its constitutive organizations. For instance, foreign policy decisions made by the White House administration are viewed as the consequence of interaction between its agencies such as the Department of Defense, the State Department, and the members from the Intelligence community. These departments employ standard operating procedures or routines that are fixed (Allison concedes that organizational change is possible. Organizations undergo change in response to budgetary changes or drastic failures in the operating environment). These agencies (or organizations), operating within their capacities and constraints provide outputs which in turn shape government behavior.

S. Sudharsan (✉)
Georgia Institute of Technology, Atlanta, Georgia, USA
e-mail: ssudharsan3@gatech.edu

 
M. E. Kosal (ed.), *Technology and the Intelligence Community*, Advanced Sciences and Technologies for Security Applications,
https://doi.org/10.1007/978-3-319-75232-7_4

remain as to why the organizations involved produced a "success" in the case of the Corona program that underwent development during the Cold War period and a so-called "failure" in the case of the Future Imagery Architecture (FIA) program that was proposed in response to terrorist threats in the 1990s. Which aspects of organizational process explain the difference in the outcome of the two programs? Who influenced outcomes in these organizations? And how did the organizational, technical, and security environment interact during the periods of study?

This chapter unpacks the influence of organizational processes on technology development for intelligence gathering. By tracing events leading up to the development of the Corona program during the period 1946–60 and the FIA program during 1999–2005, the article engages with the debate around how organizational processes influenced outcome in the development of technology for intelligence gathering. This is carried out by examining declassified documents from the Cold War period released by the National Reconnaissance Organization (NRO), and the Central Intelligence Agency, held in online repositories of the National Archives and Records Administration (NARA) and National Security Archive in George Washington University. In the absence of declassified documents on the FIA program, publicly available information, discourses by analysts, and comments from members of the Defense and Intelligence community are utilized to arrive at an explanation for the outcome of the FIA program.

The two case studies together help to draw a picture of the changes in technical, operational, and policy environment (in the United States and globally) that accompanied the development of satellite imagery technology for intelligence gathering. Analysis of the cases sheds light on the role of organizational leadership in covert technology development for intelligence gathering. It reveals that support of top leadership in constitutive organizations is a necessary if not sufficient condition for successful development of programs. Moreover, delineation of the technical, organizational, and policy domains during the periods leads to the conclusion that leaders have a role to play not only in development of technology but also to be cognizant of the need to restructure internally in response to changes across these domains.

## 2 Organizational Process and Technology Development for Intelligence Gathering: Concepts and Hypotheses

The organizational process framework includes several approaches to analyzing decision-making by firms and organizations (Allison 1971). A complete discussion is beyond the scope of this paper. However, a few important aspects of this model are worth noting. One, organizational outcome depends on rational choice considerations of risks and benefits. However, owing to limitations in the form of lack of access to complete information, organizational choice is constrained by the phenomenon of "bounded rationality" (Simon 1997). Second, the organizational process model assumes that organizations operate based on certain standard operating procedures and established routines that channel decision-making in specific ways. However, drastic changes in budgetary conditions and external environment may lead to organizational change. The discussion that follows evaluates arguments that utilize an organizational process model to explain differences in the outcome of the Corona and the FIA programs.

Dennis D. Fitzgerald, who was Deputy Director of the National Reconnaissance Organization beginning in 2001, considers a rational choice model in his account of events surrounding the Corona program. He approaches decisions made during this time in the context of tolerance to risk (Fitzgerald 2005b). He argues that tolerance for risk in the technical, operational, and financial realms in the National Security Council and the White House administration at the time resulted in continued support for the Corona program. He places the source of this tolerance for risk in the administration's uncertainty over Soviet missile capabilities during the Cold War period and claims that the question of "Bomber Gap" and "Missile Gap" reigned high in the Eisenhower administration.[2] The U-2 overflights were the primary source of intelligence during the 1950s, and the covert flights across Soviet landscape were politically risky. This in turn led to persistence over development of the Corona program. The administration accepted failure as an integral part of the endeavor and enabled the program managers to plough on despite a string of initial technological failures (Hall 1997). On the other hand, during the period 1999–2005, the NRO defined risk more narrowly in terms of program cost. As a result, fiscal constraints drove the FIA program, rather than the need to expand the boundaries of technology for intelligence needs (Fitzgerald 2005b).

The above discussion indicates that organizational definitions of risk shaped the choices made by the White House Administration during the development of the Corona and the FIA programs. Risk perception constrained the pathways available for decision-making and created a certain bounded rationality, which limited choices for the organizations involved. Other authors posit that organizational changes in the NRO determined the success or failure of reconnaissance programs.

The NRO, which was established as a covert organization in the 1960s to undertake satellite reconnaissance for intelligence gathering during the Cold War, was not a streamlined organization. During its initial years, it was organized as programs A, B, and C representing the Air Force, Navy, and the CIA, respectively. It's argued that competition between these three agencies for control over the program was conducive to the successful development of the Corona satellite Kohler (2005).

By the time that the FIA program was conceived, the internal organization of the NRO had encountered much change. While it continued to retain its hybrid identity, the previous division into programs A, B, and C had been abolished. Personnel were disengaged and shuffled across departments. It is posited that this resulted in poor accountability (Kohler 2005). Fitzgerald (2005a, b) and Taubman (2007) also argue that the organizational tolerance for risk led to greater connections between industry

[2] Bomber Gap and Missile Gap refer to belief during the Cold War period that the Soviet Union had outstripped the United States in the number and superiority of its aviation and missile development.

and leadership. This brought greater support for R&D activities and, consequently, contributed to the success of the Corona program. They claim that the cost-consciousness in the NRO during the 1990s did not allow for such partnerships, with repercussions in the form of breakdown of the FIA program.

The above account of the development of satellite reconnaissance programs lays out arguments in favor of the organizational process framework to study decision-making. They explicate how Defense and Intelligence agencies perceived threats to national security, and they argue that it resulted in routines and operating procedures that accorded a high tolerance for risk during the development of Corona satellites and a low tolerance during the development of the FIA program. From this, two types of hypotheses become evident for further testing, namely, that the Eisenhower Administration was more tolerant towards risk during the Cold War period (hypothesis 1), and that the NRO's evolution into a cost-conscious organization led to a low tolerance for risk in the later FIA program (hypothesis 2).

To test these hypotheses, this chapter takes a bureaucratic politics approach. This approach is chosen based on narratives on the development of the Corona program from declassified documents. Taking Allison (1971)'s discussion of bureaucratic politics as a point of departure, this article attempts to draw a picture of the role of top leadership that were involved in the development of the Corona and the FIA programs. This is in contrast to the black-box approach of the organizational process model that provides an output based on pre-determined operations and established organizational routines. Under the bureaucratic politics model, outcome is a result of competing interests and influence of officials involved in the decision-making process. This article considers the role of those in leadership positions who were directly involved in the development of technology for intelligence gathering. Through process tracing, the next section gives an overview of the development of Corona and FIA programs paying attention to the actors involved.

## 3 Corona and FIA Satellite Photo-Reconnaissance Programs: Case Studies

The development of satellites carrying a camera for reconnaissance purposes came during a crucial period in American history. The United States and the Soviet Union were involved in a Cold War. Tensions ran high between the two countries as the Soviet Union had tested an intercontinental ballistic missile (Johnson 2016), and the United States had no way of tracking their capabilities. Human intelligence regarding Soviet missile capabilities was minimal owing to the remote location and restricted access. Balloons in the early 1950s and U-2 reconnaissance aircraft in 1956 began to provide some intelligence. However, the limitation of area covered by aircraft reconnaissance, limitation of fuel, and finally, the shooting down of a U-2 by the Soviets sounded the death knell for this mode of intelligence gathering.

The Corona satellites (then code-named Discoverer) for photoreconnaissance of Soviet missile sites are said to have revolutionized intelligence collection in the United States. The images obtained from these satellites are credited with equipping the United States with the ability to monitor the Soviet Union's weapons capabilities through satellite imagery. The completion of Strategic Arms Limitation and Anti-Ballistic Missile treaties between the United States and the Soviet Union in the 1970s is attributed to the development of these satellites (Day et al. 1998). These satellites have since provided intelligence for monitoring compliance with international treaties, notable among them being the Nuclear Non-Proliferation Treaty.

The FIA program was conceived in 1999 as a covert, ambitious project to meet the needs of the much-changed threat landscape dominated by terrorist groups in the twenty-first century. It was conceived as a replacement for the Cold War era photo-reconnaissance satellites. A fleet of smaller-sized satellites that would enable frequent monitoring of targets over a larger geographic area and with a higher resolution to identify what are called as "entity-level" targets such as people and vehicles was proposed. However, contrary to the Corona program, FIA has been called "the most spectacular failure in the history of the intelligence community" (Taubman 2007). The Corona and the FIA satellite programs share several similarities. First, the two programs were conceived as the Defense and Intelligence agencies and policy-makers were grappling with the need to combat the dominant national security threats of their times, the Cold War and the War on Terror respectively. Second, both began as classified programs managed by National Reconnaissance Office (NRO), together with Advanced Research Projects Agency (ARPA), which was later renamed as Defense Advanced Research Project Agency (DARPA). Third, both programs faced several obstacles during the initial years of development. The first thirteen missions of the Corona program failed due to various technical glitches, while the FIA program was dismantled in 2005. Both the Corona and FIA programs overran budgetary allocations. Despite these similarities, one program is described as a revolutionary undertaking and a success, whereas the other is considered a failure. The next section explores this divergence further.

## 3.1 *Case 1: Development of the Corona Program (1946–1960)*

The Corona program began as a covert satellite-based photo-reconnaissance program, initially managed by the CIA and the Air Force. It was code-named as Discoverer, and it was publicized to the scientific community and the public as a program undertaken for space exploration. This section begins with a brief description of the technology and traces how this program came about in the first place. It identifies the most important factors that enabled the development of technology for intelligence gathering and explores the technical, operational and policy environment and associated lessons. The analysis that follows tests the following hypotheses:

H1: The White House Administration was more tolerant towards risk during the Cold War period
H2: The NRO's evolution into a cost-conscious organization led to a low tolerance for risk during development of the FIA program

The now-declassified history of Corona satellites released by the NRO in 1995 is utilized extensively for this account with corroboration against the CIA's account and other sources wherever possible.

The Corona satellite carried a camera that captured pictures from space as it passed over denied areas of the Soviet Union. Film from the camera was returned in a capsule. A parachute deployed to slow the descent of the capsule, and a C-119 J aircraft, and later JC-130 aircraft, seized the returned film in mid-air. The first thirteen missions were not successful, each due to a technical glitch in the system. However, the fourteenth mission carried out in August 1960, three months after the downing of the U-2 overflight of the Soviet Union, was successful, and sixty-four Soviet airfields and twenty-six new surface-to-air missile sites were revealed (Greer 1973).

The ideas behind the Corona satellite program originated in the years immediately following the end of WWII. RAND Corporation first drafted a report on satellite based photo-reconnaissance in 1946. In a report titled, "Preliminary Design of an Experimental World-Circling Spaceship," which was initiated by the US Air Force to compete with the Navy's activities in this area, a preliminary study of the feasibility of space technology was conducted. Its potential applications in guiding missiles, observation, and communication to meet the strategic interests of the United States were identified in a subsequent report in 1947 (Davies and Harris 1988). In 1953, the Air Research and Development Command were assigned management of the program. In the absence of cameras with sufficient resolution and the ability to retrieve them for analysis by photo interpreters, focus in the early 1950s was still on the use of balloons and bombers for reconnaissance. Eventually, the program was transferred to the Western Development Division headed by Major General B.A. Schriever. System studies were conducted, and Maj. Gen. Schriever's team proposed development of Weapon System 117 L to launch satellites into orbit in 1955. Contract to begin building the satellite system was awarded in 1958 (Greer 1973).

Approval for funding was slow to materialize during this time due to competing programs and interests. Funding was mostly concentrated on the Vanguard research satellite program and U-2 reconnaissance aircraft. In 1957, only \$13.9 million of the initially proposed \$39 million was sent to Corona program and this was later scaled down to \$24 million (Perry 2012). It is argued that the "Space for Peace" sentiment prevalent during the time discouraged the White House from approving funding for the covert reconnaissance program. International law regarding space overflights was not firmly established, and extant diplomatic skirmishes over previous covert aircraft overflights also contributed to indecisiveness to approve photo-reconnaissance satellites by the White House. Maj. Gen. Schriever faced other obstacles to securing funding during this time. Decision-makers in the Defense

Department considered it to be a preposterous proposal and a waste of time. Defense Secretary Charles Wilson and the Deputy Secretary of Defense, Donald A. Quarles, who had previously served as Assistant Secretary of Defense for Research and Engineering and then Secretary of the Air Force, were not supportive of the program. As Air Force Secretary, Quarles proposed that the department should focus on development of low-risk technologies called for cuts to so-called "non-critical" defense spending (Perry 2012). There was lack of support by leadership in the Office of the Secretary of Defense and the Department of Air Force.

However, Schriever continued to champion the program. He secured the support of retired Colonel Richard Leghorn, who had previously served as Chief of Reconnaissance Systems Branch in the Air Force and later in the Pentagon. Together, they advocated for investment in satellites by arguing that satellites would be much harder to take down compared to aerial overflights (Perry 2012). Leghorn's involvement and his support of reconnaissance for gathering peacetime intelligence on the Soviet Union's strategic assets gained the support of Dr. James R. Killian, Jr. who was the President's Science Advisor at this time and R.M. Bissell, who was Assistant to the Director of Central Intelligence (DCI), Allen Dulles (Davies and Harris 1988).

However, continued lack of support from the Pentagon leadership led Schriever to consider other avenues. He proposed that the United States create a cover story for the development of WS-1117 L as a scientific satellite under the International Geophysical Year or the "Space for Peace" program. This gained support from several influential people, including Dr. Edwin Land, a MIT Professor who was also on the Presidents Board of Consultants; Major General O.J. Ritland, who was Vice Commander of the Air Force's Ballistic Missile Division; Allen Dulles;[3] Dr. Killian; and Major General Andrew Goodpaster, Jr. who was serving as White House Staff Secretary and Defense Liaison Officer during this time having previously obtained a PhD in International Affairs from Princeton. Several of these people had already worked together in the development of U-2 reconnaissance aircraft. This may have been a contributing factor for their decision to work with each other in the satellite reconnaissance program. After ensuring buy-in from decision-makers and prominent people in leadership position with access to the administration, the Corona satellite program was finally assigned the highest priority by the White House administration in February 1958 (Perry 2012).

The Corona program emerged as a jointly managed program between the Department of Defense and the CIA (Perry 2012). The new Defense Secretary Neil McElroy established Advanced Research Projects Agency (ARPA), which was used to manage the satellite effort. The Ballistic Missile Division of the Air Force provided hardware and personnel for launching and tracking satellites, and the CIA defined and funded covert program objectives. Lockheed was chosen as contractor for integrating the system for launch, General Electric designed a capsule for recovery

[3] Even though DCI Allen Dulles was a proponent of using human intelligence, he supported both the U-2 and the satellite reconnaissance programs. While the reason is not directly evident, it may be speculated that it has to do with the difficulty of obtaining intelligence from the Soviet Union, and support of both of these programs by R.M. Bissell, who was Assistant to the DCI.

of the film payload, and Itek Corporation, which was spun out of Boston University and had Col. (ret) Leghorn on its team, developed the camera. The CIA managed the collection of intelligence and analysis through National Photographic Information Centers, which were previously used for analysis of photos from U-2 reconnaissance aircraft, and dissemination to decision-makers (Perry 2012).

The first thirteen missions between February 1959 and August 1960 did not succeed. In the fourteenth mission, payload film from space was returned, and the capsule was recovered successfully. This came at a time after the shooting down of U-2 and the capture of pilot. The camera had photographed 1.5 million square miles of the Soviet Union and East European countries. Between 1960 and 1972, 145 launches of Corona were carried out with various improvements to camera resolution and image retrieval technology (Davies and Harris 1988).

## 3.2 Case 2: The Future Imagery Architecture Program

Conceived to replace the outdated Cold War Era imagery satellites, the FIA program was designed to meet the intelligence needs of the military in a much-changed national security landscape, namely a threat environment dominated by non-state actors such as terrorists rather than a dominant state adversary such as Russia. This required additional capabilities for continuous surveillance rather than reconnaissance as in the Cold War era. Therefore, the conceptual design for the FIA program included launching many smaller-sized satellites intended to vastly increase the collection of imagery. It also required "entity level" tracking capabilities such as tracking people and vehicles rather than missile silos, spotting railroads, or roadways on fixed sites. Therefore, new requirements included satellites equipped with cameras of higher resolution (Day 2009). Moreover, the satellites combined both optical and radar capabilities in their design to gather intelligence through cloud cover during the day and at night (Flaherty and Hess 2007).

In the absence of declassified documents on this program, the decision-making process is reconstructed by studying publicly available accounts of the program and unclassified Congressional reports. The FIA program arose out of prodding by the House Permanent Select Committee on Intelligence (HPSCI), and a report was submitted by DCI John Deutch in 1996 to evaluate the feasibility of shifting to new imaging architecture. The report prepared by the Congressional Small Satellite Review Panel proposed the utilization of a large number of smaller-sized satellites that would be less expensive and also provide greater robustness and flexibility in operation by reducing the risk of being taken down and ability to replace immediately in case of failure. The report argued that the proposal would result in the incorporation of evolving technological advances from private sector on a staggered basis in subsequent satellites (House Permanent Select Committee on Intelligence 1996). The panel was headed by Robert Hermann, a former Director of the NRO, and included former NSA Director Lew Allen, former NRO Director Martin Faga, retired US Air Force General Larry Welch, scientists Dr. Sidney Drell and Dr. Ed Stone, and experts from industry. A later panel convened through the Defense Science Board Task in 1998 recommended that in addition to the imaging needs, the FIA program should include certain revolutionary aspects required by DARPA such as moving target indicator and rapid revisit image coverage (Office of the Secretary of Defense 1998).

The NRO worked with National Imagery Mapping Agency (NIMA) and six contractors to arrive at budget estimated at five billion dollars. Expecting the program to be the most expensive in the history of the intelligence community, Congress placed caps on funding in order to make sure that the program, which involved developing not only the satellites but also the supporting ground infrastructure and further integrating these systems in order to provide timely information, did not overrun proposed costs (Hess 2008). Boeing secured the contract by bidding against Lockheed, who had so far supplied all of the imaging spacecraft in (Taubman 2007).

However, by 2002, several issues began to arise. Boeing seemed to be far behind in its contract, and it began to be expected that the program would outrun its five billion dollar budget in 2002. Additional funding of 600–900 million dollars was added to bring the program back on track. In May 2003, a Joint Defense Science Board and Air Force Scientific Advisory Board concluded that primacy of cost considerations, erosion in government capabilities, and authority of program managers led to failure of space systems acquisition program. More specifically, the task force concluded that the FIA program was "technically flawed and significantly underfunded" (Berkowitz 2011).

By 2005, cost overruns had reached five billion dollars over the initial funding allocation of five billion dollars (Taubman 2007). Defense Secretary Donald Rumsfeld and Director of National Intelligence John Negroponte cancelled the program with Boeing, and components of the contract were distributed between Lockheed and Boeing. Lockheed focused on the development of electro-optical satellite and Boeing on radar (Hess 2008).

## 4 Bureaucratic Politics: Role of Organizational Leadership in Determining Outcome

The development of the Corona program indicates the prominent role played by leadership in championing the program, overcoming obstacles, and gaining buy-in from policymakers. Several people including Major General Schriever, Dr. Edwin Land, Colonel Leghorn, Dr. J.R. Killian, and R.M. Bissell seem to have been instrumental in securing buy-in from Secretary of Defense McElroy, CIA Director Allen Dulles and finally, the White House Administration. At this point, it is important to question whether in the absence of support from the above people, the Corona program would have moved forward. It may be argued that the shooting down of U-2 would have forced the administration to consider satellite photo-reconnaissance as a last ditch effort. However, it may not have been at the stage of development that it had achieved, in the absence of support from the above group of people. Moreover, process tracing in the previous section shows that Major General Schriever did not have adequate support to fund the development of the program from then-Air Force Secretary D.A. Quarles, or from then Secretary of Defense Charles E. Wilson, initially. Considering that the launch of Sputnik did not change their position on this issue, it is fair to conclude that in the absence of others championing the program, the Corona program may have stalled at a much earlier point of time.

How does organizational leadership explain failure of the FIA program? In other words, was there a lack of support from leadership among stakeholder organizations such as the CIA, NRO, the Air Force, and Office of Secretary of Defense, that may have contributed to failure of the program? Owing to lack of declassified information on the program, the examination is carried out based on publicly available information, with some speculative analysis of what could have been.

The failure of the FIA program does not seem inevitable. Delving into the role of people in leadership positions and their roles in the development of the program, it becomes evident that better engagement between these leaders could have salvaged the program possibly much earlier and with lesser financial commitment. For instance, even though DCI John Deutch brought together the committee that studied the need for smaller satellites in response to a request from the House Permanent Select Committee on Intelligence in the year 1996, he does not seem to have shared the same concerns regarding the need for smaller satellites. His priority for information distribution systems is evident from his speech at the ARPA symposium in 1996. Quoting from his speech,

> "There are also many visions of what the next step should be for our satellite architecture. It is often tempting to move to the next generation before the present generation has been successfully implemented and reached the end of useful life. The danger is that we take on too much technical risk and that we embark on a program that exceeds likely available resources…I am concerned that those outside the Intelligence and Defense Communities do not appreciate the importance of continuing the current program before moving to the next generation of technology. There needs to be a stronger appreciation of the time it takes to execute a program and the importance of sustained technical attention to program success" (Deutch 1996)

This provides some indication regarding the differing visions of future intelligence needs between decision-makers. Therefore, some of the blame for the failure of the program may be attributed to the lack of organizational leadership championing the program and driving it forward with policymakers. However, beneath this layer of explanation, it behooves to raise the question, why did leaders not engage effectively for the development of this particular technology? Several instances that demonstrate the position of stakeholders are discussed here in an effort to arrive at an answer to this question.

According to Robert Hermann, a former head of the NRO who also headed the panel that recommended the development of smaller satellites, "The FIA was technically flawed and un-executable the day it was signed…Some top official should have thrown his badge on the table and screamed, 'We can't do this system at this price.' No one did." (Taubman 2007).

Second, Congress imposed a cap on funding the FIA program in order to adequately fund other activities within the National Foreign Intelligence Program (House of Representatives 2002). The conference report on Intelligence Authorization Act for FY2003, authorizing President's budget request for funding for intelligence also observes that,

> "With respect to the nation's imagery architecture, the conferees are very concerned about the viability and effectiveness of a future overhead architecture, given the apparent lack of a comprehensive architectural plan for the overhead system of systems, specifically in the area of imagery. For example, the conferees believe the administration is facing a major challenge in addressing technical and funding problems with the Future Imagery Architecture (FIA) program that could force untenable trades between critical future capabilities and legacy systems." (House of Representatives 2002) Clearly, there is a mismatch in the positions held by leadership involved in the FIA program.

The lack of confidence evident from anecdotes is manifest in a comment made by Senator Diane Feinstein as Chair of the Senate Select Committee on Intelligence in 2009, where she refers to spending on the FIA program as follows:

> "We have extraordinarily serious concerns involving the waste of many, many dollars over a period of years and are rather determined it not happen again." (Senate Select Committee on Intelligence 2011)

Clearly, there is a discrepancy between what the different stakeholders thought about prevailing technology requirements, technical feasibility of the program, and how funding was allocated. However, a clear picture has not emerged of people in leadership that championed the program as was the case in the development of Corona satellites. Moreover, with the growth in sophistication of technology and their applications, where the NRO was involved in the launching and operation of one satellite reconnaissance program during the Cold War, in 1999 the NRO came to be treated as an umbrella organization for launching and operation of satellites for many different agencies and catering to the needs of many different users. At the same time that the CIA and NRO began to work on the FIA program, DARPA proposed the Starlite radar-imaging satellite that was also supported by the Air Force. The Air Force and DARPA were interested in the more sophisticated Starlite project while the NRO and the CIA, to some extent, seemed to be favoring the FIA program. The NRO became involved with both of these projects (Office of the Secretary of Defense 1998).

With the advent of commercial satellites, the United States' policy took a shift towards encouraging the development of US companies in this domain, thereby increasing the access to commercial imagery as well as resulting in images with good resolution from non-military sources (The White House 1994). Leadership in the Defense Department did not appear distressed by this. Quoting retired Vice Chairman of the Joint Chiefs of Staff, Marine Corps General James Cartwright, "The government's satellites are better, but the question is, What do you need? Most studies show that about 90 percent of what the military needs can be solved with commercial" (Risen 2012).

The above discussion illustrates a fundamental discord between the Congress' vision, the CIA's requirements, and the Defense Department's needs. In contrast to Congress which focused on intelligence collection systems, DCI Deutch demonstrated a preference towards better analysis and dissemination and going even further, the Defense Department focused on obtaining real-time data in its theater of operations.

## 5 Conclusions

The cases of Corona and FIA satellite programs lead to three conclusions. On the first hypothesis, that the Eisenhower Administration was more tolerant towards risk during the Corona program than the FIA program, the study reveals otherwise. The discussion reveals that leadership from participating agencies is required to champion a program in order to see it to completion. The Corona case study is evidence that the role played by leaders and decision-makers in the Air Force, CIA, and scientific advisors was crucial in the eventual approval by the White House. Consider the first hypothesis that the Administration had a high tolerance for risk during the Corona program. Certainly, once the White House approved the Corona program, it was tolerant of the initial string of mission failures between 1959 and 1960. However, the events leading up to it portray a different story. As seen in the discussion of the Corona program earlier, lack of confidence about performance of technology and uncertainty in international law about space overflights across territories of other countries resulted in some hesitation on the part of the administration to approve of the Corona program. Eventually, as the program gained support from people in leadership positions across the Defense and Intelligence communities namely, US Air Force Major General B.A. Schriever, Assistant to DCI R.M. Bissell, Dr. Edwin Land in the President's Board of consultants, Dr. Killian on the administration's Science Advisory Board, and later DCI Allen Dulles, and Secretary of Defense McElroy, they were able to convince the White House to support the program. Without the role played this motley group of people, the Corona program may not have commenced.

The second hypothesis that there was a low tolerance for risk in the NRO stemming from changes in its organizational process is not supported by the study either. Tolerance for cost overruns was low in Congress. This is evident from the funding cuts imparted to these programs by Congress. This is explained through anecdotal evidence in previous sections. However, a larger problem is identified in this study based on bureaucratic politics, and that is a lack of common vision among decision-makers. The role of organizational leadership is not immediately evident in the case of the FIA program, but it should not be ignored. Paying attention to how leaders from stakeholder organizations did *not* act also provides some insights into their importance for the development of the program. Even though the FIA program was proposed in response to threats in the form of terrorist attacks, and was managed by the NRO, CIA, and Air Force similar to the Corona program, the leaders in these organizations did not share similar goals. Reflecting the arguments posited by a bureaucratic politics model, the study also reveals that there were divergent views about the usefulness of the program among top leadership in the Defense and Intelligence communities and in Congress. Similarly, while access to leadership and competition among agencies were all important during different points of time, they do not completely explain the outcome of the two programs. The study also finds that rational choice is a valid framework in which to consider the cases of Corona and FIA programs. However, rationality can be extended to the level of the bureaucracy (organizational leadership in this study), in terms of interests of leaders that influence decision-making.

There are several limitations in this study, which ought to be noted. The article is constrained by lack of declassified information on the FIA program. Is organizational leadership more important than organizational process in determining outcomes? While the study tests an alternative approach, namely, bureaucratic politics to study technology development for intelligence gathering, it is limited by the number of cases under consideration. There are several covert technology development programs for intelligence gathering via satellite reconnaissance, as well as through other modes of intelligence collection. A broader sample of cases would enable scholars to go beyond cases that are tied to the NRO and test generalizability of findings.

## References

Allison, G. (1971). *Essence of decision* (1st ed.). Boston: Little, Brown and Company.

Berkowitz, B. (2011). *A brief history of the national reconnaissance office*. Chantilly, Virginia: National Reconnaissance Office.

Davies, M., & Harris, W. (1988). *RAND's Role in the Evolution of Balloon and Satellite Observation Systems and Related US Space Technology* (pp. 6–8). Santa Monica: RAND.

Day, D. (2009). *The space review: Gum in the keyhole*. [online] The Space Review. Available at: http://www.thespacereview.com/article/1400/1. Accessed 12 May 2016.

Day, D., Logsdon, J., & Latell, B. (1998). *Eye in the sky* (1st ed.). Washington, D.C.: Smithsonian Institution.

Deutch, J. (1996). The future of the National Reconnaissance Program. In *In: ARPA Tech '96 18th Systems & Technology Symposium*. Atlanta: Defense Technical Information Center. http://www.dtic.mil/docs/citations/ADM001169 [Accessed 4 May 2016

Fitzgerald, D. (2005a). Commentary on "the decline of National Reconnaissance Office" – The NRO leadership replies. *Journal of the Discipline and Practice, 2005*(*U1*), 45–50.

Fitzgerald, D. (2005b). Risk Management and National Reconnaissance From ·the Cold War Up to the Global War on Terrorism. *Journal of the Discipline and Practice, 2005*(U1), 9–18.

Flaherty, A., & Hess, P. (2007). *US plans new multibillion dollar spy satellites program* (p. 1). Associated Press.

Greer, K. E. (1973). Corona. *Studies in Intelligence*. [online] Supplement (17) (pp. 1–37). Available at: https://nsarchive2.gwu.edu/NSAEBB/NSAEBB90/dubious-13b.pdf. Accessed 4 May. 2016.

Hess, P. (2008). *Congress cancels novel satellite program* (p. 1). Associated Press.

House of Representatives. (2002). Intelligence Authorization Act for Fiscal Year 2003. [Conference Report] Federation of American Scientists, 2002 Congressional Reports. Available at: http://fas.org/irp/congress/2002_rpt/hrpt107-789.html [Accessed 4 May 2016].
House Permanent Select Committee on Intelligence. (1996). *Report by the DCI's Small Satellite Review Panel.* [online] Federation of American Scientists. Available at: http://fas.org/irp/offdocs/smallsat.htm [Accessed 4 May 2016].
Johnson, K. (2016). Corona and Spy Satellites. [online] The Cold War Museum. Available at: http://www.coldwar.org/articles/60s/corona.asp [Accessed 4 May 2016].
Kohler, R. (2005). One Officer's Perspective: The Decline of the National Reconnaissance Office — Central Intelligence Agency. *Journal of the Discipline and Practice, 2005*(U1), 35–44.
Office of the Secretary of Defense. (1998). *Defense Science Board Task Force on Satellite Reconnaissance.* [online] US Government Publishing Office. Available at: https://www.gpo.gov/fdsys/granule/FR-1998-01-23/98-1581/content-detail.html [Accessed 4 May 2016].
Perry, R. (2012). *History of satellite reconnaissance.* Chantilly, VA: Center for the Study of National Reconnaissance.
Risen, J. (2012). A military and intelligence clash over spy satellites. *The New York Times*, p. A13.
Select Committee on Intelligence. (2011). Report of the Select Committee on Intelligence. [online] Senate Report 112–3 from the Government Publishing Office. Available at: http://www.intelligence.senate.gov/publications/report-select-committee-intelligence-covering-period-january-3-2009-january-4-2011
Simon, H. A. (1997). *Administrative behavior: A study of decision-making processes in administrative organizations* (4th ed.). New York: The Free Press.
Taubman, P. (2007). Death of spy satellite program, lofty plans and unrealistic bids. *The New York Times*, 11.
The White House. (1994). Presidential Decision Directive/NSC-23. [Memorandum] George Washington University, National Security Archive, NSAEBB404. Washington, D.C.

# Exploring the Contributing Factors Associated with Intelligence Failures During the Cold War

**Jenna K. McGrath**

**Abstract** The goal of this chapter is to assess "intelligence failures," as defined by existing literature, in order to determine whether the failures can be attributed (at least in part) to a failure or inadequate use of technical-based or non-technical based intelligence gathering methods. Two case studies are considered in this analysis: the placement of Soviet missiles in Cuba in 1962 and the collapse of the Soviet Union. The results of the analysis conclude that while there are some technology-based intelligence failures evident in each case study, the bulk of the failure can be linked to an oversight or misuse of non-technical intelligence gathering methods. However, most striking is the pre-existing notions within the United States intelligence community that can be linked to how and why the community overlooked critical human, cultural, economic, and ethnic-based intelligence. Therefore, the analysis concludes with the suggestion that the two case studies are considered "intelligence failures" due to the internal failures of the intelligence community. Biases and pressure to deliver conclusions according to the political administration's preferences continue to impact the analyses coming from the US intelligence community. How the US intelligence community should address this failure remains unclear and requires future research that would be strengthened by the inclusion of additional case studies.

**Keywords** Intelligence failures · Cold War · Case study analysis · Process tracing

## 1 Introduction

When considering the intersection of technology and intelligence gathering techniques and how each influences the other, it is evident that technological advancements will continue to shape the nature of the intelligence community and how the

---

J. K. McGrath (✉)
Georgia Institute of Technology, Atlanta, Georgia, USA
e-mail: mcgrathj@gatech.edu

 
M. E. Kosal (ed.), *Technology and the Intelligence Community*, Advanced Sciences and Technologies for Security Applications,
https://doi.org/10.1007/978-3-319-75232-7_5

various agencies within the community process and analyze data. However, it is not exactly clear how vital the role of non-technical intelligence gathering methods remain in light of the advancing technology. Is there a risk of being over-reliant on technology-based intelligence gathering, where satellite images, big data, and drone survallience overshadows human intelligence? These questions are evolving and cannot yet be answered in full. However, understanding how technology has evolved, merged, and co-existed with traditional forms of intelligence gathering methods can provide valuable insight when thinking about these types of questions from a historical standpoint. Therefore, the Cold War era, from 1947 to 1991, lends itself to be an ideal time period to examine due to both the major technological breakthroughs during that time period, as well as the heavy reliance and resources put into the intelligence community in the United States.

This chapter specifically seeks to examine where in the intelligence gathering process the so-called "intelligence failures" during the Cold War originated from. Are "intelligence failures" attributed to insufficient or broken technology-based intelligence gathering methods (such as signal, image, and measurement intelligence)? Or are these "failures" a result of inadequate or insufficient intelligence gathering methods that do not rely on technology (such as human, cultural, and ethnic intelligence)? Two case studies of "intelligence failures" will be evaluated in this paper, starting with the placement of Soviet missiles in Cuba in 1962 and ending with the collapse of the Soviet Union in 1991. It should be noted that this chapter is not trying to argue for or against categorizing these two case studies as "intelligence failures" and instead accepts that this is how existing literature describes the two events (Betts 1978; Diamond 2008; Knorr 1964). The goal is instead to determine what, if any, of the intelligence gathering processes employed during the two events might have contributed to the ultimate intelligence failures.

To answer this, first brief background information about the intelligence gathering techniques is provided (both technical and non-technical) and relevant information regarding the American intelligence community during the Cold War is discussed. The next section will describe the two hypotheses being evaluated, followed by the method of analysis (process tracing). The following results sections will first present the results examining the placement of Soviet missiles in Cuba as a case study, second present the results from the Soviet Union collapse case study, and third, an aggregate analysis of the two case studies. In light of the aggregate of the case study results, I present a third hypothesis for intelligence failures during the Cold War and re-assess the aggregated results. The subsequent discussion considers how the final conclusions have already impacted the intelligence community and what they might mean for the future of the intelligence community.

## 2 Background

Different agencies within the intelligence community tend to focus on a specific intelligence gathering method (often referred to as INTs) more so than others but in general, intelligence gathering methods can be utilized across agencies (Kennedy 2008). Technology-based intelligence gathering methods include signals intelligence (SIGINT), telemetry intelligence (TELINT), imagery and geospatial intelligence (IMINT and GEOINT), measurement and signature intelligence (MASINT), and open source intelligence (OSINT). The National Security Agency mainly relies on SIGINT methods, including the subsets of communications and electronics intelligence (COMINT and ELINT, respectively) to monitor movements of aircrafts, ships, and other military aspects, as well as voice, email, telephone, and satellite data (Kennedy 2008). Intelligence from IMINT and GEOINT methods is gathered primarily through the National Geospatial-Intelligence Agency and analyzed via aerial surveillance and images, usually obtained through satellites, aircrafts, or drones (Kennedy 2008). Agencies within the Department of Defense, such as the Defense Intelligence Agency, obtain weapons testing and capability data through TELINT methods and MASINT information ranging from biological to nuclear, seismic to acoustic signals data to track and/or locate targets (Kennedy 2008). Lastly, many agencies use OSINT to gather information from publically available resources, such as the internet, newspapers, radio, and unclassified documents (Kennedy 2008).

Non-technical intelligence gathering methods include human intelligence (HUMINT), ethnic intelligence (ETHINT), and cultural intelligence (CULINT). The agencies that depended most on these forms of intelligence during the Cold War included the CIA and the FBI, though as stated previously, the INTs are not limited to one specific agency. Intel gained through HUMINT sources includes information from spying operations, information from political, academic, or other foreign sources within the target region, and even information gathered from refugees or defectors of the region (Kennedy 2008). Information from ETHINT requires a detailed understanding of how various societies, tribes, and groups function and relate to each other. From an outsider perspective, ETHINT can be especially difficult to obtain because the societies needing to be observed may be "invisible to us unless we are specifically looking for them; they come in forms with which we are not culturally familiar" (Renzi 2006). Though CULINT can be more straightforward in data presentation, such as economic and demographic data, training is also still required to understand how and why the culture being studied behaviors and acts a certain way, such as in social and political decisions (Coles 2006).

Intelligence failures can be linked to a variety of causes, such as a failure to deploy, utilize, or analyze the appropriate intelligence gathering methods. However, intelligence failures are also dependent on how the decision makers tasked with understanding the incoming intelligence ultimately act. As Betts writes in his 1978 paper examining intelligence failures, strategic intelligence is the "acquisition, analysis, and *appreciation* of relevant data" (Betts 1978). Betts notes that the intelligence community functions under a decentralized model, where each agency has its own focus and operation model that includes the procurement, analysis, and critical assessment of the intel. As with any model of operation, there can be breakdowns, errors, and failures within any part of the system. Considering both technical and non-technical intelligence gathering methods, how might the various intelligence gathering methods play a role in an intelligence failure?

## 3 Hypotheses and Methodology

To try to answer the above question, the following two hypotheses will be tested, keeping in mind that the goal of the analysis is to determine how intelligence gathering methods might have played a role in two pre-established "intelligence failure" case studies.

Hypothesis 1: Cold War "intelligence failures" are attributed to a technology failure within the intelligence gathering method.
Hypothesis 2: Cold War "intelligence failures" are attributed to a non-technological failure within the intelligence gathering method.

When examining both hypotheses, the failure relating to technical and non-technical intelligence gathering methods can occur during collection, the analysis of the information, or in the decisions and conclusions made regarding the intel. For technical intelligence gathering methods such as SIGINT, MASINT, TELINT, OSINT, IMINT, and GEOINT, a failure might be a breakdown or inadequate use of the technology, or a misunderstanding of the intelligence gathered. For HUMINT, CULINT, and ETHINT intelligence, a failure could be in the lack of deployment, emphasis, or execution of one of these methods, as well as a misunderstanding of the intel.

Testing the two hypotheses requires a qualitative assessment, rather than using a quantitative analysis, such as statistics, due to the lack of data available to adequately assess the research question at hand. For this chapter, process tracing is used because it offers a set structure to follow for the analysis of a research question based on historical information, particularly one involving case studies. Here, the informative article by David Collier of the University of California, Berkley is used as a guide to process tracing (Collier 2011).

The key steps for answering a research question with process tracing is to first identify the research question, paying attention to a specific field, region, or period of study. In this case, how does the changing make-up of intelligence gathering methods (technical versus non-technical) impact the intelligence community? To narrow the scope of the research, as recommended in the process tracing model, this field of study is limited to the US intelligence community during the Cold War. The next step in Collier's guide calls for testable hypotheses, with the hypotheses here are aimed to address how intelligence gathering methods played a role in previously determined "intelligence failures." To assess the hypotheses, process tracing recommends using case studies as an example in order to emphasize the details of the analysis, provide a narrative, and to provide a timeline of events.

Given the larger scope of the research question, the testable hypotheses, and the case studies, process tracing attempts to find a causal mechanism to determine and explain results. However, without quantitative data, it is difficult to draw concrete conclusions and reject or accept the null hypothesis. Therefore, a variety of conclusions are acceptable for process tracing, including: confirmation of the hypothesis; affirmation that the hypothesis is relevant, but not necessarily conclusive; an elimination of the hypothesis; and a weakened, but not yet eliminated, hypothesis (Collier 2011). The last step in process tracing is to suggest other causal factors or challenge the results of the analysis. For this chapter, this last step is critical in gaining a better understand as to how and why intelligence gathering methods (both technical and non-technical) played a role in the two intelligence failure case studies.

## 4 Case Studies and Results

Borrowing from statistical modeling, the following two case studies will be considered individually first, but then aggregated as well. In order to gain a comprehensive understanding of the scenarios surrounding each case study, both successes and failures relating to technical and non-technical intelligence gathering methods are considered. However, this analysis is not trying to prove or disprove the classification from existing literature that these case studies are "intelligence failures."

### *4.1 Case Study I – Soviet Missiles in Cuba*

The first case study investigated in this analysis is the Soviet placement of missiles in Cuba. This "intelligence failure" came on the heels of the 1961 failed Bay of Pigs operation. In the end of August and early to mid-September of 1962, there was a noticeable uptake in Soviet activity in Cuba, including military ships, military personnel, and new construction happening on the island (Knorr 1964). Around this time, US intelligence community began receiving reports that the Soviet Union had brought missiles to Cuba. Despite this information, it was not until October that reconnaissance aircraft began to fly over Cuba to look for any signs of the missiles. On October 14th, the reconnaissance flights confirmed Soviet missiles as well as the sites around the area where the missiles were going to be deployed (Knorr 1964). Table 1 provides a summary of the results of this case study, including the consideration of both successful and unsuccessful technical and non-technical intelligence gathering methods, the details of which are explained below.

An external investigation, known at the Stennis Report, into the missile crisis a few years later sought to determine why the US intelligence community was slow to act on confirming the presence of missiles on the island, despite the visible

**Table 1** Results of Case Study I – Soviet Missiles in Cuba

| | H1. Technology-based methods | H2. Non technology-based methods |
|---|---|---|
| Intelligence Success | Reconnaissance aircraft confirmation of missiles (10/14/1962) (IMINT) | Sightings and reports of the missiles were at least reaching the intelligence community (HUMINT) |
| | Confirmation of missiles being offensive before they were operational | |
| Intelligence Failures | Failure of NSA's SIGINT methods to detect any movement/presence of the missiles prior to the U2 reconnaissance images | IC was questionable of Cuban refugee and exile reports of missiles (HUMINT) |
| | IC didn't consider the increase in Soviet military personal, aircrafts, and weaponry in Cuba a threat for approximately a month (09/1962–10/14/1962) (IMINT, GEOINT, SIGINT) | IC had preconceived ideas about what was feasible for Soviet military policy, and missiles in Cuba was inconceivable (CULINT). Therefore, IC did not "give proper weight" to the situation (Knorr 1964) |

increase in Soviet activity in Cuba at that time, as well as the first-person reported sightings or knowledge of missiles. One issue was the intelligence community's hesitance to investigate the reported sightings or first-hand knowledge of the missiles. The Stennis report revealed, however, that it was not entirely unreasonable for the intelligence community to be slow acting in response to this information and there were actually a number of reasons for being wary. First, the reports of missiles came from Cuban refugees and exiles, some of whom might have had ulterior motivations for reporting such a sighting, such as the hopes of spurring conflict between the US and the USSR. Second, the intelligence community was questioned the accuracy of the reports given that the descriptions of the missiles varied across reported sightings. Furthermore, the intelligence community questioned that civilians would have the prior knowledge to accurate recognize military missiles (Knorr 1964).

The larger issue with the intelligence community's failure to act in this scenario is instead tied to pre-conceived notions the agencies had for what the Soviet Union may or may not do in regard to Cuba. It has been suggested that the US did not anticipate the USSR taking such an aggressive stance in Cuba (Westad 2000). The Stennis report concluded that the intelligence community's main constraint in adequately analyzing the information gathered was that it thought it was not possible that the USSR would consider putting missiles in Cuba. Meaning the placement of missiles went against the intelligence community's ideology and expectations of how the USSR would behave (Westad 2000). The intelligence community had to carefully balance their reactions to incoming data in order to not overreact and therefore become less sensitive to warning signs, but also still adequately consider viable threats (Betts 1978; Knorr 1964). This case study indicates that the US intelligence agencies underreacted given the incoming data (and reports suggest there might have been too much information to sort through in a short period of time) (Betts 1978).

This case study isn't made up entirely of failures, however. Once the US decided to investigate the incoming intelligence further, the reconnaissance aircraft image-based intelligence data proved invaluable and accurate indicate missiles and the areas where they would be placed on the islands (Knorr 1964). The analysis of the images also enabled the US to determine that the missiles were present in Cuba before they became operational (Knorr 1964). Though these successes in technology intelligence gathering techniques are commendable, it should also be noted that the other methods, such as SIGINT, did not pick up any information about the transport of the missiles to Cuba, a process that takes some time and planning.

## 4.2 *Case Study II – The Collapse of the Soviet Union*

There was an abundance of information being collected by the US intelligence community during the Cold War, particularly in the later years. As a result, the intelligence agencies, such as the CIA and NSA, had a difficult time analyzing and responding appropriately to much of the information being gathered (The National Security Archive 2008). Despite the influx of information, the intelligence communities were very successful in their analysis of understanding the extent and details associated with Soviet military weapons and capabilities (Diamond 2008). Furthermore, in terms of successes related to intelligence in the Cold War, it must be recognized that the national security community's goal of preventing a nuclear war between the US and USSR, as well as slowing the spread of Communism, was realized (Diamond 2008).

Intelligence agencies weren't the only interested party in assessing the threat of the Soviet Union; academics, journalists, and researchers studied the USSR as well. During the final decades of the Cold War, the popular opinion amongst the intelligence agencies was that the Soviet Union was strong and getting stronger (Diamond 2008). While some independent researchers agreed, others were starting to point out discrepancies in the gathered data that suggested weakness, rather than strength.

For example, Lieutenant General William E. Odom, the Senior advisor on President Carter's National Security Council, and his boss, Zbigniew Brzezinski, both studied and wrote about the weakening Soviet economy from 1969 all the way through the late 1980s (Diamond 2008). Similarly, Andrei Amalrik and Randall Collins wrote about ethnic conflicts and separatists being angry with the Soviet Union in the late 1960s and early 1970s (Hopf and Gaddis 1993). In 1970, Amalrik wrote that a growing part of the population was becoming disenfranchised with the Soviet Union's communist system and wanted democratic reforms, as did the elites (Diamond 2008).

Other researchers were finding that there was an increase in ethnic and cultural clashes in the Soviet Union during the later years of the Cold War. These clashes, along with other demographic information regarding an increase in alcoholism amongst adults, an increase in infant mortality, and a decrease in life expectancy, were indicators (based on non-technical intelligence) that the Soviet Union was struggling internally (Collins 1995; Diamond 2008). Moreover, the Soviet Union was isolated from the rest of the world, in part by its own doing. This isolation resulted in the Soviet Union being separated from the growing capital coming from western nations and their allies, such as Japan, and this was having detrimental impacts on the Soviet economy (Westad 2000). Despite the lagging economy, the USSR was still increasing military spending; an anomaly seemingly missed by the US intelligence community as an indicator of a weakening nation (Diamond 2008).

This, however, does not quite account for the oversight associated with highly publicized and surprising remarks made by Gorbachev in December of 1988. On December 7th, Gorbachev said in an address to the United Nations that the USSR would be reducing their military by half million troops over the next two years, especially in Eastern Europe. The reduction was in an effort to improve international relations and reduce militarization in the area. This announcement took the newly inaugurated US president, George H.W. Bush, and his administration by surprise. The CIA apparently had provided no warning of Gorbachev's announcement, only learning that he was going to be making statements regarding troops in Europe from a Washington Post article earlier that day (Diamond 2008). This was particularly embarrassing for the CIA given that the director, Robert Gates, had stated two months earlier that "the dictatorship of the Communist Party remains untouched and untouchable" (Diamond 2008). However, the embarrassment was short-lived because in 1989, less than a week after the Berlin Wall came down, the CIA issued a memo stating that they were concerned about a Warsaw Pact attack (Diamond 2008).

On December 25, 1991, Gorbachev officially announced that the USSR was dissolving. This announcement left the US government and intelligence agencies blindsided, with many being angry that the CIA did not anticipate the collapse. After all, the CIA's main purpose and more than half their budget for the past several decades was focused on assessing the Soviet threat (Diamond 2008). Some even called for the demolishment of the CIA, such as like Senator Daniel Patrick Moynihan, a Democrat from New York who became an outspoken critic of the CIA (Diamond 2008) (Table 2).

## 5 Aggregated Results

When considering the initial hypotheses in light of the results of the two case studies, a few key conclusions can be drawn about technology-based intelligence failures compared to non-technology-based failures.

For the first case study, Soviet missiles in Cuba, there were indeed technical intelligence method failures, such as detecting the movement of the missiles to Cuba prior to their arrival. However, it was a technological success that the reconnaissance aircraft detected the missiles before they were operational. Therefore, the hypothesis that this "intelligence failure" can be attributed to a failure in technical intelligence gathering techniques cannot be confirmed. Conversely though, the

**Table 2** Results Case Study II – Collapse of the Soviet Union

| | H1. Technology-based methods | H2. Non technology-based methods |
|---|---|---|
| Intelligence Success | IC was very successful/accurate at understanding the extent of Soviet's weapons & capabilities (SIGINT, MASINT, IMINT, GEOINT) | There was never a nuclear war between US & USSR |
| | | Expansion of communism was slowed/contained |
| Intelligence Failures | IC too focused on assessing military capabilities, troop movements, weaponry counts, etc.<br>Even after Gorbachev said there would be troop reductions (especially in Europe), and Berlin wall came down, CIA still feared a Warsaw pack attack | Didn't connect economic implications of decreasing GDP & increasing military spending (CULINT) |
| | | Didn't consider increasing regional & ethnic conflicts as a problem (ETHINT, CULINT) |
| | | Demographers pointed to increase in infant mortality, increase in alcoholism, and decrease in life expectancy as indication of trouble (CULINT) |
| | | Didn't consider Gorbachev a revolutionary figure before or after he rose to power (HUMINT) |
| | | CIA disregarded Soviet elite's unrest & call for reforms (HUMINT, CULINT) |
| | | Didn't consider Gorbachev's policy decisions as a whole – More so individually instead (economic reforms; giving increasing amounts of power to regional leaders) (CULINT) |

hypothesis that the "intelligence failure" is directly related to non-technical intelligence gathering methods can be confirmed. This is predominantly based on the failure of the intelligence community to put together multiple pieces of information to conclude that the abnormal increase in Soviet activity in Cuba, plus first-person accounts of the missiles, meant that it was a possibility that the USSR was acting in an unanticipated manner.

In the second case study, the collapse of the Soviet Union, the US intelligence community demonstrated that the intelligence gathering methods based in technology were advanced, accurate, and reliable. However, the intelligence community's focus on the technology to provide them with all relevant information on the USSR's activities was misguided. Therefore, as with the previous case study, the hypothesis that this "intelligence failure" can be attributed to a failure of technology is weakened, but not eliminated. Similar again to the previous study, the hypothesis that the "intelligence failure" can be attributed to inadequate use of non-technical intelligence methods can be confirmed. This is because the intelligence community, particularly the CIA, failure to connect important demographic, cultural, ethnic, and economic information with unrest in the USSR.

**Table 3** Aggregated Results from Case Study I and II

| | H1. Technology-based failures | H2. Non Technology-based failures |
|---|---|---|
| Case study 1: Soviet missiles in Cuba | Though there were some technology failures, IMINT did successfully detect offensive missiles before they were operational. | Relative dismissal for HUMINT sources, though slightly understandable. More so, a failure to adequately consider CULINT. |
| | Result: Hypothesis is relevant, but can't be confirmed | Result: Confirms the hypothesis |
| Case study 2: Soviet collapse | IC was successful in their use of tech-based methods, but was arguably too focused on this. | IC (especially the CIA) failed to connect economic data, demographic data, cultural/ethical unrest with a larger problem within the soviet union. |
| | Result: Hypothesis is weakened, but not eliminated | Result: Confirms hypothesis |

When viewing the case study's results together, it becomes clear that as technology and its capabilities improve, it is important for the intelligence community to continue to emphasize non tech-based intelligence in order to adequately understand and assess the entire scenario. The intelligence community must understand why the adversary is making these decisions and then respond appropriately, even if the activities are not what were anticipated (Table 3).

## 6 A Revised Hypothesis for Cold War Intelligence Failures

Given the results of the aggregated analysis of the two case studies, a common theme appears: pre-existing notions of how and why the Soviet Union will react was a predominant issue for the US intelligence community. Therefore, perhaps the biggest contributing factors to these "intelligence failures" was the internal and external structure and culture of the US intelligence institution/organizational. In both case studies, the intelligence community (in particular the CIA) were unwilling to consider alternatives outside their pre-existing scenarios and philosophical ideas of the Soviet Union's policies. This is more of an internal failure within the community rather than a non-technical-based intelligence method failure. With this in mind, a new hypothesis perhaps more accurately explains these "intelligence failures."

*Hypothesis 3: Cold War "intelligence failures" are attributed to institutional failures within the US intelligence community.*

Considering this revised hypothesis paired with existing research on the subject, it becomes evident that the intelligence community was indeed treating the Soviets like a threat, but it was also doing this in part because the agency was under a lot of pressure. This pressure came from superiors and leadership within the agency as well as from the presidential administration. Throughout the Cold War, the intelligence community maintained straightforward analyses that were continuing to view the USSR as a threat but not suggesting anything radically out of the ordinary in regard to this threat (Diamond 2008). The agencies, such as the CIA, knew what the White House wanted and expected to hear in reports and briefings, but what they wouldn't want to hear was also known. It would be a tarnish to the CIA's reputation to be alarmist and predict something perceived as counter-intuitive (like a breakup of the USSR) only to be incorrect (Betts 1978).

In the 1980s, it was even becoming difficult for academics to publish research that went against the dominant thinking and the intelligence community's analyses (Diamond 2008). There had been discussion, studies, and suspicion from a few academics, politicians, and even some within the intelligence community that had suggested the USSR might be running into trouble. The main problem, as said by then-director of the CIA's Soviet Analysis Office, Douglas MacEachin, is that the CIA was treating the USSR "almost entirely as a threat rather than as a 'political entity' with vulnerabilities that could lead to the kind of transformation that Gorbachev was bringing about" (Diamond 2008). Though this quote paints the CIA as misguided about the Soviet threat, the CIA was actually being rather level-headed in their assessments when compared to the Pentagon's Defense Intelligence Agency (DIA), which had inflated the perceived Soviet threat over the years, however it is suspected this was a method mostly to obtain a larger defense budget (Diamond 2008).

In addition to overestimating the Soviet threat, the US intelligence community continued to be flawed in the sense that they were reluctant to deviate from a predetermined path regarding the Soviet Union's anticipated actions, responses, and policies. This is evident in the Cuban missile case study, as well as the collapse of the Soviet Union case study. In the case of the latter, the intelligence community believed that the communist system was just fundamentally flawed and would collapse on its own eventually though not any time soon (Diamond 2008). While they were correct in predicting the flaws a communist system might have within the USSR, the intelligence community failed in understanding Gorbachev as a revolutionary figure. The intelligence community didn't expect Gorbachev to make the decisions he did in trying to change the communist system and share the power of the USSR with the regional leaders, which then allowed the regional leaders to have enough power that the resulting breakup of the USSR was rather smooth.

Robert Gates, Director of the CIA during that time, admitted that he was slow to respond to the unanticipated policy decisions being made by Gorbachev (Diamond 2008). In 1986, Gates did try to open "new lines of inquiry" as to what was happening within the USSR, but by then it was too late and the breakup of the nation was imminent. Gates conceded that the CIA wasn't open to enough alternative scenarios during the late 1970s and through the mid-1980s, and this contributed to why the agency seemed to miss many important indicators of Soviet unrest. Furthermore, the agency was too focused on Soviet military and weaponry capabilities that the economic and cultural aspects of the nation were overlooked, including the new reforms Gorbachev began to make when he took power (Diamond 2008).

Given this assessment into the underlying reasons for the "intelligence failures" during the Cold War, the revised hypothesis seems to be strengthened, that "intelligence failures" at least in part can be attributed to internal failures of the intelligence community. However, a further analysis and case studies should be considered from this time period before a definitive confirmation and other conclusions are made.

## 7 Impacts on the Intelligence Community Going Forward

Given the assessment of intelligence failures during the Cold War, it is necessary to question whether the intelligence community's structure allows for a non-biased analysis of gathered information, regardless of whether the information was obtained through technical or non-technical means. Will the current administration always have an overarching agenda that influences how the intelligence community analyzes data and makes conclusions?

Existing literature seems to suggest that this has been and will continue to be an issue for the intelligence community. Garicano and Posner suggest in their 2005 paper that there is a "year man" mentality in the intelligence community where the intelligence agencies deliver the kind of information they know their superiors want to and expect to receive (Garicano and Posner 2005). This is in line with the assessment made by Betts almost three decades earlier. Betts argued that intelligence failures can be a result of *failure in perspective*, meaning that although there will be mistakes and errors in any model, misguided efforts and motivations can increase opportunities for failures. And unfortunately, just one mistake or mis-assessment in the intelligence community can sometimes have very grave consequences (i.e., nuclear war) (Betts 1978).

Hence, despite the variety of intelligence being gathered and even an appropriate analysis of the information, the intelligence community can still draw incomplete and incorrect conclusions from the data. This can be attributed to a bounded rationality within the internal structure of the intelligence community, where the organization relies upon existing rational behavioral models and pre-existing ideas of how the adversary may act and react to a given situation (Simon 1997). Unfortunately, if an adversary does not act in a rational way, or if the rational behavioral models have a fundamental flaw, an organization can be surprised and caught off guard (Knorr 1964; Simon 1997).

If the intelligence community's internal operating systems and structures are flawed, does this put the nation at risk for additional intelligence failures? Existing literature suggest that an organizational structures based on a hierarchical balance can lead to more problems with providing non-biased analysis and decision making (Garicano and Posner 2005; Meyer and Rowan 1977). This is evident in the conclusions made in the 2002 National Intelligence Estimates of Iraq's Continuing Program of Weapons of Mass Destruction. The National Intelligence Estimates report concluded that aluminum tubes photographed by the intelligence community were centrifuges to be used for nuclear weapons, despite the Department of Energy rejection of this hypothesis stating that the tubes "were not well suited for a centrifuge application" (Garicano and Posner 2005).

## 8 Conclusions

As this study of "intelligence failures" during the Cold War has demonstrated, the US intelligence community has been subject to suffering a number of failures in their intelligence gathering process. The failures can be linked to technical failures of their information gathering methods, but more likely the failure comes from an underestimate or a lack of emphasis on non-technical intelligence gathering methods, such as from cultural, human, and ethical sources. Most pertinent to the analysis, however, is the suggestions of an alternative hypothesis to attribute to why these intelligence failures occurred. This new hypothesis suggests that it is internal failures of the intelligence community's structure and bounded rationality that leads to biased analysis.

The intelligence community has a culture predisposed to following the lead of the current administration and providing intelligence analysis and conclusions that meet the expectations of the administration. Unfortunately, this issue on predetermined notions is not new. As Betts states in his 1978 paper, "The use of intelligence depends less on the bureaucracy than on the intellects and inclinations of the authorities above it" (Betts 1978). This issue has persisted beyond the years of the Cold War and was observed again as the US entered the Iraq War. How the intelligence community should set about addressing the continuing problem within their structure remains unclear and requires further analysis. Intelligence failures do not have to define the agencies, but an acknowledgement of where and how the failures came to be can help set the stage for internal improvements.

## References

Betts, R. K. (1978). Analysis, war, and decision: Why intelligence failures are inevitable. *World Politics, 31*(1), 61–89. Web.

Coles, J. (2006). Incorporating cultural intelligence into joint doctrine. *Joint Information Operations Center*, 1–13. http://www.au.af.mil/info-ops/iosphere/iosphere_spring06_coles.pdf

Collier, D. (2011). Understanding process tracing. *Political Science and Politics., 44*(4), 823–830.

Collins, R. (1995). Prediction in macrosociology: The case of the soviet collapse. *American Journal of Sociology, 100*(6), 1552–1593.

Diamond, J. (2008). *The CIA and the culture of failure: US intelligence from the end of the cold war to the invasion of Iraq* (pp. 19–45). Stanford: Stanford Security Series. Print.

Garicano, L., & R. A. Posner. (2005). Intelligence Failures: An Organizational Economic Perspective. Working paper no. 5186. London: Center for Economic Policy Research. Discussion Paper Series. Industrial Organization and Public Policy, 6. <www.cepr.org/pubs/dps/DP5186.asp>.

Hopf, T., & Gaddis, J. L. (1993). Getting the end of the cold war wrong. *International Security, 18*(2), 202–210.

Kennedy, R. (2008). Of knowledge and power: The complexities of National Intelligence. Praeger Security International. 8–9. Print.

Knorr, K. (1964). Failures in National Intelligence Estimates: The case of the Cuban missiles. *World Politics, 16*(3), 455–467.

Meyer, J. W., & Rowan, B. (1977). Institutionalized organizations: Formal structure as myth and ceremony. *Journal of Sociology, 83*(2), 340–363.

Renzi, F. (2006). *Networks: Terra incognita and the case for ethnographic intelligence*. Military Review (September–October) (pp. 16–22).

Simon, H.A. (1997). "The psychology of administrative decisions." Administrative Behavior. Ch. 5.

The National Security Archive. (2008). "National Security Agency releases history of cold war intelligence activities." National Security Archive Electronic Briefing Book No. 260. http://nsarchive.gwu.edu/NSAEBB/NSAEBB260/

Westad, O. A. (2000). The new international history of the cold war: Three (possible) paradigms. *Diplomatic History, 24*(4), 551–565.

# The Dragon Lady and the Beast of Kandahar: Bush and Obama-era US Aerial Drone Surveillance Policy Based on a Case Study Comparison of the 1960 U-2 Crash with the 2011 RQ-170 Crash

**Leah J. Ruckle**

**Abstract** Little is known about current surveillance policy regarding unmanned aerial vehicles (UAVs), also known as aerial drones. Surveillance objectives, technologies, and policies are largely classified; however, occasionally there are glimpses into the programs. The 2011 RQ-170 crash in the Islamic Republic of Iran is one such glimpse. Upon comparison, that crash shows a number of similarities with the 1960 U-2 crash in the USSR in terms of the international political environment prior to the incident, the US agencies involved, the functional characteristics of the aircraft, the surveillance target, and the functional means of downing the aircraft. Fortunately, many details concerning the CIA U-2 program and crash are now declassified. By using the U-2 program as an analogy, this chapter explores a number of hypotheses concerning current US aerial drone surveillance policies.

## 1 Introduction

The 1960 downing of the U-2 aircraft in the USSR and the 2011 downing of the RQ-170 in Iran were both highly publicized exposures of American aerial surveillance programs during their respective eras. The two incidents, despite being separated by over half of a century, feature a number of similarities. These similarities include the international political environment prior to the incident, the US agencies involved, functional characteristics of the aircraft, the surveillance target, and the functional means of downing the aircraft.

The similarities between the two incidents suggest that American approaches to aerial surveillance have changed very little since the Eisenhower administration. The technology has certainly evolved, with the development of stealth technology,

L. J. Ruckle (✉)
Georgia Institute of Technology, Atlanta, Georgia, USA
e-mail: leruckle@gatech.edu



M. E. Kosal (ed.), *Technology and the Intelligence Community*, Advanced Sciences and Technologies for Security Applications,
https://doi.org/10.1007/978-3-319-75232-7_6

more advanced sensors, and unmanned aerial vehicles, but the policies and uses of these vehicles do not appear to have not appreciably changed.

Based on these similarities and the corresponding hypothesis that US aerial surveillance policy has not changed significantly over the past five decades, additional hypotheses can be drawn about the current US policies concerning unmanned aerial surveillance by drawing analogies to past aerial surveillance policies. These hypotheses concern topics such as the targets of US aerial surveillance drones, program oversight, program origin, and the impact on next-generation technology development.

## 1.1 Basis of Case Study Comparison & Fundamental Hypothesis

Surveillance programs, like many intelligence-gathering missions, are secretive enterprises by their nature. The sole object of the mission is to gather information without being detected by the target. In order to make this possible, as much information about the program, including the technology, budget, program management, surveillance targets, and operators is as hidden and obfuscated as possible. The United States' current covert aerial surveillance program is no exception.

However, at some point, details about surveillance programs and technologies are often revealed. This can occur either through voluntary disclosure, which usually occurs after the technology has outgrown its useful life, or through involuntary disclosure, usually in the form of discovery by the target.

Currently, there is a paucity of information on US UAV reconnaissance policy. The vast majority of information concerning policy focuses on the use of UAVs such as the MQ-1 Predator and the MQ-9 Reaper, to execute targeted attacks in the Middle East, Afghanistan, and Pakistan. However, even the disclosure given on these programs has attracted criticism for being far too scant (Stohl 2014, 2016). There is even less on information on surveillance use of UAVs.

However, there is more information concerning the US's Cold War aerial reconnaissance programs as a result of program declassification. The CIA's U-2 program, which ended in 1974, is one of these programs, and it is arguably the most notable. There are many valuable resources available, including previously classified documents, from which a detailed account of the program can be drawn. By comparison, information about current drone programs must, by and large, be gleaned from clues left in newspaper articles, anonymous interviews, FOIA requests, and budgetary disclosures.

This information about the U-2 program is valuable, not only in its own right, but also in the context of the current opaque nature of US UAV surveillance policy. This work compares the U-2 and RQ-170 programs on the basis of a number of categories: aircraft functional characteristics, program security and oversight, involved government agencies, missions prior to and after the crash, political environment prior to and after the crash, means of aircraft downing, and impact on future technology development. Some topics, such as the aircraft functional characteristics, political environment prior to and after the crash, and means of downing[1] are relatively well known.

On the other end of the information spectrum, there is almost no information concerning topics such as the program origin or aircraft missions prior to or after the crash. The information that is available is sometimes little better than speculation by journalists and bloggers who specialize in secretive military aviation programs. Officially, the US government has said very little about the aircraft, its use, or the crash in Iran. Statements from government sources have either been vague or anonymous. The lack of information is not particularly surprising - a stealth program becomes much less effective if everyone knows about it, and the intelligence community is particularly unforthcoming about its endeavors.

However, this makes the known information all the more important. The best illuminated topics concerning the RQ-170 revolve around the December 4, 2011 crash in Iran. Using the crash as a focal point, the similarities to the U-2 program, which also suffered a crash within the first decade of its use, come into sharper focus. For instance, first, we see that both aircraft were brought down by functionally similar technologies – namely recently-developed "hot topic" technologies of their day. Second, just prior to each crash the US was faced with a looming and unresolved question regarding the target country's nuclear-related technological advancement,[2] and in both cases the target was a state popularly considered, at the time, to be America's greatest enemy. Third, in both cases, there were strong indicators that countermeasures had caught up and that the aircraft were vulnerable. Finally, if the scope of the comparison is enlarged to other relatively well-known program facts, both programs were/are jointly administered by the US Air Force and the Central Intelligence Agency (CIA).

These similarities in the two programs, in addition to the thirst for aerial surveillance intelligence products which was awakened by the U-2 program but not deterred by its crash, support this chapter's fundamental hypothesis. That hypothesis, stated in the introduction, is that today's RQ-170 aerial surveillance program is functionally similar enough to the CIA U-2 program that the older, now- declassified, program can be used as an analogy to make better-educated hypotheses regarding lesser-known aspects of the RQ-170 program such as the missions prior to and after the crash, program origin, and impact on future technology development.

[1] There is still some debate, however, whether Iran successfully brought down the drone by hacking the GPS system or whether the drone malfunctioned at an inopportune time.

[2] In the case of the USSR, this was the development and alleged massive production of Intercontinental Ballistic Missiles (ICBMs) capable of carrying nuclear warheads to the United States, which was referred to as the:Missile Gap." In the case of Iran, it was the International Atomic Energy Agency (IAEA) report that Iran's nuclear program was showing signs of a weapons program.

## *1.2 Outline*

This chapter is arranged roughly chronologically in order to facilitate the reader's understanding of each program's "story." First discussed are the aspects that are more or less time-independent: aircraft functional characteristics, involved governmental agencies, and program security and oversight. These are followed by the pre-crash topics, such as the program origin, early missions, and pre-crash political environment. Finally, there are the crash-related and post-crash topics such as the means of downing, post-crash political environment, and impact on future technological investment.

Organizing the topics chronologically does have the disadvantage of placing many of the topics supporting the fundamental hypothesis (i.e., those directly related to the crashes) later in the discussion. However, this is offset by the enhanced clarity provided by a chronological approach since the logical decision-making progression throughout the programs is more readily apparent with this approach.

# 2 General Program Characteristics

## *2.1 Aircraft Functional Characteristics*

Surveillance aircraft, by design, must elude detection by enemy radar, or, failing that, at least be capable of flying through a hostile airspace without being brought down by enemy actions. The U-2 and RQ-170 both accomplish these aims using the "stealth" technology of their era and share a number of functional similarities. Although the two aircraft appear physically different, they are both designed to achieve the same functional capabilities. Namely, they are both (1) difficult to detect (by the means of their era), (2) have long endurance, (3) carry a sensor package capable of spying on the intended target, (4) are unarmed, and (5) feature a "careful" pilot selection.

### 2.1.1 U-2

The U-2 Dragon Lady is a high-altitude, unarmed, manned reconnaissance aircraft built by Skunk Works at the Lockheed Aircraft Corporation. It was designed by the legendary aircraft designer Clarence "Kelly" Johnson in 1954. It achieved first flight in 1955 and its first operational flight in 1956. In an effort to hide the fact that it was used for reconnaissance, the "U" designator for "Utility" was chosen, though neither the Soviets nor many others were fooled.

The aircraft was designed to fly with at high altitude (over 70,000 feet) for long periods of time.[3] This high cruising altitude was designed to be above Soviet radar ceilings and above Soviet air defenses, thus rendering the aircraft effectively stealthy, or at least difficult to shoot down.[4] In a further effort to avoid detection, missions were carried out in radio-silence, particularly while conducting Soviet overflights.

Its sailplane-like aerodynamics (high aspect ratio wings) and extremely low-weight construction give it a long endurance, about 12 hours, which is about the limit for human pilots, particularly on an aircraft as tricky to fly as the U-2. The high endurance enabled deep overflights of the USSR and long missions aimed at gathering large quantities of intelligence.

The aircraft, in a break with previous military aircraft specifications, had no guns and was quite fragile. Its sensor package was a high-altitude camera designed specifically for the program. It carries only single pilot, who is also responsible for manning the cameras. During the Cold War, overflights of the USSR and other politically sensitive countries were done by pilots who had officially "retired" from the USAF and were employed by the CIA, making them civilians rather than military personnel. This was done in an effort to reduce the political impact if a pilot was captured in enemy territory.

### 2.1.2 RQ-170

The RQ-170 Sentinel is a stealthy, unarmed, unmanned reconnaissance aircraft also built by Skunk Works at Lockheed Martin. Its body shape, which is similar to the much larger B-2 stealth bomber, strongly suggests that it is a stealthy aircraft, but very little information about the technical specifications are known.

Military aviation experts estimate there to be about "twenty or so" aircraft (Axe 2012). In 2009, aviation expert Bill Sweetman from *Aviation Week* dubbed it "The Beast of Kandahar" when the first photographs of the drone, taken in 2007 in Kandahar, Afghanistan, were first released by the French magazine *Air & Cosmos* (Hambling 2009).

The "RQ" designation, if representative, indicates that the UAV does not carry weapons and only conducts surveillance operations. According to the US Air Force, it is flown by the 30th Reconnaissance Squadron (United States Air Force 2009).

In an article on the RQ-170 after it was officially revealed by the US Air Force, Fulghum and Sweetman note, "visible details that suggest a moderate degree of

[3] Research done for the program in pilot life support systems and pressure suits would prove to be valuable in the space program.

[4] Unfortunately, the estimate of Soviet radar technology capability was based on the false assumption that the Soviets were still using American-built radars acquired in WWII. In reality, the Soviets had made advances in their radar technology since that point and their radar ceilings were above the U-2 cruising altitude. As a result, from early in the program, the Soviets detected the U-2 overflights, but they lacked the capability of downing the aircraft until several years later with advances in their Surface-to-Air Missile (SAM) technology.

stealth (including a blunt leading edge, simple nozzle and overwing sensor pods)" (Fulghum 2009). Additionally, they note, "if it is a high-altitude aircraft, it is painted an unusual color - medium gray overall, like a Predator or Reaper, rather than the dark gray or overall black that provides the best concealment at very high altitudes" (Fulghum 2009).

Its sensor package is also largely unknown, but it, at the very least, has video recording equipment, as evidenced by Iran's release of the video in their possession, It probably also features sensor packages aimed at detecting chemical and physical signals related to nuclear and weapons development and electronic signal intelligence.

The drone probably has moderate endurance, though the exact specifications are unknown. Unlike a satellite, UAVs can loiter over an area for long periods of time, thus gathering large amounts of data. This can be extremely useful in gathering data when the target is adept at hiding critical information while surveillance satellites, which fly on predictable paths, pass overhead. Additionally, unlike a maned aircraft, there is less political liability in the event of failure since there is no pilot for whom negotiations would need to be made, nor is there any risk that a downed pilot may be coerced to reveal classified information.

## 2.2 *Security and Level of Security*

Unsurprisingly, security is and was *extremely* high for both of these programs. For the U-2 program, very few people even knew of the program existed, and President Eisenhower was directly involved in its management. Security was so high that the intelligence gathered by the U-2 program was so closely guarded that many CIA analysts did not even know that it existed, and thus it was not figured into calculations and predictions. Given the similarly high security of the RQ-170 program, it is difficult to say much more about the security measures other than the fact that most information about the aircraft is either speculation or released by the Iranian government.

### 2.2.1 U-2

Security, which was handled by the CIA, an already secretive agency, was extremely high surrounding the U-2 program, even by CIA standards. From the very initiation of the program, extraordinary measures were taken to ensure that word of the aircraft's existence or missions were never leaked to the public.

Richard Bissell, the head of the program, paid for the airframes from the Contingency Reserve Fund, a source normally used for covert activities.[5] He also made the U-2 program, Project AQUATONE, completely self-sufficient within the

[5] See the Oversight section for more information about the Contingency Reserve Fund.

Agency. It did not need to communicate with or rely on any of the Agency directorates. Bissell reported directly to DCI Dulles or DDCI General Charles Cabell, who then reported directly to President Eisenhower.

Flight testing was done in a brand new secret test location called "The Ranch" which was handpicked by Kelly Johnson and his chief test pilot, Tony LeVier.[6] When the aircraft were later put into service, the pilots, who were recruited by the CIA out of the USAF,[7] had to deploy without their families to overseas bases. Occasionally, mission plans called for a takeoff from an air strip other than the main bases.[8] For this, the CIA had a mobile unit that could quickly set up, prep the pilot, launch the aircraft, and leave within 24 hours.

The program also had its own compartmentalized place in the CIA's Photo Intelligence Division (PID), which developed the large reels of film that were brought back from every mission and provided photo interpretation. The photographs that were taken by the U-2 were also strictly controlled, which limited their use by CIA analysts who often did not realize that the intelligence existed.

This level of tight security remained high for the entire life of the U-2 program while it was under the control of the CIA. There was some reorganization and increased public knowledge of the program after the 1960 crash, but overall, the CIA still kept the details of the program closely-guarded secrets.

### 2.2.2 RQ-170

While it is difficult to prove a claim based on a lack of information, the scarcity of information on the RQ-170 does reasonably prove that this program is kept highly confidential. There has been very little public acknowledgment of the program from the government except for the occasional anonymous source speaking to a newspaper or a USAF redaction mishap involving Freedom of Information Act (FOIA) documents (Axe 2013a). The information that is available is largely from open source documents, conclusions drawn by independent aviation experts, and statements and information released by the Iranian government following the 2011 crash.

This high level of secrecy is to be expected. A stealthy reconnaissance aircraft loses much of its advantage or utility if it is highly publicized. The whole point of

[6] This happened to be Area 51 in Utah. The number of UFO reports in the area greatly increased in the area during U-2 flight testing. At the time, the U-2 had a silver finish that would catch the light of the sun, particularly in the early evening hours, and since no one believed that manned flight was possible at the U-2's altitude, airline pilots would often report these sightings as UFOs. According the Air Force's Operation BLUE BOOK which kept track of UFO sightings, more than half of the reported UFO sightings during the 1950s and 1960s were U-2 or A-12 OXCART sightings. (Pedlow and Welzenbach 1998)

[7] Even the recruitment was secretive. Meetings and interviews often took place at motels and off-base locations.

[8] The main bases were Weissbaden in West Germany (Detachment A), Incirlik airbase near Adana, Turkey (Detachment B) and, later, the Naval Air Station in Atsugi, Japan (Detachment C).

reconnaissance is to be secretive. Like the U-2, it is expected that the RQ-170 would be a closely-guarded secret, though the *degree of high* security could be debated between the two programs.

The RQ-170 program is secretive, but it does not appear to be *as* secretive as the U-2 program considering the occasional vague mention in anonymous interviews and the UAV's brief profile on the US Air Force website. These minor differences ultimately do not amount to much since the essentials of the RQ-170 program – the missions it flies, the detailed technological specifications, details of program management, etc. – are still largely unknown.

The small differences in security speak more to a change in the times than of the program itself – in the 1950s, nobody considered that the US would build and fly a plane like the U-2 so the fact of its very existence had to be tightly controlled. In current times, nobody is surprised that the US employs these kinds of aircraft and so while the UAV is not actively publicized, occasional vague mentions are condoned.

## 2.3 Oversight of the Program Prior to the Crash

The U-2 program prior to the 1960 crash had extremely limited oversight from Congress - two senators and two members of the House of Representatives. All flights were personally approved by President Eisenhower after conferring with a small number of high-ranking advisors. All of this was in an effort to preserve the high-security surrounding the program. After the crash, the National Security Council took over executive branch oversight of the U-2 program, and it is likely that that policy has continued to the RQ-170 program, at least for highly-sensitive targets. The level of Congressional oversight of the RQ-170 program is unknown, though if the "killer drone" policy is any indication, the oversight is minimal.

### 2.3.1 U-2

By 1956, the U-2 project was nearing its first operational flight. The project was still kept highly secret within the executive branch. However, in the interest of disclosure of the U-2 program to Congress, DCI Dulles briefed Senators Leverett Saltonstall and Richard Russell, who were both ranking members of the Senate Armed Services Committee and the subcommittee on the CIA. He also briefed Representatives John Table and Clarence Cannon from the House Appropriations Committee on the advice of the Senators.

Until the crash in 1960, these four Congressmen was the full extent of formal disclosure to Congress. This limited disclosure was aided by the Central Intelligence Act of 1949 which allows the DCI to use confidentially spend funds without the direct approval of Congress. Specifically,

> The sums made available to the Agency may be expended without regard to the provisions of law and regulations relating to the expenditure of Government funds; and for objects of a confidential, extraordinary, or emergency nature, such expenditures to be accounted for solely on the certificate of the Director and every such certificate shall be deemed a sufficient voucher for the amount therein certified. (USC 403j)[9]

Considering the extreme secrecy with which the program was handled and President Eisenhower's stipulation that the U-2 program "should be handled in an unconventional way so that it would not become entangled in the bureaucracy of the Defense Department or troubled by rivalries among the services," very few members of Congress were formally aware of the top secret program until the Soviets announced that the American plane had crashed in the USSR (Pedlow and Welzenbach 1998).

However, although there were very few people who oversaw the program, particularly in the legislative branch, each overflight underwent extreme scrutiny. Eisenhower was very uncomfortable with sending an American plane into Soviet airspace. As explained by Pedlow and Welzenbach, "The President had mixed feelings about the overflights of the Soviet Union. Aware that they could provide extremely valuable intelligence about Soviet capabilities, he, nevertheless, remained deeply concerned that such flights brought with them the risk of starting a war" (Pedlow and Welzenbach 1998).

As a result, Eisenhower stipulated that he had to personally approve every overflight. He would go over each mission in great detail, even as far as redrawing flight plans. The President would only approve an overflight if it was absolutely necessary and even then it was done with reticence. Sometimes approval only came hours before the scheduled takeoff time.

This dilemma led him to the Open Skies proposal in 1955. The U-2 program was not yet ready, but it was on track for use in the near future. Eisenhower reportedly said about the Open Skies proposal, "I'll give it one shot. Then if they don't accept it, we'll fly the U-2" (Pedlow and Welzenbach 1998). Soviet Premier Nikita Khrushchev rejected the offer which would have given the Soviets access to US airfields to conduct aerial photography of US military installations in exchange for the same right in the USSR.

Eisenhower eventually agreed to the overflights under the assurances that Soviet radars were likely to not be sophisticated enough to detect the aircraft at such a high altitude, and if the plane was detected, it would not be tracked. Additionally, the CIA assured the President that even if an aircraft was lost, it could not be traced back the United States since an aircraft would be completely destroyed after falling from such a high altitude. At the very least, the pilot would not survive.[10]

---

[9] Central Intelligence Agency Act of 1949, 50 USC. 403j.

[10] Pilot survival was an important part of the political calculation since a captured pilot could be used as political leverage and would present a security risk. However, neither the assurance of effective stealth nor the assurance of complete aircraft destruction would turn out to be true. The Soviets had been improving their radar systems since WWII, but the Americans assumed that the

### 2.3.2 RQ-170

On account of, and probably to also maintain, the high secrecy surrounding the program, very little is known about the oversight process of the aerial surveillance program prior to (or since) the 2011 crash. The Obama administration publicly released some information and promised to end the "culture of secrecy about the use and oversight process of UAV strikes to kill overseas Islamic extremists" (DeLuce 2016). However, this is public information release was only focused on UAV strikes, not UAV surveillance, and criticized for still being too meager.

The Obama administration came under heavy scrutiny from independent entities regarding the transparency of the counterterrorism drone strikes (DeLuce 2016). The Stimson Center, an independent nonpartisan think tank specializing in global security and economic issues, and others, heavily criticized the Obama administration for the lack of clarity pertaining to UAV strikes (Stohl 2014).

Senator Diane Feinstein stated in an interview with *The Hill*, "Right now it is very hard [to oversee] because it is regarded as a covert activity, so when you see something that is wrong and you ask to be able to address it, you are told no" (Munoz 2013). So while the Senate Intelligence Committee has some oversight power, the decisions and power is held completely within the Executive branch, particularly the CIA, but there is "an absence of knowing who is responsible for [those] decisions" (Munoz 2013).

This however, is all in regards to the use of drones to carry out extrajudicial killings overseas. Considering that the drone attacks are much more public, harder to conceal, and there is very limited legislative branch oversight on that program, there is probably even less Congressional oversight on the surveillance drone program. The surveillance drone program is even more secretive than the attack drone program, thus promoting a need-to-know culture that makes oversight, a sharing-based activity, very limited.

Most likely, as with the U-2 program, the oversight process is confined to the executive branch, particularly the White House, National Security Council (NSC), USAF, and CIA. It is conceivable that, like the U-2 program, Presidents George W. Bush and Barack Obama required their direct authorization for overflights of militarily sophisticated and politically sensitive countries, such as China, Iran, or North Korea. As with the U-2, the reins were most likely looser concerning less technologically sophisticated countries or countries with a lower political risk. At this point in Donald Trump's presidency, it is too early to make policy conclusions.

---

USSR was still operating the same radars. In regards to the state of the aircraft and the pilot after a crash, the unusual aerodynamics of the U-2 made the aircraft usually fall in a flat spin. This meant that crashed airframes were usually found to be largely intact following a crash, even from operational altitude.

## 2.4 *Management and Agencies Involved*

In both the U-2 and RQ-170 programs, the need for intelligence gathering from an aircraft-based platform brought together the US Air Force and the Central Intelligence Agency. In the case of the U-2, the CIA was hesitant to involve itself with aircraft, a distinctively USAF role, but the need for "civilian" pilots and CIA intelligence gathering and analysis led to a joint, and nearly equal, involvement in the program from the two organizations. In the case of the RQ-170, it is known that both the USAF and CIA are involved in the program, with the USAF taking ownership of the aircraft (unlike the U-2 program), but the degree of cooperation is not known. Given that the intelligence community has grown since the days of the U-2, there are likely other interested agencies, such as the National Geospatial-Intelligence Agency (NGA) and Director of National Intelligence (DNI), even if they are not directly responsible for the program management.

### 2.4.1 U-2

Overall, although the U-2 was technically a CIA project, as stipulated by President Eisenhower, the USAF was very involved. "As [head of the program] Richard Bissell later remarked about the U-2 project, 'The Air Force wasn't just in on this as a supporting element, and to a major degree it wasn't in on it just supplying about half the government personnel, but the Air Force held, if you want to be precise, 49 percent of the common stock'" (Pedlow and Welzenbach 1998).

The CIA paid for the aircraft and was responsible for the cameras, security, film processing, and arrangements on bases overseas. The USAF was responsible for pilot selection and training, weather information, mission planning, and operational support. However, the CIA also had input in pilot selection and mission targets. Additionally, the USAF rerouted some of the Pratt & Whitney J57 engines from its other aircraft programs so that the CIA would not have to buy them, thus raising suspicion (Pedlow and Welzenbach 1998).

### 2.4.2 RQ-170

Lines of responsibility are not made public in the RQ-170 program, but like the U-2 program, it is publicly acknowledged that the USAF and the CIA are working together on the missions. However, unlike the U-2 program, the RQ-170 program is officially controlled by the USAF. The aircraft are operated by the 30th Reconnaissance Squadron and remotely flown from Creech Air Force Base in Nevada (United States Air Force 2009). The CIA however, also plays a role in the program. The extent of this role is not clear. Taking clues from the U-2 program and the CIA's role in the IC, the CIA likely performs intelligence analysis and makes requests for mission targets based on intelligence needs and priorities.

In the U-2 program, Richard Bissell, the head of the program, created the Ad Hoc Requirements Committee (ARC) to set intelligence gathering requirements for the U-2 and other methods. The ARC had representatives from the Army, Navy, Air Force and CIA. In 1957, the National Security Agency (NSA) began sending a representative, and in 1960 the State Department, began sending a representative as well.

Presumably the mission planning for the RQ-170 is similar to that of the U-2. There may not be a formal committee, but the fact remains that the stealthy drone can produce intelligence that is highly sought by a number of organizations within the executive branch and that intelligence is otherwise very difficult to obtain. In fact, the same interested parties would be expected: Air Force, Navy, Army, CIA, NSA, State Department, in addition to additional agencies that have been added since the 1960s such as the National Geospatial-Intelligence Agency (NGA) and the Director of National Intelligence (DNI).

From the information that is available publicly, it is clear that the USAF and the CIA are working together on the RQ-170 program. It would be surprising if these additional military branches, departments, and agencies were not a part of the program as well, even if their role is just to advise and make requests for mission targets.

## 3 Pre-Crash Topics

### *3.1 Origin of the Program*

It is generally true that aircraft programs, being expensive investments, are not initiated without a distinct need. The U-2 and RQ-170 programs are no exceptions. For both aircraft, the international political environment likely had a large impact on the creation and implementation of the program. In the case of the U-2, it is now known that the U-2 design would have be scrapped if not for the "Bomber Gap" debate. In the case of the RQ-170, the origins of the program are unknown, but its estimated first use around the time of the 2003 US invasion of Iraq suggests that the aerial surveillance capability gap formed by the retiring of the SR-71 in 1998 and post- 9/11 political environment played a large part in the successful launch of the program.

#### 3.1.1 U-2

Prior to WWII, peace time reconnaissance was a relatively rare practice. However, in the years following WWII, the Iron Curtain formed and Americans became increasingly alarmed that the USSR was an aggressive nation, particularly after the Soviet-backed North Korean invasion of South Korea in 1950.

The tight security and secrecy of the USSR made intelligence gathering very difficult in Soviet Bloc countries. Many traditional forms of intelligence gathering were blocked, dangerous, or afforded only outdated information. Additionally, increasing Soviet aggressiveness in its air defense policy made aerial surveillance flights dangerous.[11]

In response to the strategic need for intelligence in the face of an extreme challenge in obtaining that intelligence, Lt. Col. Richard Leghorn, the head of the Reconnaissance Systems Branch of the Wright Air Development Command suggested the development of a high-altitude aircraft. He felt that a high-altitude aircraft was the solution since it would probably fly above the range of the Soviet radar, and, even if detected, it would be out of the range of Soviet air defense capabilities.

Although his initial idea was rejected, the idea of a high-altitude aircraft gained traction in the USAF. Requests for proposals of a high-altitude photoreconnaissance aircraft were sent to several small aircraft companies.[12] Lockheed Aircraft Corporation, although not formally contacted by the USAF submitted an unsolicited design by Kelly Johnson in March 1954.

The initial military response to the Lockheed design, then called CL-282, was unfavorable. The uniformed officials did not like that the plane was unarmed, fragile and did not come close to meeting military aircraft specifications. As General LeMay of the Strategic Air Command said, "[I am] not interested in a plane that has no wheels or guns." [13] (Pedlow and Welzenbach 1998).

However, several of the USAF civilians were interested in the design and turned to the Central Intelligence Agency (CIA) in attempt to inspire interest, but were met with very little success. The Director of Central Intelligence (DCI) at the time, Allen Dulles, preferred traditional, HUMINT-based espionage, and he believed that the CIA should not interfere with a USAF project.

As a result, the design was rejected in June 1954 and lay quiet for a few months. Meanwhile, in 1953, the USSR had detonated a lithium deuteride hydrogen bomb only nine months after the first US hydrogen bomb test. Additionally, a Top Secret RAND report in early 1953 warned that the Strategic Air Command's (SAC) US bases were vulnerable to a Soviet long range bomber attack. In 1954, a new Soviet bomber capable of carrying a nuclear bomb from Russia to the United States, the Mysishchev M-4 Bison, was revealed, thus kicking off the "Bomber Gap" debate.

In mid-1954, President Eisenhower instructed James Killian, the President of MIT, to form a panel of experts to investigate offensive, defensive, and intelligence

---

[11] The USAF and US Navy flew RB-47 s along the Soviet border in the early 1950s. When a radar gap was detected, the aircraft would dart in and dart back out. Soon these gaps were filled and these relatively low-flying aircraft were at high risk for being shot down as the Soviets became more aggressive in defending their airspace.

[12] The USAF believed that a small company would give the project more time and effort than a larger company.

[13] The initial design for the U-2 had no wheels. The aircraft was designed to land on its belly. Later, this was changed to a two-wheel ``bicycle" configuration.

capabilities. This panel was called the Technological Capabilities Panel (TCP). One group of the panel, called Project Three, focused on intelligence. When Project Three found out about the Lockheed CL-282 design, they were very excited about its prospects and, using Killian's connections in the White House, pitched the idea directly to the President.

Eisenhower, although reluctant to sign off on a project that would result in overflights of the Soviet Union, approved the program on the basis that intelligence was direly needed to resolve the alleged Bomber Gap. The President agreed with the recommendation of the Project Three panel that the CIA should be in charge of the program to lower the risk that a detected overflight would provoke a war. After the President signed off on the program, the CIA and USAF also agreed to the project.

Kelly Johnson promised that the first aircraft would be flying by August 1955, just eight months after the start of production. He accomplished this feat on time and under budget. According to the original operational plan, organizing, building, and deploying the CL-282 design would take 20 months and cost $22 million (FY1954) (Pedlow and Welzenbach 1998).

The first operational flight occurred on June 20, 1956 when a U-2 overflew Poland and East Germany. The first Soviet overflight had to be delayed until after the USAF delegation to the Moscow airshow left the USSR, and occurred on July 4, 1965, with a second occurring the next day. Those two flights overflew Leningrad and Moscow[14] respectively.

### 3.1.2 RQ-170

Considerably less information is known about the origins of the RQ-170. It is designed and built by Lockheed Martin Skunk Works, the same organization that designed and built the U-2. However, a number of facts about its origin are still unknown – it is not even clear exactly when it was first flown either in testing or operationally. According to some estimates by journalists specializing in military aviation programs, the aircraft "entered service some time before the 2003 US invasion of Iraq" (Axe 2014).

This timing would make sense in the context of the political environment. At that point, the War on Terror had just been launched, renewed questions regarding whether Iraq had weapons of mass destruction had arisen, and the US was preparing for an invasion. A technology capable of providing high-quality data on both the battlefield environment in Afghanistan and provide photoreconnaissance of Iraq would have been very attractive.

Like the U-2, the international political environment almost certainly played a role in the development of the RQ-170 program. Aircraft are expensive investments that are not developed without a need. If the U-2 program is taken as an analogy, the program development cycle probably followed the traditional product development

[14] Reportedly, this was the first and last time a U-2 overflew Moscow.

path: (1) the impending capability gap is recognized, perhaps by a select few; (2) a proposal to bridge that gap is generated and potentially initially rejected either through lack of consumer interest, funding, or need, but ultimately; (3) the proposed program bridges the technological "Valley of Death" when a crisis in need of a quick, effective solution arises. In the case of the U-2, the crisis was the "Bomber Gap." In the case of the RQ-170, it was probably the 9/11 attacks and the following War on Terror and invasions of Iraq and Afghanistan, and the recognized impending ISR capability gap was that formed by the retirement of the SR-71 in 1998.

When put together, this forms a hypothetical story that fits well within aircraft development timelines. The U-2, with its eight month development and production timeline is an exception rather than a rule in aircraft development timelines. Current manned aircraft development periods are closer to five to fifteen years,[15] though unmanned systems are generally simpler than maned systems leading to shorter development times. Additionally, the USAF information page on the UAV hints that the RQ-170 was developed faster than an average aircraft or even an average UAV: "the Air Force's RQ-170 program leverages the Lockheed Martin Advanced Development Programs and government efforts to rapidly develop and produce a low observable UAS [Unmanned Aircraft System]" (United States Air Force 2009). Estimating of five years from development to operational flight (the U-2 took two years for that process) in 2003, that would place the start of the RQ-170 program in 1998, the year the SR-71 was retired.

Although it is not founded on any official story, following the U-2 analogy, it is hypothesized that the RQ-170 program was started in the late 1990s as the SR-71 program was coming to a close; developed for a few years but perhaps shelved; then following the 9/11 attacks, the program was put on the fast track due to an operational need, increased budgets, and renewed interested; and finally operational in 2003.

## 3.2 *Role and Missions Prior to the Crash*

The U-2 was employed on a variety of strategic and tactical photoreconnaissance missions prior to (and following) the 1960 crash. Primarily, it was used to gather intelligence on Soviet and Soviet-bloc military technology and sophistication, particularly related to nuclear weaponry and development. The RQ-170 has probably been used similarly for intelligence-gathering over Iraq, Pakistan, North Korea, and China though the information is scarce.

[15] Though there are exceptions, such as the F-35 program which has been mired by program delays and has been in development since 1993 (when it was the JAST program).

### 3.2.1 U-2

The main targets of the U-2 within the Soviet Union were the bomber force, missile sites, air defense system, submarine yards, and nuclear program (Pocock 1989). Twenty-three successful deep penetration overflights of the USSR were made by U-2 aircraft prior to the unsuccessful attempt by Francis Gary Powers on May 1, 1960. No overflights of the USSR were attempted after the crash.

In addition to overflights of the Soviet Union, the U-2 was also used to gather photoreconnaissance of other Soviet Bloc countries and monitor international incidents around the world. For example, the U-2 was used to monitor troop movements during the Suez Crisis in 1956, thus displaying its importance in tactical as well as strategic intelligence. U-2 tactical intelligence gathering capabilities was again valuable in 1958 with the US intervention in Lebanon.

### 3.2.2 RQ-170

Unsurprisingly, there is very little disclosure of the missions that the RQ-170 flies, but there are clues as to where the RQ-170 has been involved. The most notable "semi"-disclosed mission prior to the crash was the successful operation into Pakistan to kill Osama bin Laden in 2011. According to an unnamed *Washington Post* source,

> Using unmanned planes designed to evade radar detection and operate at high altitudes, the agency [the CIA] conducted clandestine flights over the compound for months before the May 2 assault in an effort to capture high-resolution video that satellites could not provide (Miller 2011a).

The unmanned stealth aircraft mentioned was almost certainly the RQ-170. No other known American aircraft fits that description in terms of abilities and unwillingness by the government to disclose the full details. That assertion is supported by unnamed sources to the *National Journal* (Ambinder 2011).

There is also evidence of the UAV's use in other countries. It was first seen in 2007 by a civilian photographer in Kandahar, Afghanistan, thus earning the nickname, "The Beast of Kandahar" (the USAF did not acknowledge the aircraft until two years later). It is therefore known that the RQ-170 was stationed in Afghanistan, but a stealth reconnaissance drone is hardly necessary in an environment with no air defenses. Other non-stealthy drones such as Predators and Reapers, while most well-known for their role in targeting missions, are also capable of reconnaissance missions. Thus, while the RQ-170 could have been used for tactical intelligence and battlefield awareness in Afghanistan, it was most likely used to overfly Afghanistan's neighbors, particularly Pakistan and Iran, which are now known overflight areas following the Osama bin Laden mission and the crash in Iran.

Thanks to FOIA-released documents with incompletely redacted footnotes, some information is known about the RQ-170's use in the Pacific (Axe 2013b). From January 15 to February 16, 2009, the 30th Reconnaissance Squadron deployed to Guam. Although the RQ-170 is not specifically mentioned, the 30th RS operates the RQ-170 and the drone probably made the journey as well. What happened in Guam is not specified either, but the island is a common take off location for US military aircraft intending to fly into airspace around China (Axe 2013b). Then, in September of 2009, the RQ-170 was deployed to South Korea, presumably to gather intelligence on the North Korean nuclear and missile programs.

Both of the deployments to presumably gather intelligence on China and North Korea were surprisingly risky endeavors. Both countries have strong air defense systems. Assuming the UAV did fly in or near these airspaces, a shoot down or crash could have resulted in major diplomatic difficulties. This shows that whoever is in charge of the RQ-170 program is, or was, very confident in the UAV's stealth technology and/or needed the intelligence gathered so badly that the political risk was taken despite the strong air defenses.

Most probable strategic targets of the RQ-170 are those types of sites targeted by the U-2: nuclear facilities, missile manufacturing sites, submarine and ship yards, air defense systems, and air bases. Additionally, the drone is probably used for tactical surveillance in the Middle East similar to the non-lethal role of the Predator and Reaper drones.

## 3.3 *Political Environment Immediately Prior to Crash*

In both political environments prior to their respective crashes, the US government was under internal pressure to confirm or refute concerns about the nuclear-related military sophistication of a state that was, at the time, considered to be America's "greatest enemy." In the case of the U-2 crash, the "Missile Gap" debate prompted President Eisenhower to sign off on the flight. In the case of the RQ-170, the IAEA reported that Iran had failed to suspend its enrichment activities and was showing signs of pursuing a nuclear weapon.

### 3.3.1 U-2

In 1959, the "Missile Gap" debate emerged. It was provoked by heavy but false Soviet propaganda about its missile program. This came after recent Soviet successes in its space program and successful intercontinental ballistic missile (ICBM) tests. On December 4, 1958, the Soviet delegate at the Geneva Conference on Prevention of Surprise Attack declared, "Soviet ICBMs are at present in mass production" (Pedlow and Welzenbach 1998). Soon after, Premier Khrushchev stated that the USSR had an ICBM capable of carrying a 5-megaton nuclear weapon 8000 miles. The Soviets continued to boast that Soviet ICBMs could hit "precisely any point," that ICBM production was in progress, and that its Army had a series of long, medium, and short range missiles (Pedlow and Welzenbach 1998).

Eisenhower was suspicious of these claims and rightfully so. The Soviets were notorious propagandists, but in the absence of hard evidence to contradict these claims, many members of the intelligence community believed the Soviet boasts. However, to make matters worse, in February 1959, Secretary of Defense Neil McElroy testified to the Senate Preparedness Investigating Committee that by early 1960 the Soviets could have a three-to-one missile advantage on the United States. However, he also stated that this advantage would be short lived because the United States was making progress with more advanced solid-fueled missiles whereas Soviet missiles were liquid-fueled. This latter detail was largely ignored and significant concern was raised over the impeding gap.

Eisenhower was still reluctant authorize an overflight to confirm or disprove these assessments since he judged that the political risk was not worth the expected intelligence. He did, however, authorize a series of ELINT-gathering missions along the border.

Concern over the potential missile gap continued to grow throughout 1959. The Soviets continued to make bold propaganda statements such as "in one year, 250 rockets with hydrogen warheads came off the assembly line in the factory we visited" (Pedlow and Welzenbach 1998). Despite these claims, Eisenhower refused to approve any overflights.

However, limited U-2 overflights in 1958 and 1959 had left the intelligence community with very little current information on the Soviet missile program. Eventually, Eisenhower agreed to allow an overflight at the encouragement of his advisors. The resulting February 1960 overflight did not reveal any missile sites, even though the mission was designed to target probable missile sites.

Despite this, the missile gap debate persisted. The Army, Navy, and CIA now doubted the Soviet missile claims based on the results of the overflight, but the Air Force still believed that the Soviets could have deployed up to 100 missiles. In order to put the debate to rest, Eisenhower agreed to allow two more overflights of the Soviet Union as long as they occurred in April 1960. The first, Operation SQUARE DEAL returned safely, but the second, delayed until May 1, 1960, Operation GRAND SLAM, did not.

### 3.3.2 RQ-170

The political environment prior to the RQ-170 crash was similar to that of the U-2 crash: tensions were high and the United States strongly desired answers to a pressing nuclear sophistication issue. In the case of the RQ-170, this took the form of nuclear weapon development in Iran. At the time of the RQ-170 crash, the United States was receiving pressure from its allies concerning Iranian nuclear ambitions. Additionally, there was high domestic concern about Iran as evidenced by a Gallup poll in 2011 which found that "Americans are most likely to mention Iran when asked which country they consider to be the United States' greatest enemy" (Jones 2011).

Then, in November 2011, the IAEA published a report enumerating concerns about the Iranian nuclear program (IAEA 2011). There was significant concern that Iran's nuclear program was not strictly for peaceful purposes. Iran had failed to suspend enrichment activities and heavy-water-related projects, and it had engaged in several activities specific to the development of a nuclear weapon. As stated in the report, "Since 2002, the Agency has become increasingly concerned about the possible existence in Iran of undisclosed nuclear related activities involving military related organizations, including activities related to the development of a nuclear payload for a missile, about which the Agency has regularly received new information" (IAEA 2011).

Tensions escalated further a few weeks later when Iranians attacked the British Embassy and a diplomatic residence in Tehran. Britain withdrew its diplomatic staff and ordered that Iran remove its diplomats from the UK (Jaffe and Erdbrink 2011). By the time that drone crashed a week later, tensions with Iran were very high across the West.

## 4 Crash and Post-Crash Topics

### *4.1 The Crash & Means of Downing the Aircraft*

The crashes of both aircraft share a number of similarities. First, in both incidents, the aircraft were downed by the new and upcoming-technology of the era. For the U-2, it was an improved Surface-to-Air Missile (SAM), and for the RQ-170, it was probably GPS spoofing, which falls under the realm of hacking and cyberwarfare.[16] Second, in both situations, a cover story was issued by a secondary organization. For the U-2 crash, a cover story was issued by NASA, and for the RQ-170, a press release was issued by the NATO-led International Security Assistance Force (ISAF). Third, there was some warning in both crashes that enemy countermeasures had caught up and the aircraft was vulnerable to attack. Finally, the aircraft was brought down by and landed within the borders of "America's greatest enemy."

#### 4.1.1 U-2

On May 1, 1960, the approval was given for Operation GRAND SLAM despite recent warnings that Soviet missile technology may have progressed to the stage of being a danger to U-2 overflights. Later that day, the U-2, piloted by Francis Gary Powers, one of the most experienced U-2 pilots, was shot down deep into Soviet territory near Sverdlovsk. The aircraft was shot down using recently-developed

[16] There is some debate over whether Iran successfully hacked the UAV or whether the aircraft malfunctioned. The author believes that the evidence is strong enough to believe that Iran successfully hacked the drone.

SA-2 Soviet surface-to-air missiles (SAMs) at an altitude of 70,500 feet (Pedlow and Welzenbach 1998).

Powers managed to survive the crash and was promptly captured by Soviet forces. He subsequently endured an extensive interrogation and was eventually sentenced to "Ten years of confinement, with the first three years to be served in prison" (Powers 1970). His aircraft also managed to survive largely intact, including the valuable camera and film detailing the photo intelligence that he had gathered to that point.

Premier Khrushchev, in attempt to ferret out more information out of the United States government, announced that an American spy plane had been shot down over the USSR, but no mention of the state of the pilot or aircraft was made. Initially, the US issued a cover story that the downed aircraft was a NASA high altitude weather research aircraft and that the pilot had reported difficulties with his oxygen system before going off course. Two days later, Khrushchev announced that both the pilot and the plane had survived, thus catching the United States in a lie. This led to international embarrassment for the United States, particularly for President Eisenhower who took the complete blame for the program.

The use of the SA-2 SAMs to down the aircraft is particularly important. Missiles, particularly guided missiles, were the major up-and-coming military technology of concern at that time. In fact, the mission was approved for the expressed purpose of investigating Soviet ICBMs, a much larger type of missile.

### 4.1.2 RQ-170

The RQ-170 was also downed by the up-and-coming, emerging military capability of the time: cyberwarfare. Although it has not been confirmed by the US government, the Iranians have credibly claimed that the UAV was brought down by Iranian electronic warfare specialists (Peterson 2011).

According to an exclusive interview conducted by *The Christian Science Monitor* with an anonymous Iranian engineer who participated in the drone operation, Iran managed to jam the communication feed that connected the UAV to the remote pilot at Creech Air Force Base in Nevada (Peterson 2011). Then, Iranian engineers spoofed the UAV's GPS system to make it believe that it was returning to its home base in Afghanistan when it was actually landing in Iran. The drone landed near Kashmar, Iran, about 140 miles from the Afghan border.

Just one month prior to the downing of the RQ-170 by the Iranians, a group of academics from ETH Zurich and University of California, Irvine published a paper at the Association for Computing Machinery (ACM) conference on Computer and Communications Security entitled, "On the Requirements for Successful GPS Spoofing Attacks" (Tippenhauer et al. 2011). In the paper, the researchers outline the necessary steps for spoofing a civilian or military GPS device – the same technique potentially used by the Iranians. An expert later commented on the paper, "It's a PDF file…essentially, a blueprint for hackers" (Russon 2015). The researchers have been criticized for not giving the military any warning of the security flaws, but ultimately, the published paper was available for everyone to view, including the US military and intelligence community. There was a warning that an exploitable bug had been found, but the US military and CIA either chose to ignore that warning or were ignorant of it.

Less than a year later, researchers at the University of Texas at Austin were able to successfully duplicate the GPS spoofing technique on a GPS-guided UAV, and a year after that demonstration, they were able to successfully spoof the navigation system of an $80 million yacht off the coast of Italy and steer it off course (Kerns et al. 2014, UT News 2013). The GPS system on a military spy plane is better than that of a yacht, but their experiment does show that GPS spoofing is possible and give credence to the Iranian claims and the Tippenhauer et al. paper.

Immediately following the December 4, 2011 downing, US officials acknowledged that an aircraft was missing near the Iranian border, but reports from Iranian state TV news and the Fars News Agency that the UAV was intentionally brought down with a cyber-attack were dismissed. In an interview with *The Washington Post*, an anonymous Pentagon official stated, "If this happened, it is a 95 percent chance that it just malfunctioned" (Jaffe and Erdbrink 2011).

As in the case of the U-2 crash, the true origin of the drone was initially obscured and a cover story was released. CIA officials declined to comment on the crash (Miller 2011b). Instead, the NATO-led International Security Assistance Force (ISAF) in Afghanistan issued a statement,

> The UAV to which the Iranians are referring may be a US unarmed reconnaissance aircraft that had been flying a mission over western Afghanistan late last week. The operators of the UAV lost control of the aircraft and had been working to determine its status (ISAF 2011).

The statement is extremely vague, and although the fact that the statement came from the ISAF seems to suggest that the ISAF was responsible for the aircraft, no actual statement of ownership was made. There was no mention that the CIA was involved, and the CIA and military continued to decline to comment. However, soon after, US officials confirmed that the UAV was part of a CIA mission, but insisted that the ISAF statement was technically accurate (Starr 2011). Furthermore, US officials have continued to stress that the aircraft had not been brought down by a cyberattack.

## 4.2 *Immediate Aftermath of the Crash*

The U-2 crash was a disaster for President Eisenhower's hopes of thawing the Cold War. Premier Khrushchev cited the crash and Eisenhower's refusal to apologize as a justification for walking out of the Paris Summit. After that point, the Cold War "refroze" and continued for the next three decades, though the Soviets' previous blustery but ultimately toothless reactions to the overflights suggest that Khrushchev simply used the U-2 incident to embarrass President Eisenhower and storm out of discussions he never had much interest in. The aftermath of the RQ-170 was less dire by comparison but still not insignificant. Iran acquired the drone technology and issued a sternly-worded complaint to the UN citing the crash as a US aggression and therefore against the UN charter.

### 4.2.1 U-2

On May 16, 1960, President Eisenhower and Premier Khrushchev traveled to Paris for the scheduled summit. Khrushchev, who spoke first, read a long protest statement regarding the crash and the US infringement of Soviet rights, and he demanded an apology. Eisenhower replied that overflights were suspended but refused to give an apology. Khrushchev stormed out of the meeting, and the summit ended in a failure.

The Paris Summit had been arranged to address a number of Cold War issues, particularly the nuclear arms race and management of Berlin. There were signs that the Cold War could come to an end, and leaders were optimistic. The summit was supposed to capitalize on this feeling and make progress towards peace.

However, it is hard to pin the failure of the Paris Summit solely on the U-2 incident. The Soviets had known for years that the USA was conducting overflights within Soviet borders, but prior to May 1, they had never been able to intercept the plane. They issued complaints but little else. The downing of the U-2 and investigation of the camera film confirmed their suspicions about the program, but there was already little doubt as to the overflights' purpose.

In this light, it is suspicious that the crash of a U-2 in Soviet territory was enough to change Premier Khrushchev's mind regarding a thawing of the Cold War. It seems much more likely that the U-2 incident merely provided him an opportunity to embarrass the United States and President Eisenhower and to storm out of talks that he had little intention of taking seriously. Perhaps, some progress of could have been made at the Paris Summit, but it seems overly idealistic to believe that the following three decades of Cold War could have completely avoided had it not been for the U-2 incident.

In 1962, Francis Gary Powers, the U-2 pilot, was returned to the United States in exchange for the Russian spymaster Rudolph Abel. Upon his return home, Powers underwent a series of CIA and Congressional testimonies. Powers, although absolved of blame for the incident, faced a very negative opinion from the American public. He died in a helicopter crash in California in 1977.

### 4.2.2 RQ-170

While the US government was very quiet about the program and crash, Iran was very public about it. On December 8, 2011 the Iranian Revolutionary Guards released photos of the nearly intact downed drone surrounded by anti-American propaganda, in what appeared to be a school gymnasium (Peterson 2011). The United States requested that Iran return the drone, but the request was refused.

Iranian General Hossein Salami stated in response to the request, "No nation welcomes other countries' spy drones in its territory, and no one sends back the spying equipment and its information back to the country of origin" (CNN 2011). Aside from the request for return, President Obama declined to comment on the issue (White House Office of the Press Secretary 2011).

Iran made a formal public protest regarding the incident in a letter to the UN Ambassador Mohammad Khazaee to UN Secretary General Ban Ki-moon, the heads of the General Assembly, and the UN Security Council. In the letter he wrote,

> My government emphasizes that this blatant and unprovoked air violation by the United States government is tantamount to an act of hostility against the Islamic Republic of Iran in clear contravention of international law, in particular, the basic tenets of the United Nations Charter. The Iranian Government expresses its strong protest over these violations and acts of aggression and warns against the destructive consequences of the recurrence of such acts. The Islamic Republic of Iran reserves its legitimate rights to take all necessary measures to protect its national sovereignty (Khazaee 2011).

Tensions between Iran and the US remained high, but despite the strong words from Iran, no further actions were taken against the US, aside from periodic claims of Iranian technical prowess in learning the technological secrets of the UAV. Similar to the U-2 incident, Iran used the crash as an opportunity to embarrass the US and issue a formal complaint, but it did little else to escalate the issue.

A little over a year later, in February 2013, Iran claimed to have decoded surveillance footage from the drone and displayed it online. A year later, in May 2014, Iran unveiled a copy of the RQ-170, which they claimed to have reverse-engineered. Iran released a video of the reverse engineered UAV allegedly flying later that same year. Iran claimed it to be "faster, more fuel efficient, and even less detectable by radar than the US original" (Peterson 2014).

All of these claims have come under scrutiny. Officials from the US government, which still maintains the stance that the drone suffered a malfunction rather than being brought down by an Iranian cyberattack, dismiss these claims. Pentagon spokesman Col. Steve Warren said of the copy, "[there is] no way it matches American technology" (Peterson 2014). There are also skeptics amongst aviation experts outside of the government. David Cenciotti from *The Aviationist* notes that while the footage of the UAV flying looks authentic, the landing sequence "seems to be a bit weird... more than a UAV, the aircraft moves and reacts to the remote pilot's input as a small remotely piloted scale model" (2014).

## 4.3 Post-Crash Roles and Missions

After the U-2 crash, the overflight approval process was put under the control of the National Security Council and the program itself was merged with other "overhead reconnaissance" programs. The aircraft never overflew the USSR again, but it was used extensively during the Cuban Missile Crisis and the Vietnam War. The RQ-170 program likely underwent a similar re-evaluation of targets and, at least for the time immediately following the crash, no longer overflew Iran. However, it was (and is) probably still used over other targets such as Syria, territory seized by ISIL, Yemen, Pakistan, and the South China Sea, and less likely, but still possibly, North Korea and China.

### 4.3.1 U-2

After the crash, there were organizational changes in how overhead surveillance intelligence was handled. The Ad Hoc Requirements Committee created by Richard Bissell was merged with the Satellite Intelligence Requirements Committee to form the Committee on Overhead Reconnaissance (COMOR). Overhead reconnaissance was defined as, "all reconnaissance for foreign-intelligence purposes by satellite, or by any vehicle over denied areas, whether by photographic, ELINT, COMINT, infrared, RADINT, or other means" (Pedlow and Welzenbach 1998).

Additionally, the review structure for overflight approval was overhauled. Previously, the approval process had been unstructured, with President Eisenhower holding the final say after consulting with a very select group of advisors. After the crash, the process became much more structured. A detailed submission justifying the overflight had to be prepared by the CIA and sent to the National Security Council (NSC) Special Group for approval. The decision of the NSC was usually final although it could be taken to the President for his decision.

Soon after the crash, President Eisenhower terminated all overflights over the Soviet Union. Although he did not rule out the possibility publicly, internally it was understood that the President would no longer approve overflights. When John F. Kennedy took over the Presidency later that year, he followed in line with Eisenhower's decision, stating, "I have ordered that the flights not be resumed, which is a continuation of the order given by President Eisenhower in May of last year" (Pedlow and Welzenbach 1998).

Secretively, however, President Kennedy and the intelligence community seriously considered Soviet overflights during the Berlin crisis of 1961. Kelly Johnson wrote in his project log, "It seems that President Kennedy, who publicly stated that no U-2 aircraft would ever fly over Russia while he was president, has requested additional flights. Some poetic justice in this" (Pedlow and Welzenbach 1998). These flights never materialized and the U-2 never overflew the USSR after May 1, 1960.

The aircraft was, however, not taken out of service. It was used elsewhere for strategic and tactical surveillance over a number of countries and crises. It overflew Cuba to support the ill-fated Bay of Pigs invasion, and it overflew the country again to collect intelligence during the Cuban Missile Crisis. The U-2 was also used in Asia during the build up to Vietnam War. Thirty-six missions were flown over North and South Vietnam from 1962 until the Gulf of Tonkin Resolution in 1964 handed responsibility to the Air Force (Pedlow and Welzenbach 1998).

CIA U-2 s reentered North Vietnamese airspace in 1973 when President Nixon tasked them with monitoring the 1973 ceasefire agreement. Under the agreement,

USAF planes were prohibited, but the CIA U-2 s and pilots were technically civilian and exempt. By this time, the use of Agency U-2 s was waning, and the CIA ended its U-2 program on August 1, 1974 when it transferred all its remaining aircraft to the Air Force.

### 4.3.2 RQ-170

In December 2013, immediately following the 2011 crash, Secretary of Defense Leon Panetta told Fox News that the US would "absolutely" continue UAV surveillance flights into Iran from Afghanistan (Fishel 2011). There are three plausible reasons for why Secretary Panetta responded the way that he did: the US did not terminate Iranian overflights and he was telling the truth; the US did terminate Iranian overflights, but the Secretary could not publicly state that the overflights were terminated without implicitly giving credit to the Iranian's claims that they brought down the UAV with a cyberattack; or President Obama had not yet made the decision of whether to terminate overflights and at that present time, and the Department of Defense's strategy was to continue to perform overflights.

In the case of the first option, continuing overflights of Iran would indicate that the US was absolutely sure that the drone was not downed as a result of any Iranian efforts. If there was any chance that the UAV was brought down by a cyberwarfare unit or even tracked within the country to the point that it could be shot down with other countermeasures, it would be extremely risky to attempt another Iranian overflight. In that case, the risk-reward balance would be too unfavorable, no matter how pressing the need for intelligence on Iranian nuclear activities.

In the case of the second option, accusing the Secretary of Defense of lying when asked a direct question is not something that should be taken lightly. It is, of course possible however to rationalize why he may have lied. At that point, Iran was viewed as America's greatest enemy. If Secretary Panetta gave credit, even only implicitly, to Iran's claims that it had brought down a top secret American drone, it would have undermined the American people's faith in the military. By saying that the United States would continue to send drones into Iranian airspace, it sends the message that the United States is being proactive about concerns over the nuclear ambitions of America's greatest enemy, and Iran's claims of a successful cyberattack are bogus.

In the case of the third option, that President Obama had not yet made a decision, this is reasonably plausible. Even Eisenhower, who was nervous about overflights from the start of the U-2 program, did not terminate overflights until seven days after the incident. Even then, when Eisenhower took complete credit for the program four days later, he did not provide any assurance that overflights had stopped or would stop. When Khrushchev demanded an apology and an assurance that overflight were terminated, Eisenhower refused. It is completely possible that at the time that Secretary Panetta spoke to the press, the decision still had not been made, and in the interest of protecting the image of American military strength, he responded that the overflights would continue.

Of the three options, the first option seems the most unlikely. There is too much evidence that Iran could have the cyberwarfare capabilities that it claims. The GPS spoofing tactic was publicly published was repeated by American researchers. Additionally, Iran seems to have repeated its spoofing abilities recently when two American Navy vessels "misnavigated" into Iranian waters in January 2016 (Goward 2016).

The second option also seems unlikely, although to a lesser extent. It would be bold to lie so directly, but catching the Secretary in the lie would be a difficult task. These UAVs are designed to be undetectable and the military does a very good job in keeping their movements secret. If the UAVs never flew above Iran again, how would anyone know?

Most likely, however the last option was what actually transpired. A decision had not yet been made and the Secretary commented on policy as it stood at that time. When a decision was made, it most likely resulted in a termination of overflights, as in the case of the U-2 and the USSR. Overflights were too risky.

This does not mean, however, that the RQ-170 was taken out of service. Just like the U-2, it is likely being used to collect intelligence on other countries with a more favorable risk-reward balance. This hypothesis is further supported, though not proved, by the noted lack of information about the aircraft even to this day. It is difficult to prove the negative, but if the aircraft were scrapped following the crash in Iran (which is highly unlikely), the USAF would reasonably be more willing to publish information on the aircraft. It could be a political maneuver to hide their hand, but the safer conclusion is that the aircraft is still being used. Additionally, activity involving the 30th Reconnaissance Squadron, which operates the RQ-170, strongly suggests that the aircraft is still employed (Axe 2014).

It is difficult to say exactly where the RQ-170 is flying, but recent conflict areas are all possible. The Syrian government has some air defenses, but it is not sophisticated enough, nor currently strong enough to track and down a stealthy American drone. Other targets, particularly ISIL strongholds, Yemen, and Pakistan are likely targets. In the Pacific region, North Korea, although it boasts a strong air defense system, is a possible though risky target, particularly with its recent nuclear and missile tests. China, another potential past target, is less likely considering the tricky political relationship and the fact that Iran may have very likely sold the Chinese information regarding the downed RQ-170 (Cenciotti 2013). Still, it is possible that overflights in Chinese airspace have occurred after 2011 if the US was confident enough in the UAV's ability, particularly after the recent and continuing tensions in the South China Sea.

## 4.4 Future Aerial Surveillance Technology Development

In the case of the U-2, the reveal of US aerial surveillance efforts and technology only temporarily slowed the development of future aerial surveillance platforms. Only seven years after the U-2 crash, the U-2's successor, the A-12 OXCART, achieved its first operational flight. Similarly, although even less is known about the RQ-170's successor, RQ-180, it seems that the 2011 crash did not force the US government to halt aerial surveillance efforts or technology development.

### 4.4.1 U-2

At the time of the U-2 crash in 1960, the CIA had been engaged for two years in the A-12 OXCART program, the plane that would become the SR-71. The A-12 was the successor to the U-2, whose life flying above Soviet air defenses was soon coming to an end as the Soviets made strides in their air defense systems. The A-12 was an exceptional aircraft designed to fly at 100,000 feet at a speed of Mach 3.1.

Following the U-2 crash in May, the fate of the A-12 was under question. In a memo, General A.J. Goodpaster, Eisenhower's Staff Secretary, wrote "[President Eisenhower] said he was inclined to think that it [the A-12] should go forward, on low priority, as a high performance reconnaissance plane for the Air Force in time of war" (Goodpaster 1960). As would be expected, given the termination of overflights over the USSR, the mission for which the A-12 was designed, enthusiasm for the aircraft waned.

Still, the A-12 OXCART program survived and achieved its first operational flight in 1967. However, its use was short-lived and the entire fleet was retired just one year later in 1968. The duties of overhead reconnaissance were transferred to the USAF. The A-12 design was tweaked into a slightly slower two-seater aircraft for the USAF called the SR-71. The SR-71 was flown until 1998, when the program was terminated for budgetary reasons (Richelson 2002), All CIA management of manned aircraft programs ended in 1974 with the transfer of the Agency's fleet of U-2 s to the Air Force (Pedlow and Welzenbach 1998).

### 4.4.2 RQ-170

Although there is very little information about the RQ-170, there is even less information on the UAV's successor, the RQ-180. It is made by Northrop Grumman and according to Amy Butler and Bill Sweetman at *Aviation Week* who first announced the existence of the drone in 2013, "the aircraft will conduct the penetrating ISR [intelligence, surveillance, reconnaissance] mission that has been left unaddressed, and under wide debate, since retirement of the Lockheed SR-71 in 1998" (Butler and Sweetman 2013). Additionally, the UAV is stealthier and has a longer range than its predecessor, the RQ-170, and it is intended for use in denied airspace. As with previous programs, the RQ-180 will be jointly managed by the Air Force and the CIA.

Butler and Sweetman have a reasonably good report of the capabilities and niche of the RQ-180, but there is no indication of whether the RQ-170 crash had any impact on the decision to go forward with the RQ-180 project. According to Sweetman, the RQ-180 project was awarded to Northrop Grumman in early 2008

(Sweetman 2012). This means that the RQ-180 was already three years into development at the time of the RQ-170 crash. David Axe remarks that the existence of the RQ-180 program, "could help to explain some peculiar moves by the Air Force in 2012 and 2013. The flying branch chose to abandon all of its essentially brand-new Global Hawk UAVs while also cutting production of the smaller Reaper drones" (Axe 2013a).

This suggests a total buy-in to the RQ-180 program. Whereas the A-12 program faced some doubts about its future use and relevancy in light of the termination of Soviet overflights, the RQ-180 program does not appear to have faced these same reservations. If there were reservations about the future of the program after the RQ-170 crash, they were brief. Recall that the crash occurred in December of 2011 and the USAF was already abandoning its uncontested airspace drones in 2012. Rather than make officials cautious about performing overflights, it seems like the RQ-170 crash galvanized the Air Force's leadership's decision to invest in surveillance UAVs designed for denied airspaces.

## 5 Conclusions

In general, the secrecy surrounding intelligence gathering efforts makes it difficult to know what the current policies and practices are in place for those programs. Aerial surveillance programs are no exception to this trend. However, the little information that is available regarding current aerial surveillance shows that very little has changed over the past decades. The technology may be more advanced, but the policies that govern and guide the use of that technology have not greatly changed. From this observation we can make informed hypotheses on the current state of aerial surveillance policies based on what we know about past programs.

Using this basis, we form several conclusions about the current state of the RQ-170 program. First, the aircraft is still in service, but it is not currently overflying Iran, particularly in light of the recent nuclear deal.[17] The aircraft is overflying countries with a more favorable risk-reward balance, countries in which intelligence is needed and the risk of being downed is acceptable in light of the urgency of this need. Countries such as Syria, North Korea, and ISIL hotspots in the Middle East are good candidates.

Second, American aerial surveillance technologies are still vulnerable to up-and-coming countermeasures. Both the U-2 and the RQ-170 were downed by the "hot topic" technology of their day- missiles and cyberwarfare, respectively. It is, of course, difficult to design countermeasures against a technology that has not been invented yet, but as the surveillance technology ages and evolutionary nature of weapons development progresses, the operators of surveillance technologies must be vigilant in identifying when their surveillance technology is in danger. These

[17] It is too early in the Trump presidency to evaluate whether this policy has changed, but with President Trump's criticism of Iran and the Iran Nuclear Deal, it has possibly changed.

programs operate with extreme secrecy. That secrecy is broken with one failure. If the United States wants to improve the effectiveness of its secret surveillance programs, it needs to make a more intensive, and perhaps more pessimistic, assessment of enemy capabilities and decide whether to enter into that environment.

Finally, aerial surveillance has become an established tool in intelligence gathering. Despite the highly-public crashes that both the U-2 and RQ-170 programs suffered and despite the emergence of satellite technology, aerial surveillance continues to be an important part of intelligence gathering, particularly areas of limited access. In both programs, the crashes did not terminate development of next-generation technology, and in the case of the RQ-170, the crash seems to have reinforced the belief that a next-generation vehicle was needed. This decision making pattern, that a crash serves as a justification for further development rather than a halt on the activities, makes the second conclusion all the more important.

Aerial surveillance UAVs appears to be here to stay. However, the words of President Eisenhower are just as true today as they were in the 1950s when he said, "Well, boys, I believe the country needs this information, and I'm going to approve it. But I tell you one thing. Some day one of these machines is going to be caught, and we're going to have a storm" (Pocock 1989). During this new wave of unmanned system development, it is important to realize that lessons can be taken from mistakes and programs of the past. Additionally, it is important to remember that new technologies can be built, but countermeasures will always eventually catch up. The critical part is to realize they have caught up before the plane is brought down.

## References


Ambinder, M. (2011). The little-known agency that Hellped kill bin Laden. *The Atlantic*, [online] May 5. Available at: http://www.theatlantic.com/politics/archive/2011/05/the-little-known-agency-that-helped-kill-bin-laden/238454/. Accessed 28 March 2017.

Axe, D. (2012). 7 secret ways America's stealth armada stays off the radar. *Wired*, [online] Dec 13. Available at: https://www.wired.com/2012/12/steath-secrets/?pid=1688. Accessed 28 March 2017.

Axe, D. (2013a). Yes, America has another secret spy drone-we pretty much knew that already. *War is Boring*, [online] Dec 6. Available at:. https://warisboring.com/yes-america-has-another-secret-spy-drone-we-pretty-much-knew-that-already-41df448d1700#.er1ih0d5u [Accessed 28 March 2017]. [In-line citation: Axe 2013a].

Axe, D. (2013b). Stealth Drone's secret Pacific missions: Air force accidently cops to RQ-170's 2009 Asia tour. *War Is Boring*, [online] Dec 7. Available at:. https://warisboring.com/stealth-drones-secret-pacific-missions-3f12eee5c0d1#.a5d6utjh0 [Accessed 28 March 2017]. [In-line citation: Axe 2013b].

Axe, D. (2014). America's stealth drones stay very busy. *War Is Boring*, [online] Sep 26. Available at:. https://warisboring.com/americas-stealth-drones-staying-very-busy-f0f77011141#.ncuz-brdtu [Accessed 28 Mar 2017].

Butler, A., & Sweetman, B. (2013). Secret new UAS shows stealth, Efficiency Advances. *Aviation Week*, [online] Dec 6. Available at:. http://aviationweek.com/defense/secret-new-uas-shows-stealth-efficiency-advances [Accessed 28 March 2017].

Cenciotti, D. (2013). New photo shows that China has really copied the US RQ-170 sentinel stealth drone. *The Aviationist*, [online] Jun 2. Available at: https://theaviationist.com/2013/06/02/china-rq170-copy/. Accessed 29 March 2017.

Cenciotti, D. (2014). Iran releases first (somewhat suspicious) video of its RQ-170 stealth drone in flight. *The Aviationist*, [online] Nov 12. Available at: https://theaviationist.com/2014/11/12/video-of-flying-iranian-rq-170/. Accessed 28 March 2017.

Central Intelligence Agency Act of 1949, 50 U.S.C. §§403j (2017).

CNN Wire Staff. (2011). Obama says US has asked Iran to return drone aircraft. *CNN*, [online] Dec 12. Available at: http://www.cnn.com/2011/12/12/world/meast/iran-us-drone/ [Accessed 28 March 2017].

De Luce, D. (2016). Obama's drone policy gets an 'F'. *Foreign Policy*, [online] Feb 23.. Available at: http://foreignpolicy.com/2016/02/23/obamas-drone-policy-gets-an-f/ [Accessed 28 March 2017].

Fishel, J. (2011). Panetta says drone campaign over Iran will continue. *Fox News*, [online] Dec 13. Available at: http://www.foxnews.com/politics/2011/12/13/panetta-says-drone-campaign-will-continue.html. Accessed 28 March 2017.

Goodpaster, A. (1960). *Memorandum for Record.* NSA archive, [online]. Available at:. http://nsarchive.gwu.edu/NSAEBB/NSAEBB74/U2-06.pdf [Accessed 28 March 2017].

Goward, D. (2016). Opinion: Were US sailors 'spoofed' into Iranian waters. *The Christian Science Monitor*, [online] Jan 15. Available at:. http://www.csmonitor.com/World/Passcode/Passcode-Voices/2016/0115/Opinion-Were-US-sailors-spoofed-into-Iranian-waters [Accessed 28 March 2017].

Hambling, D. (2009). Mysteries surround Afghanistan's stealth drone (Updated). *Wired*, [online] Dec 4. Available at:. https://www.wired.com/2009/12/mysteries-surround-afghanistans-stealth-drone/ [Accessed 28 March 2017].

International Atomic Energy Agency. (2011). Implementation of the NPT safeguards agreement and relevand provisions of security council resolutions in the Islamic Republic of Iran. Available at:. https://www.iaea.org/sites/default/files/gov-2015-34.pdf [Accessed 28 March 2017].

International Security Assistance Force. (2011, December 5). ISAF releases statement on missing unmanned Aerial vehicle, Dec 5, 2011. *Resolute Support News*, [online] Dec 5. Press Release. Available at:. http://www.rs.nato.int/article/rs-news/isaf-releases-statement-on-missing-unmanned-aerial-vehicle-dec-5-2011.html [Accessed 28 March 2017].

Jaffe, G., & Erdbrink, T. (2011). Iran says it downed US stealth drone; Pentagon acknowledges aircraft downing. *The Washington Post*, [online] Dec 4.. Available at: https://www.washingtonpost.com/world/national-security/iran-says-it-downed-us-stealth-drone-pentagon-acknowledges-aircraft-downing/2011/12/04/gIQAyxa8TO_story.html [Accessed 28 March 2017].

Jones, J. M. (2011). Americans continue to rate Iran as greatest US enemy. *Gallup*, [online] Feb 18. Available at: http://www.gallup.com/poll/146165/americans-continue-rate-iran-greatest-enemy.aspx. Accessed 28 March 2017.

Khazaee, M. (2011). Iran UN letter: US drone is aggression. *The Iran Primer*, [online] Dec 9.. Available at: http://iranprimer.usip.org/blog/2011/dec/09/iran-un-letter-us-drone-aggression [Accessed 28 March 2017].

Kerns, A. J., Shepard, D. P., Bhatti, J. A., & Humphreys, T. E. (2014). Unmanned aircraft capture and control via GPS spoofing. *Journal of Field Robotics, 31*, 617–636.

Miller, G. (2011a). CIA flew stealth drones into Pakistan to monitor bin laden house. *The Washington Post, [online]*, (May 17.) https://www.washingtonpost.com/world/national-security/cia-flew-stealth-drones-into-pakistan-to-monitor-bin-laden-house/2011/05/13/AF5dW55G_story.html [Accessed 28 March 2017

Miller, G. (2011b). After drone was lost, CIA tried a head fake. *The Washington Post*, [online] Dec 6. Available at:. https://www.washingtonpost.com/blogs/checkpoint-washington/post/after-drone-was-lost-cia-tried-a-head-fake/2011/12/06/gIQAJNrnZO_blog.html [Accessed 28 March 2017].

Munoz, C. (2013). Sens. Feinstein, Leahy push for court oversight of armed drone strikes. *The Hill*, [online] Feb 10.. Available at: http://thehill.com/policy/defense/282033-feinstein-leahy-push-for-court-oversight-of-armed-drone-strikes-#ixzz2RzPHdGfL [Accessed 28 March 2017].

Pedlow, G. W., & Welzenbach, D. E. (1998). The CIA and the U-2 Program, 1954–1974. Central Intelligence Agency.

Peterson, S. F. (2011). Exclusive: Iran hijacked US drone, says Iranian engineer (video). *The Christian Science Monitor*, [online] Dec 15.. Available at: http://www.csmonitor.com/World/Middle-East/2011/1215/Exclusive-Iran-hijacked-US-drone-says-Iranian-engineer-Video [Accessed 28 March 2017].

Peterson, S. (2014). Iranians fly knock-off of US stealth drone. Did they get it right? *The Christian Science Monitor*, [online] Nov 13.. Available at: http://www.csmonitor.com/World/Middle-East/2014/1113/Iranians-fly-knock-off-of-US-stealth-drone.-Did-they-get-it-right [Accessed 28 March 2017].

Pocock, C. (1989). *Dragon Lady*. Shrewsbury: Airlife Publishing Ltd.

Powers, F. G. (1970). *Operation overflight*. New York: Holt, Rinehart and Winston.

Richelson, J. T. (2002). The U-2, OXCART, and the SR-71. National Security Archive Electronic Briefing Book No. 74.. Available at: http://nsarchive.gwu.edu/NSAEBB/NSAEBB74/ [Accessed 28 March 2017].

Russon, M. (2015). Wondering how to hack a military drone? It's all on Google. *International Business Times*, [online] may 8. Available at:. http://www.ibtimes.co.uk/wondering-how-hack-military-drone-its-all-google-1500326 [Accessed 29 March 2017].

Starr, B. (2011). Drone that crashed in Iran was on CIA recon mission, officials say. *CNN*, [online] Dec 7. Available at: http://www.cnn.com/2011/12/06/world/meast/us-iran-drone/ [Accessed 29 March].

Stohl, R. (2014). Recommendations and report of the task force on US drone policy. The Stimson center. Available at:. https://www.stimson.org/content/recommendations-and-report-stimson-task-force-us-drone-policy-0 [Accessed 28 March 2017].

Stohl, R. (2016). Grading progress on US drone policy: Report card on the recommendations of the Stimson task force on US drone policy. The Stimson center.. Available at: https://www.stimson.org/content/grading-progress-us-drone-policy-0 [Accessed 28 March 2017].

Sweetman, B. (2012). Reading secret USAF bomber, ISR Plans. *Aviation Week*, [online] Dec 3. Available at:. http://aviationweek.com/awin/reading-secret-usaf-bomber-isr-plans# [Accessed 28 March 2016].

The White House Office of the Press Secretary. (2011). Remarks by President Obama and Prime Minister al-Maliki of Iraq in a Joint Press Conference. Washington D.C.: Office of the Press Secretary, [online] Dec 12. Available at:. https://obamawhitehouse.archives.gov/the-press-office/2011/12/12/remarks-president-obama-and-prime-minister-al-maliki-iraq-joint-press-co [Accessed 28 March 2017].

Tippenhauer, N. O., Pöpper, C., Rasmussen, K. B., Čapkun, S. (2011). On the requirements for successful GPS spoofing attacks. *Proceedings of the 18th ACM conference on Computer and communications security*. Pages 75–86. Chicago: USA. Oct 17–21.

United States Air Force. (2009). RQ-170 Sentinel. *Fact Sheets*, [online] Dec 10. Available at:. http://www.af.mil/AboutUs/FactSheets/Display/tabid/224/Article/104547/rq-170-sentinel.aspx [Accessed 28 March 2017].

UT News. (2013). UT Austin researchers successfully spoof an $80 million yacht at sea. *UT News*, [online] July 29.. Available at: http://news.utexas.edu/2013/07/29/ut-austin-researchers-successfully-spoof-an-80-million-yacht-at-sea [Accessed 28 March 2017].

# Wiretapping: The Interaction of Policy and Technology

**Ben Johnson**

**Abstract** In recent history, wiretapping has provided an effective means for uncovering potential security threats, both domestically and abroad. Wiretapping as a technology developed first in the 1890's with the development of the telephone recorder, which quickly followed the invention of the telephone in 1876. Through the years, as new communications technology was developed, methods were quickly adopted to capture the information shared through the new method. The biggest challenge initially with capturing the information has been the ethical and legal battles that have ensued. The ethics and legality associated with wiretapping has been accessed and reassessed with a slew of hearings, policies, and court orders. The amount of discussion on the issue makes the controversial nature of the method clear. Later, with the invention of computers and the internet, wiretapping technology was needed to allow the government a method to maintain security, and policies shifted from a focus on the legality of wiretapping to a focus on regulating technology so the government could maintain necessary security measures.

## 1 Introduction

Through its almost one hundred year history, wiretapping policy and technology have been intertwined, with new applications of the technology enabling law enforcement and intelligence with enhanced abilities to monitor individuals and collect information on targets and suspects. At the same time, there exists a tension with privacy rights. As is often the case, new technological developments have pushed the boundaries of policy. Courts have been persuaded both for wiretapping rights as well as for privacy rights. Legislative bodies have responded attempting to balance privacy rights and needs of law enforcement and intelligence. As a result, in the following examples it is clear how executive authority has swayed over the years, moving in a cyclical pattern in response to recent events of the day. In this

B. Johnson (✉)
Georgia Institute of Technology, Atlanta, Georgia, USA
e-mail: benjohnson@gatech.edu

 
M. E. Kosal (ed.), *Technology and the Intelligence Community*, Advanced Sciences and Technologies for Security Applications,
https://doi.org/10.1007/978-3-319-75232-7_7

chapter, several policies will be analyzed, beginning with the introduction of wiretapping in the early 1900's leading up to policies passed in recent years in response to the need for additional national security measures to protect the country domestically. The policies will be analyzed in the context of the executive freedom they provided, either intentionally or unintentionally, and in the context of the interaction between the policy and the technology at the time. These two factors (current policy and technological capability) are the two driving factors for executive authority. Within this context, a number of key events are analyzed under this framework ranging from Supreme Court decisions, key choices, and interpretations by the executive branch and legislation passed through Congress.

## 2 1928 - The Beginning

As in the case of many new technologies, the policies to regulate the use of wiretapping did not exist as it gained popularity. Shortly after the invention of the telephone, the capability to listen in on phone conversations by attaching a listening piece to the proper wire carrying the phone conversation was discovered. This gave law enforcement the ability to collect evidence without being physically in the room, giving birth to the beginning of executive freedoms exercised through wiretapping.

In 1928, the Supreme Court ruled wiretapping legal without a court issued warrant in a landmark case against the bootlegger Roy Olmstead (Olmstead v. United States 1928). Despite broad political and legal authority, the executive branch was limited by the technological capability of wiretapping. Law enforcement agencies began wiretapping as early as the 1890's after the invention of the telephone in 1876. However, in the earliest years, wiretapping required inserting a literal wire in the connection between the phone and the network switches. At first, this was done by connecting a phone receiver or listening device outside a person or organization to intercept the signal leaving the building. Later, this was replaced by recording devices (How Wiretapping Works 2001). The drawback to these types of technologies was the need to either listen continuously or have a large volume of recorded conversations to sift through. Overall, during this period of time, law enforcement had almost unlimited authority to wiretap, limited more by their view on ethical wiretapping practices than an established legal framework and by the technology available at the time.

## 3 1934 - First Congressional Attempts to Limit Wiretapping

In response to the seeming unlimited legal use of wiretapping available to the executive branch, Congress stepped in to limit the use of wiretapping. The result was the 1934 Communications Act (Communications Act of 1934 1934). In the Act,

Congress intended to outlaw all wiretapping with or without a warrant. The act stipulates:

"No person receiving or assisting in receiving, or transmitting, or assisting in transmitting, any interstate or foreign communication by wire or radio shall divulge or publish the existence, contents, substance, purport, effect, or meaning thereof, except through authorized channels of transmission or reception, to any person other than the addressee, his agent, or attorney, or to a person employed or authorized to forward such communication to its destination, or to proper accounting or distributing officers…".

Here, Congress attempted to define wiretapping via public law. The executive branch and law enforcement agencies argued federal agents were not "persons" as mentioned in the act and thus were not subject to the legislation (Atkins 1977). Later, after the Senate Interstate Commerce Committee called for tougher controls, the courts ruled the intention of the act was clear, and the Attorney General at the time advised Congress the law was clear and the FBI would not continue wiretapping (Atkins 1977). Despite this agreement, within a year as tensions mounted before World War II, President Roosevelt urged Attorney General Robert Jackson to issue internal Justice Department guidelines that only eliminated disseminating contents of wiretapped messages outside the executive branch (Atkins 1977), effectively giving the executive branch authority to collect information for intelligence purposes. As war approached, Congress even considered a number of laws to legalize wiretapping, all of which were opposed by FBI Director Hoover as any redefinition of the current law and any clarification of wiretapping practices would call attention to FBI practices and limit their ability to continue operations. The FBI at the time opposed both laws legalizing wiretapping as well as laws prohibiting it. The lack of definition allowed federal law enforcement to operate in a gray area without the hindrance of additional legislative hoops (Atkins 1977).

This period of time was characterized first and foremost by Congress attempting to limit the authority of the executive branch to wiretap. Despite legislation such as the 1934 Communications Act, the Executive branch did exert independent authority, leaving this period as one with more technical capability than in previous years and an Executive Branch largely unchanged.

## 4 Leading Into World War II

Leading up to World War II, there were clear national intelligence needs. Additionally, the policies regulating wiretapping remained vague at best. At the time, the three main institutions conducting intelligence operations were the FBI, the Military Intelligence Division, and the Office of Naval Intelligence. The Military Intelligence Division and the Office of Naval Intelligence were able to operate quite independently and focused the majority of their efforts on international espionage, an area which remained largely covert. The FBI, alternatively, focused the majority of their efforts domestically where policies regarding wiretapping had a much stronger impact. Since the existing policies were non-specific, the FBI was able to use clauses in existing legislation allowing them to collect information about activities deemed "subversive" or activities with an intelligence focus. Further, the director of the FBI noted in a memorandum:

> "Under this provision investigations have been conducted in years past for the State Department of matters which do not in themselves constitute a specific violation of a Federal Criminal Statute, such as subversive activities. Consequently, this provision is believed to be sufficiently broad to cover any expansion of the present intelligence and counter-espionage work which it may be deemed necessary to carry on... In considering the steps to be taken for the expansion of the present structure of intelligence work, it is imperative that it be preceded with, with the utmost degree of secrecy, in order to avoid criticism or objections by either ill-informed persons or individuals having some ulterior motive" (Atkins 1977).

Later, as technology advanced and especially with the development of the internet, the types of policies developed differed significantly. For example, initially polices were written to legally define wiretapping, often in the context of the communication technology available at the time. Much later as seen with the passage of CALEA in 1994, the policies shifted toward a focus on the specific aspects of the technology and regulating it to ensure the government could maintain this ability. The wiretapping technology available at the time was sufficient for the level of communications technology that existed. It was not until later when telecommunications technology expanded with the development of the internet and challenged the ability of the government to collect surveillance information using current technology.

The memorandum cited above was in response to growing interest in domestic espionage. During this time period, the idea of expanded domestic surveillance was growing in Congress and in the Executive branch (US Senate 1976). Domestic and international wiretapping differ with respect to the location of the targets and the ease of acquisition. Wiretapping domestically presents an easier challenge technically because of proximity and existing regulatory authority over the infrastructure. Politically and legally, tensions exist between enabling the mission and goals of law enforcement and protecting rights of citizens within the limits of the US Constitution. Internationally, obtaining information can be a more dominant factor, since foreign nationals outside the borders of the country do not have Constitutional protections. However, the technological and political challenges are greater since there are now other governments and other infrastructure systems involved. At this time period, the policy differences between international and domestic wiretapping were relatively minor due to the Supreme Court ruling on wiretapping. In the following years, there will be a growing difference between domestic and internationally wiretapping policies and practices however.

## 5 The 1950s to 1970s

In 1967, the Supreme Court reversed their previous decision regarding wiretapping in the case Katz v. United States (Katz v. United States 1967). In this decision, the Court found that wiretapping could represent an unreasonable search and seizure and that a seizure could be defined beyond actual property. In opposition to its previous decision, the Court found there is an expectation for privacy even in public places and that the FBI technology, despite having a listening device placed on the outside of the telephone booth, was in fact an unlawful seizure without a court issued warrant. The SCOTUS ruled that there is an expectation for privacy in a phone booth, particularly as demonstrated by an effort to close the booth doors, and that the FBI did not have the authority to conduct such operations and needed additional approvals from the judicial system.

After this landmark decision, wiretapping without a warrant was ruled Unconstitutional by the Supreme Court, and ability of law enforcement to conduct surveillance operations without going through the process of obtaining a warrant was severely limited. In response, Congress passed the Omnibus Crime Control and Safe Streets Act of 1968. Included in the legislation were exceptions relating to national security:

> "Nothing contained in this chapter or in section 605 of the Communications Act of 1934 (48 Stat. 1143; 47 U.S.C. 605) shall limit the constitutional power of the President to take such measures as he deems necessary to protect the Nation against actual or potential attack or other hostile acts of a foreign power…" (Omnibus Crime Control and Safe Streets Act 1968).

In the Act, Congress further clarified what wiretapping was legally. They limited domestic wiretapping substantially and made it consistent with the Supreme Court's ruling on search and seizure.

In 1978, Congress passed the Foreign Intelligence Surveillance Act (FISA) addressing a concerns of the Executive branch including those regarding wiretapping and formalized the relationship between national security and intelligence gathering activities, including wiretapping (US Congress 1978). The Act established the FISA Court, which is a secret court, allowing the Executive branch legal authority to obtain a warrant without exposing the nature of the operations to protect national security. Before, broad exceptions had been made for matters of national security, which the Executive branch had taken to be any subversive activity that it considered could potentially damage the country. The Act was in large part a response to events brought e to light by the Church Committee and how much authority they were taking. For example, under past policies, subversive activities that could potentially be wiretapped for security reasons were classified broadly into several categories including "Maritime Industry," "Activities in Organized Labor Organizations," "Activities in Government Affairs," among others (US Senate 1976).

Congress and the American public became aware of how much license the Executive branch was taking in a scandal involving wiretapping that originated outside of the intelligence community. Through what has become known as the Watergate Scandal, it was discovered that the Nixon administration had been wiretapping the headquarters of the opposition Democratic Party. Shortly thereafter, the House judiciary committee issued articles of impeachment against President Nixon due to efforts to cover-up the activities.

The Watergate scandal and the surrounding response is the first time the Executive branch was significantly curtailed in their wiretapping operations. Another takeaway from this scandal is the limitations of the technology available at the time. In the Watergate scandal, the perpetrators were caught inside the office of the Democratic National Committee (Craig 2012). It was later discovered they had installed listening devices into the phone of two prominent members of the Democratic Party (Pear 1992). While the technology for the listening devices was well developed, it was not perfect. Problems in one of the listening devices is what created the need for the second "burglary" to replace the faulty device, during which the burglars were caught.

In 1975, Congress held the Church Committee Hearings, a series of Senate investigations into CIA and FBI intelligence operations. In these meetings, it became clear the Executive branch was willing to conduct operations Congress deemed far outside the scope of national security. The Executive branch had wiretapped hundreds of people without court ordered warrants including leaders such as Martin Luther King, Jr., students groups, and a US Congressman (Markels 2005).

## 6 Shifting Technology Needs: The 1980s and 1990s

Advances in technology often necessitate a need for policy change. As new communication technologies are developed, old legislation is too narrow, with wording to specific to old technologies, to cover newer technologies serving the same or very similar functions. As an example, Congress passed the Electronic Communications Privacy Act in 1986 (Electronic Communications Privacy Act 1986), expanding the scope of the Federal Wiretap Act of 1968 (also known as the Omnibus Crime Control and Safe Streets Act of 1968), which was written to cover only physical telephone lines. The new policy updated the policy to cover electronic communications via computers and addressed electronically stored communications.

In 1994, with the passage of the Communications Assistance for Law Enforcement Act (CALEA), there was a shift in the wiretapping policies. Unlike previous policies, technological support for law enforcement and the executive branch was required by the industry. The Act "[r]equires a telecommunications carrier to ensure that its equipment, services, or facilities that provide a customer or subscriber with the ability to originate, terminate, or direct communications are capable of: (1) isolating and enabling the Government, pursuant to a court order or other lawful authorization, to intercept all of the subscriber's wire and electronic communications…" (Communications Assistance for Law Enforcement Act 1994). In essence, the law requires telecommunications providers to maintain the ability to wiretap specific phones so that the government with a valid warrant will be able to wiretap a given target. In the past, law enforcement hypothetically had the necessary technology to do any of the wiretapping they needed. With the advancement of telecommunications technology, the intelligence agencies and law enforcement no longer possessed the technology necessary to ensure wiretapping was available to them as an option when needed. In interest of national security, this legislations was passed making sure the technology was in place so law enforcement has the means it needs to intercept communications.

In 1998, wiretapping via computer networks was first used in a criminal case by the Department of Justice. An Argentine computer hacker was extradited and pled guilty based on evidence obtained by the first wiretap of a computer network. The wiretap was obtained legally using technology available via the internet. Keywords the hacker used as file names and specific targeted IP address were used to create a profile of the hacker to obtain the warrant (Rindskopf 1998). Technologically, this represents an important step in the prosecution of wiretapping offenses, particularly as it relates to international cybercrime. Before this case, warrants had to be obtained for individual people or individual phone numbers known to be used by people who are targets of law enforcement investigations. However, in the case of cybercrime it is very easy to disguise identities, necessitating the use of more general profiles for obtaining the warrant.

## 7 Post 9/11 Policies

On September 11, 2001, America experienced one of the most devastating terrorist attacks on domestic soil. The nation's response to this action was strong, and law enforcement found they wanted additional capability than was available to them domestically at the time. To give law enforcement the tools necessary to better complete their jobs, Congress passed the USA Patriot Act of 2001, which is an acronym for the longer, more descriptive name "Uniting and Strengthening America by Providing Appropriate Tools Required to Intercept and Obstruct Terrorism Act of 2001." The Patriot Act gave law enforcement officials far greater authority to conduct surveillance operations, such as collecting foreign intelligence information on those in the US and roving wiretaps, which follow a specific target as opposed to a particular line or phone. Through a combination of technological developments in the ensuing years and the use of CALEA to ensure law enforcement's access, law enforcement and the executive branch did not need any new technologies to quickly respond and increase their surveillance operations. They had the technical abilities and only needed to obtain the necessary legal authority to act more efficiently.

The Protect America Act of 2007 further codified the differences between domestic and international surveillance. The Act modified the FISA such that "nothing in the definition of electronic surveillance under section 101(f) shall be construed to encompass surveillance directed at a person reasonably believed to be located outside of the United States" (Protect America Act 2007). Over the years, FISA had been modified and with the world growing ever flatter, the Executive branch needed to ensure it had the capabilities to monitor foreign subjects and that the Constitutional protections on individuals were given to Americans without restricting the use of technology to monitor around the globe.

## 8 Analyzing the Shift in Technology

Over time, policies shifted from limiting the use of technology in the 1934 Communications Act and the Foreign Intelligence Surveillance Act to requiring technology in the form of hardware or software wiretapping mechanisms to enhance government surveillance capabilities and ensure the government had access to the communications networks in case of a national security matter. Recently, there has been another shift in technology that has been seen in the policies as well, although not stated explicitly. One of the biggest limitations on surveillance operations is the political authority required to conduct those operations. With the advent of new technology, specifically big data analytics technologies, restrictions on political authority impose significant barriers to conducting surveillance. In the surveillance world, big data is meaningless without the authority to collect and analyze large quantities of data. Restricting the possible sources to only those persons for which a court issued warrant has been collected severely limits the technology, if not eliminating its use entirely.

First, as communications technologies become more complicated and more far-reaching, the federal government needs assistance from the companies developing these capabilities to keep up. With regards to internet wiretapping capabilities in proposed legislation in 2010, law enforcement and national security officials mention "that their ability to wiretap criminal and terrorism suspects is 'going dark' as people increasingly communicate online instead of by telephone" (Savage 2010). The scope of these technologies created an intelligence-gathering problem for intelligence agencies. However, with the political authority to collect the data, the intelligence world is opened to the tremendous possibilities to use cutting edge data technologies for intelligence gathering and analysis.

Looking at the emergence of big data and analytics capabilities, this begs the question about what technologies are being used in surveillance operations and what capabilities exist to conduct those operations. Big data technology needs storage and infrastructure. The NSA has established a new data center in Utah in 2015. The cost of the data center has been estimated at $2 billion with estimates for NSA spending on data gathering around $5 billion (Bamford 2012; Brown 2015).

The exact nature of the technology available to the intelligence community is not clear, but there are hints as to what some of the capabilities may be. Then-President Bush described the intelligence situation at the time, saying "We use FISA still.... But FISA is for long-term monitoring....There is a difference between detecting so we can prevent, and monitoring. And it's important to know the distinction between the two....We used the [FISA] process to monitor. But also....we've got to be able to detect and prevent" (Drum 2005). The quote provides insight into thinking for the future of intelligence. To the former President's point, there is a difference between the technologies used for these situations. For monitoring, once a problem has been identified, the FISA process is available. To detect and prevent situations from occurring, you need to be ahead of the terrorists, often before a problem is known that can be monitored.

After the Patriot Act was passed, there was a period of time when the NSA was using a combinations of policies to pursue mass surveillance within the United States. In the name of preventing domestic terrorism, they were storing vast numbers of American's phone records without any sort of court warrant. The technology needed to use and process this kind of data requires extensive data mining capabilities. Once revealed, this type of capability domestically was changed through the passage of the USA Freedom Act, which left in place the majority of the surveillance options for law enforcement but made sure to eliminate the mass data collection by the NSA domestically (Mascaro 2015). This limit, however, does not apply to the use of these capabilities internationally on foreign subjects.

Another hint at the desired surveillance capabilities of the intelligence community comes in the form of a Defense Department research program called Total Information Awareness, which was developed by the Total Information Awareness Office developed at DARPA (Ron Wyden Senate Office 2003; Stokes 2006). The program was designed to take advantage of developments in computing technology and data analytics, which was meant to reduce the number of people needed to manually analyze data, but rather, computers could be used to sift through large amount of data in the form of phone calls, text messages, and other types of electronic data with the goal of identifying patterns related to terrorism. If used to its fullest potential, a program like this could be used to discover a trove of useful information. Technology alone, while it can be effective when used, will not solve the intelligence gathering problem in relation to wiretapping.

## 9 Conclusions and Thinking Forward

Technology is one effective means for intelligence gathering. With the proper authority and legislation passed to allow wiretapping, the technologies used are capable of collecting vast quantities of data. Law enforcement and intelligence authorities have had capable technologies for wiretapping on parties of interest. Domestically, those technologies have been capable enough that through much of history, the technology was not the limit on the ability of the government to collect intelligence. Rather, the limitations on the intelligence community domestically have been the efforts to balance of citizen's privacy rights and national security interests, as manifested through the ability to obtain the proper warrants through the judicial system. Historically, the Executive branch has taken any authority it has been granted. Once discovered, Congress and the court system then restores a more appropriate balance of power regarding domestic surveillance. International surveillance is mostly open to the technologies available to the intelligence community. The technologies have been enabled through the passage of legislation such as CALEA, which ensures the executive branch has the technical ability to wiretap communications.

The power of the executive branch to wiretap is a combination of the legal authority to conduct surveillance operations and the technical ability of the intelligence community to collect and decipher information. The technical capability of surveillance has grown substantially over the years, beginning with a hard "wiretap" placed on the phone lines. Now, legislation (e.g., CALEA) has enabled the Executive branch access to communications, when allowed through the judicial process, and big data analytics technologies may enable the ability to analyze huge portions of data.

In the early years, the Executive branch had virtually unlimited permissions when the Supreme Court initially decided wiretapping did not require a court warrant. Later that decision was reversed, but not until Congress passed legislation outlawing any form of wiretapping with or without a warrant. In times or warfare or terrorism, the Executive branch is granted greater freedom to operate independently. In more peaceful times, Congress will pull back the reigns on the Executive branch.

Overall, technologies have affected the policies developed through CALEA and the mandate that the Executive branch have the technical capability to wiretap even if they have not received court approval. Additionally, policy decisions have affected the technologies available for surveillance operations and what can be done with those technologies. Clearly, there is a strong interaction between the technologies available for surveillance and the policies regulating it. One thing we know for certain, however, is that technology in and of itself is not sufficient to solve the surveillance problem. There are other factors which must be considered. Lastly, it is also clear the large impact technology has had on wiretapping policies. When advanced technologies did not exist, policies regulating wiretapping had a lesser impact. Now, with increasingly advanced technologies and data analytics capabilities, those policies regulating surveillance activities are critically important to maintain the balance of power and will continue to be as technology advances farther into the future.

## References

Atkins, D. (1977). Wiretapping legislation: Past and present. *First Principles: National Security and Civil Liberties, 3*(1), 1–6.

Bamford, J. (2012). The NSA is building the country's biggest spy center (Watch what you say). *Wired.* https://www.wired.com/2012/03/ff_nsadatacenter/. Accessed 2 Dec 2017.

Brown, M.S. (2015). NSA mass surveillance: Biggest big data story. *Forbes Magazine.* http://www.forbes.com/sites/metabrown/2015/08/27/nsa-mass-surveillance-biggest-big-data-story/#15974adc9acc. Accessed 3 May 2016.

Communications Act of 1934. (1934). http://www.criminalgovernment.com/docs/61StatL101/ComAct34.html. Accessed 28 Apr 2016.

Communications Assistance for Law Enforcement Act. (1994). United States.

Craig, S. (2012). The bartender's tale: How the watergate burglars got caught washingtonian. http://www.washingtonian.com/2012/06/20/the-bartenders-tale-how-the-watergate-burglars-got-caught/. Accessed 2 May 2016.

Drum, K. (2005). What is the NSA up to? *The Washington Monthly*. http://www.washington-monthly.com/archives/individual/2005_12/007812.php. Accessed 3 May 2016.

Electronic Communications Privacy Act. (1986). US Department of Justice. https://it.ojp.gov/privacyliberty/authorities/statutes/1285. Accessed 3 May 2016.

How Wiretapping Works. (2001). How stuff works. http://people.howstuffworks.com/wiretapping1.htm. Accessed 28 Apr 2016.

Katz v. United States. (1967). US, vol 389. Supreme Court.

Markels, A. (2005). Timeline: Wiretaps' use and abuse. *NPR*. http://www.npr.org/templates/story/story.php?storyId=5061834. Accessed 2 May 2016.

Mascaro, L. (2015). Congress' passage of NSA bill will rein in surveillance, a first since Sept. 11. *Los Angeles Times*. http://www.latimes.com/nation/la-na-senate-advances-nsa-20150602-story.html. Accessed 3 May 2016.

Olmstead v. United States. (1928). US, vol 277. Supreme Court.

Omnibus Crime Control and Safe Streets Act. (1968). United States.

Pear, R. (1992). 2 decades after a political burglary, the questions still linger. *The New York Times*. http://www.nytimes.com/1992/06/15/us/watergate-then-now-2-decades-after-political-burglary-questions-still-linger.html. Accessed 2 May 2016.

Protect America Act. (2007). United States.

Rindskopf, A. (1998). Argentine computer hacker agrees to waive extradition and returns to plead guilty to felony charges in Boston. *DOJ*. https://fas.org/irp/news/1998/05/arditasnt.htm. Accessed 3 May 2016.

Ron Wyden Senate Office. (2003). *Wyden Calls for Congressional Oversight*. Accountability of Total Information Awareness Office. Ron Wyden Senate website: https://www.wyden.senate.gov/news/press-releases/wyden-calls-for-congressional-oversight-accountability-of-total-information-awareness-office. Accessed 25 Apr 2016.

Savage, C. (2010). US tries to make it easier to wiretap the internet. *The New York Times*. http://www.nytimes.com/2010/09/27/us/27wiretap.html?pagewanted=all&_r=0. Accessed 3 May 2016.

Stokes, J. (2006). TIA (aka Topsail) unveiled: the real scope of the NSA's domestic spying program. http://arstechnica.com/uncategorized/2006/05/6813-2/. Accessed 25 Apr 2016.

US Congress. (1978). Foreign intelligence surveillance act of 1978. *Public Law*, 95–511.

US Senate. (1976). Supplementary detailed staff reports on intelligence activities and the rights of Americans. Book III. Final report of the select committee to study governmental operations with respect to intelligence activities.

# Nuclear Counterproliferation Intelligence

**Abdalla Abou-Jaoude**

**Abstract** Spying on nuclear weapons programs can be particularly challenging for the international community. A wide range of sophisticated technologies need to be employed to gather information, while analyzing complex technical programs isn't without its own challenges. From the review of the US intelligence efforts on the Iraqi, Iranian, Syrian, and North Korean program, it is found that responses from states to apparent violations of the NPT are not always rooted in accurate intelligence analysis. Furthermore, responses are often counterproductive and embolden a state's resolve to pursue nuclear weapons. Lastly, intelligence agencies can have distorted views of programs due to a range of political, cultural, bureaucratic, or organizational factors. This goes to demonstrate that technologies cannot reduce the reliance on sound analyses and that the 'human-in-the-loop' will always maintain a crucial role in the investigation of these types of programs.

## 1 Acronyms

**Acronyms used in the Chapter and their definitions**

| | |
|---|---|
| AFTAC | Air Force Technical Applications Center |
| CIA | Central Intelligence Agency |
| CMO | DIA's Central MASINT Office |
| COMINT | Communication Intelligence |
| CTBTO | Comprehensive Nuclear-Test-Ban Treaty Organization |
| DCI/DNI | Director of Central Intelligence/Director of National Intelligence |
| DIA | Defense Intelligence Agency |
| DOE | Department of Energy, Office of Intelligence and Counterintelligence |
| DPRK | Democratic People's Republic of Korea |

(continued)

A. Abou-Jaoude (✉)
Nuclear & Radiological Engineering Program, Georgia Institute of Technology, Atlanta, Georgia, USA
e-mail: aj.abdalla@gatech.edu



M. E. Kosal (ed.), *Technology and the Intelligence Community*, Advanced Sciences and Technologies for Security Applications,
https://doi.org/10.1007/978-3-319-75232-7_8

| | |
|---|---|
| HEU | High-Enriched Uranium |
| HUMINT | Human Intelligence |
| IAEA | International Atomic Energy Agency |
| IMINT | Imagery Intelligence |
| IMS | International Monitoring System |
| INR | State Department Bureau of Intelligence and Research |
| LEU | Low-Enriched Uranium |
| LWR | Light Water Reactor |
| MASINT | Measurements and Signatures Intelligence |
| NIE | National Intelligence Estimate |
| NIMA | National Imaging and Mapping Agency |
| NPT | Non-Proliferation Treaty |
| NRO | National Reconnaissance Office |
| NSA | National Security Agency |
| NSG | Nuclear Supplier Group |
| SISMI | Italian Military Intelligence and Security Service |
| UNSC | United Nation Security Council |
| UNSCOM | United Nation Special Commission on Iraq |
| WINPAC | CIA's Weapons Intelligence, Nonproliferation, and Arms Control Center |
| WMD | Weapons of Mass Destruction |

## 2 Overview of Nuclear Programs

This section provides an overview of the selected nuclear programs in Iraq, Iran, North Korea, and Syria. The four cases are taken to be representative of future trends, with 'rogue' states developing secretive programs to gain leverage in the international community. Due to the multiple developments in the Iraqi program, it is divided in this section between the pre-1991 and the post-1991 phase. The overview emphasizes both historical developments as well as intelligence gathering efforts. Table 1 summarizes the main types of responses in each case (whether military strikes or broad diplomatic agreements). It is noted that a coupling of diplomatic and military responses is not observed. A country is more likely to suffer additional military attacks having been the victim of strikes earlier, i.e., Iraq. Similarly, even when diplomatic efforts fail, the international community has preferred to keep resorting to this option, i.e., Iran and North Korea. This aspect of interventions is discussed in further detail in Sect. 3.3.

**Table 1** Overview of responses on the selected nuclear programs. Note that the list of diplomatic responses is not exhaustive

| Military Responses | Diplomatic Responses |
|---|---|
| Iraq | North Korea |
| – Operation Opera (1981) | – Denuclearization agreement (1992) |
| – Persian Gulf War (1991) | – Agreed framework (1994) |
| – Iraq War (2003) | – Statement of principles (2005) |
| Syria | Iran |
| – Operation Orchard (2007) | – Paris Agreement (2004) |
| | – Joint Comprehensive Plan of Action (2015) |

## 2.1 *Iraqi Program Pre-1991*

Iraq's nuclear program can be traced back to the mid-1950s. In 1959, the country obtained two small 2 MW research reactors from the USSR. Following the Baathist coup, the program accelerated in the mid-1970s (Braut-Hegghammer 2011). This began fueling international concerns, most notably Israeli fears that Iraq was attempting to pursue the plutonium path to the bomb. The Iraqi government began negotiations with France in 1974 to acquire a larger research reactor and with Italy for reprocessing laboratories. Israel's main fears originated from claims by Saddam Hussein regarding "the first Arab attempt at nuclear arming" and senior officials quoted saying, "if Israel owns the atomic bomb, then the Arab must get a bomb" (Smith 1978). These concerns later grew after Iraq refused to switch the fuel for the research reactor it was receiving from France from highly-enriched to low-enriched uranium (HEU to LEU) fuel.

In reality, the program started out as relatively open-ended, with the main objective being to acquire know-how in various aspects of the nuclear fuel cycle (Iraqi Atomic Energy Commission 1992). Key members of the program even admitted they struggled to identify the real intentions of the political leadership. It wasn't until 1975, that Saddam Hussein is believed to have finally given the order for the establishment of a program. Even then, the endeavor was still abstract, without any dedicated infrastructure or budget. Following the agreements with France and Italy, Iraq acquired the basic capability to reprocess small amounts of plutonium. By 1979, the country could reprocess around one fuel assembly at a time from one of the earlier research reactors it had obtained from the USSR (Braut-Hegghammer 2011).

International concerns began to join Israeli ones following the agreement with France to provide Iraq with the two Osirak reactors. Iraq originally requested a graphite reactor, but was turned down because it was deemed obvious that this couldn't be used for any power or research purposes (Richelson 2007). This gravely worried the Israeli intelligence services, which began a systematic assassination and sabotage campaign (Corera 2006). The Mossad leaked information about the reactors to media outlets such as the *London Daily Mail* to stir up international sentiment (Ford 2004). The agency took more aggressive actions in 1979 with the detonation of a facility building the reactors in France, only two days before their shipping date. In 1980, an Egyptian physicist known to have been working on the program was assassinated (Corera 2006). Meanwhile, the 1979 Iranian Revolution had in fact, substantially disrupted the program with many members arrested or dismissed on suspicions of supporting Shiite movements opposed to the regime. The program appeared to be in standstill in the background of the Israeli attack in 1981 (Braut-Hegghammer 2011).

The main concerns with the Osirak reactor were twofold: (1) HEU fuel could be diverted for a uranium weapon, and (2) plutonium could be bred in the reactor, then extracted in a reprocessing plant. Even though the French offered guarantees on how much HEU would be available to the Iraqis at any one time and that the reactor was to be under IAEA safeguards, this did not alleviate Israeli concerns. Much later, a *Nature* study would go on to show how the design itself was relatively non-ideal for plutonium production even (Wilson 1983). It appears that the Israeli intelligence community had incorrectly assumed the same design specification as a typical French Osirak reactor, which was not the case (Government of Israel 1981). While the reactor could be modified to enhance plutonium production – by either replacing assemblies with natural uranium or using the neutron hall – US estimates indicated that it would take Iraq between 10 to 30 years to divert enough plutonium using routine operations (US Congress 1981).

Regardless of these subtleties, the Israeli Air Force launched an attack on June 7, 1981. Eight F-16 jets were dispatched, evaded Iraqi antiaircraft defenses, and destroyed the reactor core and the buildings nearby (Black and Morris 1991). The next day, the Israeli government acknowledged its role in the attack and provided justification for the strike (Government of Israel 1981). The attack brought the expected UN condemnation.

In the aftermath of the strike, several key scientists of the Iraqi program began pressing the leadership to double down on the developments. The program was to continue covertly and would shift from the reprocessing route to the enrichment route (Stone 2005). Relying mostly on satellite imagery, the CIA determined that the strike was a significant setback to the program. In a report, the Director of Central Intelligence (DCI) assessed that it would take Iraq several years before the country could rebuild capacity (Director of Central Intelligence 1981).

Between 1982 and 1987, at least six secret weapons laboratories were established. The program investigated calutrons, thermal diffusion, and centrifuge enrichment. Calutrons were considered obsolete at the time, and this was one of the main reasons they were 'missed' by intelligence agencies. The program did not completely abandon reprocessing, and a separate laboratory was focused on developing that type of capability (Davis and Kay 1992). Significant assistance came also from a wide range of international companies. Interatom, a subsidiary of Siemens was an important contributor. A Japanese firm provided high speed video testing that could be used for implosion studies. Yugoslavia actively assisted in the management and procurement of material to complexes. Even the US assisted the Iraqi program following their defeat in the Iran-Iraq war. The Commerce Department authorized sales of up to $1.5 billion of sensitive technology to Iraq (Hedges and Cary 1991).

Iraq attempted to intentionally mislead intelligence efforts on their program. Having been assisted by the CIA with satellite images during the Iran-Iraq war, it knew what signals the agency looked for (Woodward 1986). As such, all buildings in the complex were not suggestive of what was inside, and they were designed to prevent any material from leaking out. Most facilities had dual applications, e.g., civilian reactor operation, manufacturing window frames etc. The program also tried to convey that what they were doing was of limited importance by not heavily guarding facilities and only using light fences. Knowing that US agencies would look for electricity diversion patterns, the country even built an independent power plant and buried the cables (Burrows 1994).

By 1988, the CIA became increasingly aware of proliferation efforts and started raising concerns. However, even as far as 1991, doubts persisted on whether Iraq was close to building a weapon. High-ranking officials believed this was still around 10 years away (US Congress 1992). In the lead up to the 1991 invasion, intelligence-gathering efforts were intensified. Resources started being diverted from Eastern Europe following the fall of communism. Foreign intelligence provided crucial assistance, with some even indicating that Iraq might be able to obtain the bomb within a year (Hertz 1990). In the end, it wasn't the nuclear line that was crossed in the summer of 1990, but another: the border with Kuwait. This prompted the US to launch an international coalition to invade Iraq in what became known as the First Gulf War. Military operations began in January of 1991 and ceased only a month later.

## 2.2 *Iraqi Program Post-1991*

In the aftermath of the First Gulf War, the UN Security Council demanded that Iraq disclose all nuclear related activities and submit them to inspection. A subsequent UN Resolution ordered the destruction of all biological, chemical, nuclear, as well as ballistic technology in the country. The IAEA was instructed to carry out immediate inspections and to designate suspect undeclared facilities. The first inspection team arrived on May 14 1991.

During the inspection process, Iraq thrived to be as uncooperative as possible. The administration went as far as setting up an entire agency with the sole purpose of feeding misinformation and concealing as much as it could about the program. Nevertheless, the IAEA inspections were slowly able to roll back the nuclear program. With assistance from US intelligence agencies (mostly in the form of satellite intelligence), the inspectors uncovered enriched uranium, separated plutonium, calutron technology (which they were surprised to find), and reports on important facilities. Following threats from President Bush to pursue further military action, Iraq was pressured to reveal even more of the program. Upon discovering the scale of the Iraqi program, the UN passed Resolution 707 condemning the breach of the Non-Proliferation Treaty (NPT) and Iraq's failure to meet its obligations. By the end of 1991, uranium enrichment and processing facilities were all destroyed. The joint

IAEA and US efforts to unravel the program were proceeding faster than Iraqi attempts to conceal it (Richelson 2007).

Despite progress made, the IAEA conceded that the extent of the program came as a shocking surprise (Smith 1991). Around mid-1992, activity could still be detected around some of the sites, and this lead a 1993 assessment to claim that 'Iraq sill presents a threat' (US Congress 1993). As Richelson (2007) points out in his book, while Iraq had lost most of its facilities, it maintained one key asset: skilled personnel with expertise. This meant that the country could rebuild its program much faster than any other non-nuclear nation. Inspections were uninterrupted throughout 1995, at which point the UN Security Council concluded that the essential components of the program had been identified and destroyed; any remaining uncertainties were judged to be mostly detail (IAEA 1995). In 1998, Saddam Hussein announced that the country would cease all cooperation with the UN Special Committee on Iraq. However, by that point, the IAEA assessed that Iraq did not appear to retain any practical capability (Richelson 2007).

Nevertheless, by the dawn of the new century, the US became increasingly concerned about a resumption of the Iraqi program. Several defectors approached the CIA between 2000 and 2001 claiming to be working on a renewed program; these claims could not be validated. The intelligence agencies began taking a more serious look at the matter following a report from the Italian intelligence agency, SISMI, claiming that Niger was planning to ship several tons of uranium to Iraq. The report asserts that negotiations had been underway since 1999 and were concluded in 2000 (Hersh 2003). A second report from SISMI in 2002 provided additional information, including the alleged text of the accord – which would later be discovered to be a forge (Richelson 2007). The CIA dispatched an agent to the country to discreetly verify the story by interviewing officials. He saw no evidence of a sale, strongly questioned the authenticity of the signatures, and even the possibility of conducting such a sale covertly (Hersh 2003). Despite of this, the 'Niger intel' would prove to be one of the main pillars upon which the CIA based its assessment of the Iraqi program.

The second pillar was the now infamous, 'Aluminum Tubes.' The information here was more robust than the Niger case. The intelligence community uncovered that Iraq was trying to purchase high strength aluminum tubes that were controlled items by the Nuclear Suppliers Group (NSG). Agencies detected a shipment of 2000 tubes leaving China to Hong Kong, on route to Iraq via Jordan. The information was obtained from the Australian intelligence agency, and is believed to have contained a combination of intercepted faxes, emails, and telephone conversations relating to the deal. The CIA allowed the material to be shipped and subsequently seized it as soon as it arrived in Jordan (Richelson 2007, Australia Broadcasting Corporation 2003). This would prove to be the foundation of the CIA's case against the Iraqi program.

Supplemental intelligence was also obtained regarding the program but was largely unfounded, uncertain or invalidated. For instance, the US intelligence communities obtained reports that scientists were being reassigned to reconstruct the nuclear program, but other reports indicated that the majority of scientists had either retired or left Iraq by this point (US Congress 2004). Due to the level of conflicting information and uncertainties, different agencies drew varying conclusions about the Iraqi program at the time. This pitted the CIA and DIA against the State Department's Bureau of Intelligence and Research (INR) and the Department of Energy's (DoE) office of intelligence (hereof referred to as DoE). The CIA's Weapons Intelligence, Nonproliferation and Arms Control Center (WINPAC) acknowledged that there might be other application for the tubes, but its analysts strongly believed they were most likely intended for centrifuges. WINPAC claimed these would fit well for use in the 'Zippe centrifuge' design of the 1950s, and noted Iraq's known tendency to revert to relatively outdated technologies (such as calutrons) to avoid detection. The DoE countered, based on analysis from technical experts, that these tubes would be inefficient for use in centrifuges and that Iraq had used even more advanced material in the 1980–1990s. Expert analysis concluded that Iraq would need twice the number of centrifuges that it had in the 1990s; several thousand would need to be built, something no proliferator had ever done before. The agency also noted that Iraq had previously obtained similar tubes for their rocket program, and they believed that this was the most likely intended application. The INR on the other hand, questioned the CIA's analysis of the Niger data, noting that France had direct control over the country's uranium industry and would have been aware of a potential deal (US Congress 2004).

In parallel to intelligence debates on Iraq, the US administration started leaning towards military intervention. In his 2002 State of the Union address, President Bush rebuked Iran, Iraq, and North Korea for posing a threat to international security. During that year, it became increasingly clear that the administration was willing to go to war with Iraq, accusing it of violating the terms of the 1991 ceasefire agreement. The rhetoric against Iraq gradually escalated and assertions about intelligence proofs appeared more certain. As the debate was taken to the UN, many allied countries to the US backed the accusations on the Iraqi programs through their intelligence analyses. Evidence was provided to the IAEA to make the case for an intervention, but the agency concluded that the Niger intel was clearly a forgery. With the US threatening to take unilateral action, the UN passed Resolution 1441, declaring that Iraq had breached existing resolutions and called on it to provide a declaration of any renewed activity. Inspectors were allowed into the country in 2002, but they found no substantial evidence of revived activities. Despite Iraq submitting the requested declaration, the US administration found it to be unsatisfactory. After weeks of unsuccessful diplomacy following the opposition of France, Germany, and Russia; the US and UK concluded that the Security Council would not approve further action against Iraq. On March 17, 2003, the US gave Iraq an ultimatum ordering Saddam Hussein to leave the country or face military action. Following his refusal, military operations began on March 17 and were deemed 'completed' less than one month later (Richelson 2007).

With the fall of the regime, intelligence agencies believed it would be easier to search the entire country without obstruction. The opposite proved to be true, with suspects intimidated and threatened by militants, while the survey teams were frequently attacked during inspections. The subsequent chaos in the country rendered the Iraq Survey Group's task all the more difficult. Even though the team consisted of over 1300 individuals, it was not able to identify any stockpiles of WMD-related program activities that were allegedly concealed to the 2002 UN inspections (Richelson 2007). In a final three-volume report in 2004, the head of the survey team concluded that while Baghdad did maintain interest in nuclear weapons and tried to maintain its capabilities, there were no signs that it had revived its program. The report also invalidated claims regarding the aluminum tubes and the Niger uranium (Duefler 2004).

## 2.3 *Syrian Program*

Information on the Syrian program has been relatively scarce. This is mostly attributed to how recent military development against it were, and to the chaos resulting from the raging civil war that has kept the IAEA from accessing crucial sites. The country signed the NPT in 1968 and started discussions with the IAEA about developing nuclear power around the mid-1970s. After many unsuccessful attempts at finding contractors to build reactors, the country was able to ink a $100 million deal with Argentina in 1990 (Kessler 1990). The deal was vetoed shortly after however, under alleged pressure from the US and Israel. A similar deal with India was also shelved in 1991 after significant US pressure (Nuclear Threat Initiative 2016c). Finally, in 1991, China agreed to build a research reactor that would be fueled with HEU. It went critical in 1996 and was placed under IAEA safeguards (IAEA 2008a).

Around this time, the CIA began actively monitoring the Syrian program. In 1996, the agency reported that the program was still very rudimentary, and there was no clear evidence of attempts to proliferate (Deutsch 1996). The US did remain cautious and pressured Russia into halting deals signed with Syria. In 2004, a CIA report alleged that the A.Q. Khan network provided Syria with nuclear information and equipment (Tetrais 2004). Prior to 2007, there appears to have been limited concern from US intelligence agencies, except in terms of known chemical weapons programs (Nuclear Threat Initiative 2016c).

On September 6, 2007, Israel launched an airstrike in the northeastern region of Dair Alzour. A facility was destroyed and believed to have housed a partially completed graphite reactor, capable of producing enough plutonium for 1–2 bombs per year (IAEA 2008b). Israeli authorities maintained silence over the issue, and even Syria stated that it was an unused military building. Despite these assertions, Syria leveled what remained of the site and built over it before any external access was ever granted. In the midst of wild speculations from the media on the nature of the site, in April 2008, the US released satellite intelligence of the site prior to the airstrike. The facility appeared strikingly similar to the Yongbyon production reactor, suggesting North Korean assistance in its development (Albright and Brannan 2008). In light of the allegations, the IAEA was finally provided unrestricted access to the site on June 2008. The agency reported that the size and water pumping capacity fits with the reactor allegations made. Inspectors also found traces of natural uranium particles at the site (IAEA 2009a). Syria quickly retorted that the particles had derived from Israeli munitions, although this was thought to be very unlikely.

After further satellite intelligence and analysis, the IAEA adopted a resolution in June 2011 accursing Syria of being in non-compliance with its obligations under the signed safeguards agreements. However, with the start of civilian uprising in the country, China and Russia were opposed to any action against the country in the UN Security Council. The Syrian dossier has been postponed ever since, in light of the deteriorating security situation. In February 2013, it is believed that Syrian rebels – including Al Nusra affiliates – took control of the Alzour facility. They were then overtaken by Islamic State (IS) troops in 2014. The terrorist organization apparently tried to excavate the site in search of radioactive material (Nuclear Threat Initiative 2016c). In January 2015, *Spiegel Online* published an article alleging that the Syrian government was still trying to develop nuclear material at a secret facility near the Lebanese border (Follath 2015). However, an assessment in June 2015 stated that satellite imagery did not support the assertion (IHS Global Limited 2015).

## 2.4 Iranian Program

The Iranian nuclear program was first established in 1957. A research reactor was obtained from the US in 1960. The program's ambitions grew in the 1970s with the signing of an agreement with a German company to build two reactors at Busheher in 1974 (Albright and Sticker 2015). Around that time, it appears that the Shah wanted to keep the option of building nuclear weapons open. According to a 2003 interview with *Le Figaro*, the head of the Iranian nuclear program during the 1970s stated that a special research team was tasked with investigating the weapon option (Lorieux 2003).

Following the Iranian Revolution of 1979, all nuclear efforts were suspended due to opposition of the Ayatolla Khomeini, who deemed it 'un-Islamic' (Nuclear Threat Initiative 2016b). Nonetheless, efforts were restarted in the wake of the Iran-Iraq war. Demands for a weapon were bolstered by tensions with the US and growing evidence of an Iraqi program. It is believed that the final decision to start a weapons program was made in 1984. The centrifuge development efforts began the following year, with significant assistance from the A.Q. Khan network. The country then proceeded to take deliberate steps to build up its nuclear capacity. In addition to enrichment, it developed uranium mining and milling infrastructures, as well as an indigenous heavy-water reactor capacity (Albright and Sticker 2015).

In the 1990s, US intelligence agencies began monitoring the Iranian program more closely. A detailed report from 1991 concluded that Iran was actively trying to acquire the bomb (Sciolino 1991). The US blocked many sales of nuclear goods to Iran from China and Argentina during that time. In 1993, reports surfaced that Iran was recruiting former nuclear engineers from the USSR and may have even purchased warheads, the latter of which was an overestimation (Richelson 2007). Intelligence agencies then received reports that Iran was trying to purchase HEU from Kazakhstan in 1994; in response, the US preventatively purchased all the material and transported it out of the country before Iran could gain access to it (Gertz 1994). In the face of these proliferation attempts, all US firms were banned from doing any business with Iran, while Russia, China, and the Czech Republic were pressured to refrain from any nuclear sales to the country (Richelson 2007).

Despite some intelligence successes, the CIA and NSA admitted in 1999 that they had low confidence in their ability to monitor the Iranian program and that they couldn't rule out the possibility that weapons capability was already acquired (Risen and Miller 2000). In 2002, satellite images of the Natanz (enrichment) and Arak (reactor) facilities were made public (Albright and Sticker 2015). As concerns continued to grow about the program, the IAEA was allowed access to the Natanz facility in 2003 where they uncovered the centrifuge program (Richelson 2007). With evidence on the program mounting, negotiations started with Western countries, and the Paris Agreement was signed in November 2004. Tehran agreed to temporarily suspend its enrichment activities and to seek-out a long-term diplomatic solution (IAEA 2004).

Diplomatic progress broke down in 2005, with Iran deciding to resume enrichment. The IAEA adopted a resolution stating the country was in non-compliance with safeguards agreement, and the US imposed further bans on doing business with Iranian entities. In response to continuing Iranian defiance, the UN Security Council passed Resolution 1696 in July 2006 demanding a halt to nuclear activities, banning international transfer of any nuclear material to the country, and freezing of assets of organizations associated with the program. Iran continued to defy UN resolutions and pursued enrichment activities, leading to additional sanctions on the country (Nuclear Threat Initiative 2016b). Around 2008, the US and Israel are believed to have lead cyberattacks against the Iranian program via the now infamous computer virus Stuxnet (Sanger 2012). This was in parallel to an alleged Israeli campaign to assassinate Iranian scientists (Sinha and Beachy 2015).

With tensions continuing to rise and multiple talks ending in failure, the US, France, and UK declassified substantial amounts of intelligence on the Iranian program in 2009. They also disclosed a secret underground enrichment plant in Fordo (DeYoung and Shear 2009). In the face of these continued violations by Tehran, the US Congress imposed sanctions on all foreign companies dealing with the Iranian oil industry (Reuters 2009). The IAEA was finally given access to the underground facility near the end of 2009, where it confirmed that it can house additional centrifuges (IAEA 2009b). The UN Security Council imposed additional sanctions in 2010, while negotiations continued to stall with Iran. Following the release of another damning IAEA report, the UN approved further sanctions in November 2011, as well as a series of unilateral actions. With speculations growing of a possible Israeli strike on Iran, the international community became more forceful in its imposition of sanctions. That same year, the US Congress also enacted an amendment sanctioning any financial institution for processing Iranian oil transactions. The European Union followed suite by freezing all assets of the Central Bank of Iran and phasing-out Iranian oil imports (Nuclear Threat Initiative 2016b).

With economic pressure on Iran mounting, Washington and Tehran held back-channel negotiations between 2009 and 2013. The election of the moderate Hassan Rouhani to the presidency paved the way to the Joint Plan of Action in November of 2013. The agreement placed a freeze on Iranian nuclear activities and introduced greater transparency. This was followed by the Joint Comprehensive Plan of Action, known in the US as "the Iran Deal," which was signed in July 2015. Iran agreed to limit is nuclear program and give enhanced inspector access to facilities in return for sanction relief (Albright and Sticker 2015).

## 2.5 North Korean Program

North Korea is the only one of the four cases that can be deemed a 'failure' of the nonproliferation regime. The country acquired material for bombs and conducted six underground tests at the time of writing. The intelligence community cannot take all the blame for missing the nuclear proliferation; multiple reports had already uncovered the scale of the program even before the tests were conducted. A combination of several factors including failed diplomacy and lack of confidence in intelligence contributed to the country being able to reach weapons capability.

The nuclear program is believed to have started as early as 1955 with a delegation being sent to Moscow and the subsequent signing of a collaboration agreement. North Korean engineers received training in the USSR as well as assistance in the construction of the IRT-2000 research reactor during the 1960s. Around that time, it is reported that the DPRK requested from Beijing to share its nuclear weapons technology following the country's successful 1964 test. However, China is believed to have refused (Bermudez 1991).

During the 1970s and 1980s, the program continued largely under the metaphorical radar of the international community. North Korea was able to acquire reprocessing technology from the USSR, construct a uranium fuel fabrication complex, another small 5 MW reactor, and started building a larger 50 MW reactor. The original IRT reactor was placed under safeguards in 1977 and the country signed the NPT in 1985 (Nuclear Threat Initiative 2016a). The CIA underestimated the DPRK's capabilities during this time (CIA 1986), but suspicions began to grow after the agency became aware that the new 5 MW reactor was completed in 1986. The country's refusal to implement NPT reforms and to allow inspector access, increased these fears. Based on its assessments, the CIA concluded in the late 1980s that the nation might be willing to risk international sanctions in order to acquire a military edge over its southern rival (CIA 1989).

These concerns lead to the announcement of President George H.W. Bush that the US would withdraw nuclear weapons from South Korea. The decision was subsequently followed by the Joint Declaration on the Denuclearization of the Korean Peninsula in 1992. The agreement bound each side not to develop or build nuclear weapons and to agree to a bilateral inspection regime (Nuclear Threat Initiative 2016a). Pyongyang signed a safeguards agreement with the IAEA that same year and six inspections were carried out between June 1992 and February 1993. The inspectors obtained access to multiple sites and documents on the program, including an admission that small amounts of plutonium had already been separated (Albright and Hibbs 1992). Following further analysis, the IAEA concluded that North Korea had separated more than the 80 g of plutonium it claimed. After multiple denied requests for access to sites, the IAEA then issued an ultimatum to the country until March 1993. In response, the DPRK announced it would withdraw from the NPT, but the threat was suspended after the US agreed to conduct high-level talks (Albright 1994). During these talks, the 5 MW reactor in Yongbyon continued operation, and technicians even removed spent fuel rods without the supervision of IAEA inspectors (Sanger 1994). The crisis was finally diffused when former US President Carter traveled to Pyongyang and reached an agreement on the outline of a deal that was later signed as the Agreed Framework in 1994 (Webb 1994). Under the terms of the agreement, North Korea froze its reactors and other related facilities. The country allowed IAEA inspectors back into the country, and promised to uphold the NPT as well as the 1992 Denuclearization Agreement. In return, the US agreed to lead a consortium to build two conventional Light Water Reactors (LWR), provide 500,000 tons of heavy oil per year, and give formal 'assurances' against the use of nuclear weapons (KEDO 1994).

While the agreement did freeze the reactor program for almost a decade, both sides remained unsatisfied; the US due to the lack of inspection on past activities, and the DPRK due to the lack of progress with the construction of the LWR (Nuclear Threat Initiative 2016a). By 1999, there was growing concern among US intelligence agencies that North Korea was attempting to develop an enrichment program. Indicators included suspicions about Pakistani assistance along with the purchase of frequency converters that could be used for centrifuges, as well as other centrifuge-related material. As evidence mounted and the CIA became more confident in its analysis, the US confronted the DPRK over the issue between 2001 and 2002 (Richelson 2007). Unhappy with the lack of progress on the matter, the Bush administration decided to halt oil shipments to the country. In response, North Korea restarted its 5 MW reactor and reprocessing plant, resumed construction of its larger reactor, and expelled IAEA inspectors (Stevenson 2002). In January 2003, Pyongyang finally withdrew from the NPT. Later that year, activity was detected around the reprocessing facilities. This provided the basis for US intelligence agency to increase their estimate of the North Korean arsenal to eight bombs. Uranium enrichment was also believed to be operational by 2007, producing up to six bombs' worth of plutonium per year (Kessler 2004).

Talks between key stake-holders stalled until September 2005 with the signature of the Statement of Principles. North Korea would abandon its nuclear program, return to the NPT, and allow inspector-access. It received in return economic aid, security assurances, electricity from South Korea, assurances on normalization of relations with the US, and the resumption of negotiations on the LWR. However, much of the detail surrounding the agreement still needed to be finalized (Kahn and Sanger 2005). Following multiple disputes, the deal remained dormant for another year. Deteriorations reached a new low point in October 2006 when North Korea conducted its first nuclear test. This led to UN Security Council Resolution 1718 imposing a wide range of sanctions on the country. Both sides reached a new agreement in February 2007, but this only lasted until March 2009. During which time, inspectors were allowed access to key facilities. In May 2009, a second nuclear test was conducted, and UN Resolution 1874 imposed further sanctions on the country. No substantial advances in negotiations occurred since. Pyongyang detonated three other nuclear devices in February 2013, March 2016, and September of that same year, followed by a more recent one in September 2017. The tests were all accompanied by the usual condemnations, and sanctions, with very little change to the program's further development (Nuclear Threat Initiative 2016a).

## 3 Intelligence Gathering, Analysis and Responses

### *3.1 The Role of Technology in Intelligence Gathering*

Gathering intelligence on nuclear program can be particularly complex and demanding, relative to other types of espionage. The main reason is that they are notoriously secretive and hard to infiltrate. Developments within nuclear programs are extremely opaque and can be exceptionally challenging to extract reliable information from. The difficulty of spying on these programs is exacerbated by the level of the threat that they pose to the international community. The stakes are incredibly high when a country is attempting to acquire nuclear weapons. Many scholars consider a successful attempt a 'revolutionary' event in international relations (Hymans 2010). The pressure on intelligence agencies to acquire 'good intel' cannot be overstated. Intelligence gathering efforts for these programs often tend to have international dimensions, more so than other forms of espionage (excluding counter-terrorism). Allied nations' agencies often collaborate when it relates to these types of threats (e.g., US, Italy, UK, and Australia on Iraq in the 2000s) and even share intelligence with rival countries (e.g., US with Russia and China on Iran and North Korea). Coordination with the UN and IAEA is also necessary to gather information or effectively respond to a violation.

**Table 2** Overview of the different espionage methods for the case studies. Note that these are based on known intelligence reports; oftentimes the means of espionage is not reported

| | Iraq-81 | Iraq-91 | Iraq-03 | Syria | Iran | DPRK |
|---|---|---|---|---|---|---|
| Satellite | ✓ | ✓ | ✓ | ✓ | ✓ | ✓ |
| Spy planes | ? | ✓ | ✓ | ? | ✓ | ✓ |
| Drones | | | | | ✓ | ✓ |
| COMINT | ✓ | ✓ | ✓ | ? | ✓ | ✓ |
| MASINT | | ✓ | | ✓ | | ✓ |
| IAEA | ✓ | ✓ | ✓ | ✓ | ✓ | ✓ |
| Defectors | | ✓ | ✓ | | ✓ | ✓ |

In this section, the different sources of intelligence gathering tools are discussed with an emphasis on technology-based intelligence. As Table 2 shows, a wide range of different tools have been used. From a first glance, it appears that IMINT-based intelligence has been overwhelmingly favored by the intelligence community. Satellite imaging stands out as one of the preferred espionage tool. Due to the technical complexity of nuclear programs, MASINT plays an important role as well. Another characteristic of note is that the longer spying efforts last (Iran and North Korea), the more tools the intelligence agencies are likely to call upon as they become increasingly desperate for information.

### 3.1.1 Human Intelligence

As previously mentioned, nuclear programs tend to have the highest level of secrecy surrounding them. Managers go to extreme lengths to keep programs secret, with even high-ranking officials not being made aware of them oftentimes. This makes it exceedingly difficult to infiltrate such programs, and there are only few cases of spies penetrating them. One such instance is Taiwan in the 1970s and 1980s, but there are no known occurrences with the four programs considered in this case study (Richelson 2007). Nevertheless, intelligence agencies are still able to acquire substantial HUMINT via defectors and IAEA inspectors.

The majority of governments in the cases studies were authoritarian and even brutal regimes. It is not surprising therefore that scientists and officials involved in the nuclear programs defected to western countries. In the case of Iraq, the country led a crackdown on its own nuclear scientists following the 1979 Islamic Revolution in Iran. Many important figures were accused of supporting Shiite movements, were imprisoned and even tortured. Even Jaffar, the equivalent of an Iraqi Oppenheimer, was placed under house arrest during those turbulent times. Iraqi defectors greatly assisted the CIA with unravelling the extent of the nuclear program in the aftermath of the 1991 invasion. Two of Saddam's sons-in-law, holding important administrative role defected in 1995. One of them had previously served as minister of defense (Cockburn and Cockburn 1999). In the case of North Korea, defectors in 1991 help bring to the attention of the CIA that their country was actively developing a nuclear program (Sanger 1991). In 1997, high ranking officials even warned the CIA that their country already possessed nuclear weapons and planned on using them (Gertz 1997). With Iran, the steady stream of information that the CIA and NSA were obtaining from defectors led the Revolutionary Guard to lead a crackdown on employees, with at least two being arrested (Charbonneau 2004).

While defector intelligence can prove invaluable, agencies cannot rely solely on it. Defectors have a tendency to embellish their involvement in the hopes of appearing important in the eyes of their host. They can sometime feed misinformation purposely or unintentionally. Iraqi defectors to the CIA between 2000 and 2001 wrongfully claimed that they were working on a renewed nuclear program (Miller 2001). Generally, intelligence agencies prefer to rely on more trusted eyes-on-the-ground. This tends to be IAEA inspectors that are given access sensitive facilities.

Intelligence agencies have referred to IAEA inspector reports in their assessments of the nuclear program in all of the cases considered. Inspectors provided testimonies of what they observed and photographs of facilities, as well as intelligence in the form of MASINT. The latter type of intelligence will be discussed in the next section.

### 3.1.2 Technology-Based Intelligence

Due the difficulties of obtaining traditional human intelligence (HUMINT), intelligence agencies must rely increasingly on technological means to assess nuclear programs. This section provides a non-exhaustive overview of different tools employed by intelligence agencies. As highlighted in Table 2, the vast majority of intelligence is in the form of IMINT. Satellites are the primary mean of providing images, but spy planes are deployed when 'close-up' photography is required. MASINT also plays a crucial role because it is among the most irrefutable sources of intelligence. Nuclear facilities and tests emit key radionuclide signatures that are not observed in nature.

Imagery intelligence is obtained from a wide array of tools. In the case of Iraq in 1990s, image data collection was done via: 3 KH-11 satellites, 1 Onyx radar imagery satellite, 2 signal intelligence satellites, 2 U-2, and 1 SR-71 aircraft (Richelson 2007). IMINT provide vital information on the layout of suspected sites, surrounding human activity, and helps identify new facilities. In the case of the North Korean reactor, satellite images of plumes of steam allowed the CIA to know when the facility restarted or was shutdown (Sanger 2003). The main limitation is that IMINT can be easily contestable; agencies prefer to draw conclusions with a combination of different types of intelligence. The National Reconnaissance Office (NRO) is generally in charge of gathering IMINT. The National Security Agency (NSA) is tasked with gathering COMINT. While this type of intelligence has played a lesser role in the case studies, it is nonetheless important. US embassies are often hosting NSA communication eavesdropping operations, notably in the case of Iraq (Richelson 2007). Specific instances of interceptions include communications between Iran and China on the sale millions of dollars' worth of hydrofluoric acid, which could be used to produce fissile material (Gellmand and Pomfret 1998).

MASINT, on the other hand, is a significantly more complicated and subtle type of intelligence. It can take many shapes or forms; whether identifying nuclear tests, detecting radioisotopes, monitoring electricity diversions, or even flight travelling patterns. Nuclear test detection is handled by a range of US institutions, key of which is the Air Force Technical Application Center (AFTAC). These signals include seismic measurements, radionuclide detections, light flashes, as well as acoustic signals in both air and sea. Similar capabilities are also part of those of the International Monitoring System (IMS) of the CTBTO (CTBTO 2016). Radionuclide signals can additionally be used for other types of intelligence gathering. The US placed Krypton-85 detectors at the Russian embassy in the DPRK and flew them in planes to monitor reprocessing activities (Risen 2003 and Gertz 2003). IAEA inspectors gathered a wide range of MASINT using a variety of tools, such as gamma detectors (for uranium enrichment), Cerenkov detectors (for plutonium), wall and soil radioisotope detectors, and alloy detectors (for metals used in nuclear activities). While these types of MASINT tend to be considered irrefutable, the experience with Syria proved otherwise. IAEA inspectors detected traces of uranium at a suspected facility, which proved, in their opinion, that Syria was operating a covert reactor. The Syrian government on the other hand, retorted that these particles originate from the Israeli munitions used in the airstrike (IAEA 2009a). MASINT can also be obtained from subtler means. For example, US agencies looked for large amounts of electricity diversions and monitored flights from foreign works and international businessmen that are known to be dealing with suspected programs (Richelson 2007).

## 3.2 *Data Analysis & Intelligence Agencies*

Gathering intelligence is only one side of intelligence assessments; analyses play an equally important role in the lead-up to a decision by a state. As in any other institutions, intelligence agencies are subject to organizational challenges and biases that can alter their perception of a program. Understanding the dynamics between different parts of the organization is crucial in evaluating successes and failures of the intelligence community. This section will start by explaining the different roles of the relevant agencies and how they interact with each other. An overview of different intelligence 'successes and failures' will then be provided. Lastly, distortion in the analyses will be discussed with the aim of deriving lessons learned.

### 3.2.1 From Gathering Intelligence to Making a Decision

In the four case studies, the main intelligence agencies charged with analyzing the data gathered were the CIA, DIA, (Department of Defense's intelligence agency), DoE's Office of Intelligence (referred to as DoE for simplicity), and INR (State Department's agency). Due to the high level of technicality associated with nuclear programs, intelligence gathering relies heavily on nuclear experts. In the case of Iraq in the late 1980s, a task force of over 80 individuals at Oak Ridge National Laboratory was created with the sole objective of analyzing intelligence data and help direct efforts (US Congress 1992). These scientists and engineers tend to be based at national laboratories (e.g., Oak Ridge, Los Alamos, Lawrence Livermore) and operate under the directive of the DoE. The four main agencies do not always receive intelligence directly; gathering efforts usually go through other bureaus such as the NSA (COMINT), the NRO/NIMA (IMINT), and AFTAC/CMO (MASINT). After drawing their conclusions, the four main intelligence agencies draft reports for the leadership and provide recommendations. These reports are usually joint efforts led by the CIA (e.g., National Intelligence Estimates, or NIE),

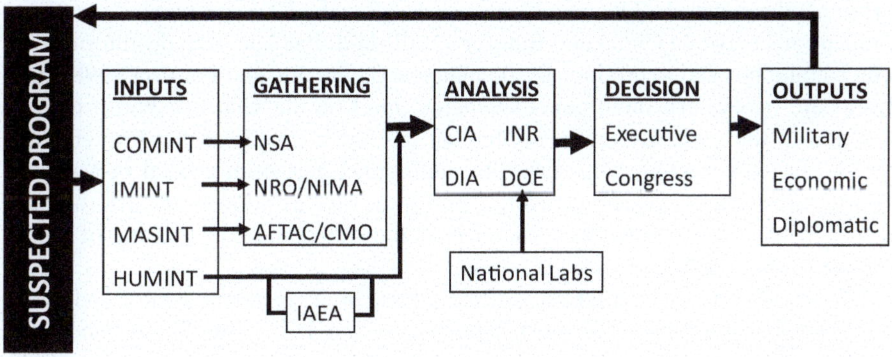


**Fig. 1** Overview of how intelligence investigations on nuclear programs take place. Note that this is relative to the US point of view, and that the IAEA only plays the role of an 'intelligence gatherer' even though the agency conducts its own analyses

but they can also be in the form of individual reports by each agency (for instance INR reports to the Secretary of State). As was previously pointed out, a good deal of crucial HUMINT intelligence in the cases analyzed is provided by an external organization, the IAEA. This makes for a curious dynamic whereby the CIA can have a tenuous relationship with the IAEA at times or a very productive one at others. All of these different interactions are summarized by Fig. 1.

### 3.2.2 Overview of Intelligence Successes and Failures

It is difficult to establish what can be deemed 'successes' or 'failures' of the intelligence community. In one sense, all cases apart for North Korea can be considered 'successes' since the countries did not acquire a working weapon. However, this perspective can be misleading as the intelligence community cannot take all the credit for stopping nations from proliferating. In this section, a more specific review of instances where intelligence helped lead to direct action is provided. This is followed by an evaluation of cases where intelligence agencies missed important aspects of nuclear programs or drew the wrong conclusions from the data. The examples listed are not by any means exhaustive, but they are intended to provide the reader with a good overview of what are the types of accomplishments and setbacks of the intelligence community.

Instances where intelligence led to specific actions against a state are numerous, notably when relating to Iran and North Korea. In 1994, the CIA received reports that Iran was trying to purchase over on thousand pounds of HEU from Kazakhstan. This prompted President Clinton to authorize a mission that involved buying all the stockpile, and flying it directly to Delaware before the Iranians could obtain it (Gertz 1994). The US also uncovered numerous deals between China, Russia, and the Czech Republic to sell Iran or North Korea sensitive material or technologies. This allowed the US to apply pressure on these countries to terminate the contracts. One notable case was when the NSA intercepted two communications between China and Iran for a million-dollar sale of hydrofluoric acid, which could be used to produce fissile material (Gellmand and Pomfret 1998). The disclosure of western intelligence agencies' intel on the secretive Fordo facility was one of the main drivers to implement the harsh sanctions that brought Iran back to the negotiating table. Satellite intelligence has been crucial in assisting IAEA inspectors in Iraq, North Korea, and Iran. Images allowed the US to know exactly when the North Korean reactor had shut-down and reprocessing of plutonium was about to begin.

On the other hand, intelligence estimates have often overestimated or underestimated nuclear programs. The most flagrant case was Iraq in the 2000s, where the CIA severely overestimated the extent of the program. Another instance was a 1992 intelligence report predicting that Iran could have the bomb by 2000 (Cordesman 2000). But intelligence agencies have also severely underestimated the development pace of some nuclear programs. The most blatant cases were Iraq prior to the 1991 invasion and North Korea during the 1980s. Scholars have often warned that external assistance to weapons program play a crucial role in their development (Kroenig 2009). Regrettably, the CIA has often missed Pakistani collaboration with countries, notably through its A.Q. Khan network. This was the case with both Iraq and Iran in the 1990s. Lastly, intelligence agencies have often made incorrect assessments from the intelligence or signals they received. Most notable was the report that Niger had sold several tons of uranium ore to Iraq, which later turned out to be forged. Another instance was in 2003 when officials believed that North Korea was bluffing when it claimed it had already reprocessed enough plutonium for five bombs (Sanger and French 2003).

### 3.2.3 Distortion of Intelligence

Many lessons can be learned from these successes and failures of the intelligence community. Providing a critical analysis of past performance can ensure the same errors are not repeated in the future. As Montgomery and Mount (2014) discuss in their review of US intelligence failures, analysis can be distorted by a range of political, cultural, bureaucratic, or organizational factors. Political factors include issues such as decision makers nit-picking from analyses or tacitly encouraging conclusions that adhere to their preferred lines. Cultural factors consist of incorrect assumptions about a very different country than the US as well as preconceived biases. Bureaucratic issues relate to inter-agency politics or delaying in the aggregation of data due to over-fragmentation of institutions. Lastly, organizational factors include the inability of separating signal from noise, the strong preference of secret over open source data, and previous failures leading analysts to forego conservative assertions. Several of these issues have been encountered during the review of intelligence efforts on the four case studies.

One of the most apparent issues of the 'behind-the-scenes' aspects of intelligence analysis is the role that disagreements play. Dividing the intelligence analysis between four main agencies inherently leads to competition and disagreements. This can be a double-edged sword: disagreement encourages skepticism, but it can also lead to biases and hardening of positions. Inter-agency disagreements were most animated during two cases: on Iraq during the 2000s and North Korea in the 1990s. In the first instance, the CIA and DIA were pitted against the INR and DoE. One group strongly believed that the Iraqis were actively rebuilding their capacity, while the other was very skeptical of the basis of these claims. In the case of North Korea, the CIA and DIA disagreed strongly with the INR over how advanced the nuclear program was. In each instance, biases in the institutions are believed to be the responsible for the divergence of opinions. With regards to the CIA, its failure to detect the scale of the Iraqi program in 1991 led it to it becoming overly pessimistic in its following analyses in the 2000s. A culture of 'better-safe-than-sorry' overtook the agency; its fear of missing the next proliferation violations led it to overestimate subsequent Iraqi, North Korean, and Iranian programs. On the other hand, because the State Department was actively advocating for diplomacy with the DPRK, this biased the INR's judgement in the 1990s. The agency was inclined to down-play intelligence on the nuclear program's progress to justify claims that there was enough time for negotiations to take place (Sciolino 1992).

Because institutions are inherently subjected to biases, disagreements should, in theory, allow for the more impartial judgements to become predominant. Conversely, this has often not been the case. Even though the CIA was incorrect in its analysis of Iraq during the 2000s, its view became the dominant one among policy makers (even within the broader State Department, which rejected its own INR's analysis). This is all the more of a concern since the agency's position can be traced back to a single analyst (known as Joe T.) in its WINPAC section. It appears that inadequate structural measures were in place and kept policy makers from being aware of the opposing views of different agencies. Joint intelligence statements, while useful in forcing agencies to converge to common conclusions, can also lead to crucial viewpoints being ignored. There additionally seems to be a trend among policy makers to favor the CIA viewpoint over others, which can be attributed to its higher recognition and prestige.

On the other hand, cultural and organizational issues discussed by Montgomery and Mount (2014) were observed in all four of the case studies. One of the main reasons that led the CIA to 'miss' the Iraqi program in the 1990s was attributed to its distraction by the Iran-Iraq war during that time (Richelson 2007). Not being able to filter out the 'noise' can make-or-break attempts to detect proliferation. Furthermore, agencies often have preconceived notions about the capabilities of a state. Understandably, the CIA wrongfully doubted the abilities of the North Koreans to develop a weapons program in the 1980s. A 1986 report claimed that previous estimates had underestimated the threat due to a belief that a country with rudimentary means could never develop the bomb (CIA 1986). Another factor in the CIA's underestimation of the 1980s Iraqi enrichment program was that they did not consider proliferation via technologies thought to be obsolete by experts

(Montgomery and Mount 2014). These preconceived notions can also have the reverse effect with agencies overestimating capabilities of states. As the Iraqi nuclear program began to be uncovered in the 1990s, the CIA was taken aback by how sophisticated were deception means and how covert the program was. This wrongfully led the agency to conclude that the same level of deception was being employed in the 2000s (US Congress 2004).

These findings show how the analysis plays such an important role in intelligence efforts. Gathering of data is not self-sufficient and agencies cannot rely exclusively on technology to gather as much signals about a program as they can. Sound judgement and evaluation is crucial in drawing the right conclusions. As is the case with any institution, intelligence agencies can often be biased in their analysis and have a tendency to become overzealous in defending their position. Overall, the scientific community with subject matter expertise should be given a louder voice in the ongoing debate regarding nuclear proliferation intelligence.

## *3.3 Counterproliferation Responses and their Effects*

### 3.3.1 Military Responses

Getting the UN Security Council to fully agree on an issue can prove to be an unsurmountable challenge. As a result, many nations have opted to take unilateral military actions to subdue a perceived threat. A case in point is Israel's decision to strike Iraq in 1981 and Syria in 2007, or the US, when it invaded Iraq in 2003. From previous sections, it is apparent that most of these interventions were not necessarily backed by confident or accurate intelligence. This lack of correlation between severity of response and confidence in the intelligence indicates that nations have other motivations that more strongly affect their decision-making process. In this section, some potential hypotheses that lead a state to pursue a military response are discussed, as well as the resulting effect of these interventions.

The most obvious reason for a state to favor military intervention is when the perceived threat to is national security is great enough to outweigh the potential fallout. This is undeniably the case with Israel during its 1981 airstrike. The government even released a statement justifying the attack and explaining why it could not allow Iraq to develop nuclear capabilities (Government of Israel 1981). These arguments were likewise used to make the case for potential Israeli airstrikes on Iran more recently (Times of Israel 2015). Another reason that can factor into a government's decision-making is its ability to keep an attack 'quiet'. This was the case with the 2007 airstrike of Syrian facilities by Israel. Both sides downplayed the incident and substantial international backlash was avoided. The ability of a suspect government to retaliate is also believed to play an important role. One of the main reasons cited for why the US invaded Iraq on proliferation grounds rather than North Korea, was fears that the latter may have already acquired the bomb (Gordon 2002).

Similarly, US military was likely very confident in its ability to successfully invade a country it had swiftly defeated just ten years earlier.

Regardless of what motivates states to revert to military attacks, the end results have often been mixed. The case for 'preemptive' strikes has always been hotly debated in the literature ever since the Osirak operation (Reiter 2005 and Raas and Long 2007). However, as Braut-Hegghammer (2011) points out, the Iraqi weapons program expanded even further *after* the strike took place. Countries usually pursue nuclear weapons in response to a perceived threat (often from another state with nuclear capability). Attacking such a country would therefore, only increase its resolve to acquire nuclear weapons. The 1981 airstrike forced the program underground, making it harder to gather intelligence on or to obstruct. Nevertheless, the First Gulf War in 1991 did contribute substantially to the dismantling of the Iraqi program. By reverting to the UN Security Council, the IAEA, and the threat of force, the US administration was able to successfully coerce Iraq into unravelling its nuclear program, which was subsequently destroyed. It appears that while military strikes are not sufficient in themselves, it is possible that they could lead to joint international efforts that are more effective. Lastly, it should be remembered that intelligence estimates are never flawless and opting for military intervention can oftentimes be baseless (e.g., the 2003 Iraq War).

### 3.3.2 Diplomatic & Economic Responses

While many academics advocate strongly for diplomatic or non-military responses to nuclear proliferation, these means aren't necessarily more effective than their counterparts (Cortright and Lopez 2005). As two of the case studies show, the record on Iran and North Korea is mired with controversies and setbacks. States that are attempting to pursue nuclear weapons often have more leverage when entering the negotiating table. At the start of discussions, they seldom have more to gain from stopping their program. For instance, US security and economic assistance to North Korea pale in comparison to the international standing the country obtains from 'going nuclear'. On the other hand, harsh sanctions alongside the right type of incentives can persuade a state to negotiate more genuinely.

The fact that negotiations were used as the primary means to block North Korea might be seen to make the case for diplomacy weaker, given that the country was able to acquire weapons. However, the recent Iran Nuclear Deal was hailed by many as groundbreaking for nonproliferation (Brown et al. 2015). It seems that the record on the effectiveness of diplomacy is at best, mixed. As the DIA feared, North Korea employed the 1994 negotiations as a means to stall (Smith 1993). The country did shut down its graphite reactor but used the reduced scrutiny to develop an enrichment program. When confronted about this and pressured to return to the negotiating table, the DPRK stalled again until it could reprocess enough fuel from its restarted reactor. Furthermore, the country always insisted on being 'bought-off' with a range of economic benefits such as oil shipments, civilian nuclear reactors, or food aid (Weisman 2003). In retrospect, it seems the country was able to outma-

**Table 3** Overview of the intelligence gathering process, and the resulting responses for each case. The H/L symbols correspond to high and low confidence in the intelligence gathered. The O/U/A symbols indicate whether the intelligence overestimated, underestimated or adequately estimated the nuclear program at the time. M/D highlight if the resulting response was military in nature or diplomatic. E/I point to if the response was effective or ineffective. Double letters are used for emphasis to compare between different cases. Finally, the verdicts indicate how correlated the responses and their effectiveness were with respect to confidence and accuracy of the intel. For instance, in the case of Iraq in 1981 ('–'), the response was ineffective and the intelligence gathered was overestimating the program. In 1991 ('luck'), the response was adequate and effective, but is attributed to chance since intelligence severely underestimated the program. With Iran in 2015, high confidence and accuracy in intelligence did lead to an effective response, but it remains to be seen if it will stand the test of time ('+?')

| | Iraq | | | Syria | Iran | | North Korea | | |
|---|---|---|---|---|---|---|---|---|---|
| | 1981 | 1991 | 2003 | 2007 | 2004 | 2015 | 1992 | 1994 | 2005 |
| Intel confidence | H | L | L | L | H | HH | L | H | HH |
| Intel accuracy | O | UU | OO | O | O | A | U | A | A |
| Response | M | M | M | M | D | D | D | D | D |
| Effectiveness | II | EE | I | E | II | E? | I | E/I | II |
| Verdict | – | Luck | – | Luck | – | +? | – | – | – |

neuver the international community by receiving some benefits, while still developing nuclear weapons. Similar conclusions can be drawn from the Iranian experience in the 2000s. The Paris Agreement struck in 2004 broke down only two years later. The country had more centrifuges after the deal than it did prior to it. It wasn't until economic sanctions started targeting the oil industry it depended on, that a more favorable climate for negotiations was reached. The election of a new president proved instrumental in striking the final agreement, indicating that more democratic regimes may be more sensitive to economic sanctions than autocratic ones.

### 3.3.3 Lessons Learned from Interventions

Two overarching conclusions on interventions can be drawn from the case studies: (1) intelligence is not the primary driver for state responses, and (2) all types of responses can be equally ineffective. Conclusion 1 explains why states have never opted for a mix of diplomatic and military responses, or why military action is pursued even if confidence in the intelligence is lacking. Conclusion 2 can be reached by having a holistic view of the different responses. While the international community might take comfort in the notion that only a single case of these 'rogue states' was able to develop a weapon, many of the successes are not intentional as Table 3 outlines. The majority of interventions can be said to have been counterproductive with the targeted country's program expanding afterwards (or in the case of Iraq in 2003, not being existent to begin with). In the few instances where interventions appear to have had a more positive impact, it can be mostly attributed to chance. In the case of Iraq in 1991, the US had little knowledge of the extent of the nuclear program, and it was not the main driver for the war in any case. In Syria, while it does seem that the country's program hasn't advanced in the aftermath of the Israeli strike, the main reason is the ongoing civil war rather than the attacks themselves. It is sensible to speculate that Operation Orchard might have had the same impact as Operation Opera in 1981, by emboldening the Syrian resolve to acquire weapons. Table 3 also illustrates the varying levels of confidence intelligence agencies had in their analysis and how accurate the analysis really was. The two worst estimates were Iraq prior to 1991 (heavily underestimated) and prior to 2003 (heavily overestimated). It is interesting to note, that even when there is high degree of accuracy and confidence in the intelligence – such as the case of North Korea – this does not always constitute enough vindication for a military strike, further reinforcing conclusion 1.

## 4 Conclusions & Future Trends

Intelligence efforts on four cases were reviewed in order to gain a better understanding of how to spy on 'rogue' states and how to effectively counter attempts to proliferate. The study emphasized on three main aspects: (1) the intelligence gathering, (2) data analysis, and (3) the subsequent responses. On the gathering front, technology plays a crucial role and can often displace the 'human' aspect. The same cannot be said about intelligence analysis, where biases and disagreements can distort reports and recommendations. Irrespective of how adequate the gathering of intelligence or how sensible was the analysis, decisions are oftentimes driven by external factors unrelated to the data at hand and the resulting intelligence.

Moving forward, technology can play an increasingly vital role in espionage activities. MASINT notably holds a lot of promise, with new technologies on the horizon, such as antineutrino monitoring of reactors, active interrogation techniques for detecting smuggled material, as well as nuclear test detection and characterization. More emphasis should be given to this form of intelligence because it is harder to refute than other, with the notable exception of the disagreement over the Syrian 2007 strike. Combinations of multiple independent sources of intel should be favored when building a case against a state.

In future intelligence assessments, human and organizational aspects must be taken into considerations more carefully. Scientists and subject-matter experts should have a louder voice in intelligence analyses (e.g., CIA agent on Niger intelligence, and DoE experts on aluminum tubes). Disagreements should be stated more clearly in joint reports to government officials, in order to emphasize the relevant uncertainties more clearly. While policy-makers tend to prefer summarized, brief assessments, 'the devil is always in the details.' More effort is needed on the political side to skeptically review the presented data. Political and organizational biases need to be carefully scrutinized in the future; here the leadership of the different agencies has a bigger role to play, notably the DNI.

Reverting to military responses does not always correlate with the level of confidence in the intelligence. Despite the aforementioned limitations, intelligence should be a bigger driver of responses. The review also shows that both military and diplomatic responses can be equally ineffective in halting proliferation, e.g., Iraq in 1980s and DPRK in 1990s. Attacking a state might embolden it in its pursuit of the bomb, while negotiations might lead it to stall and press forward with its clandestine activities. Counterproliferation responses need to be taken on a case-by-case basis (accounting, for instance, on if the country is democratic or not) and rely on well-coordinated mixes of incentives, threats, and sanctions. Unilateral action tends to be less effective than a coordinated international response; it decreases the probability of reaching a more long-lasting solution. Decision-makers should leverage the international consensus on nonproliferation and be willing to collaborate more closely with rivals during responses.

Although the international community has had many shortcomings, the case for strengthening institutions such as the IAEA has never been stronger. Owing to its neutrality, the agency can provide crucial 'eyes-on-the-ground' that greatly expand the understanding of programs. Tending towards more open counterproliferation intelligence can benefit the nonproliferation regime. It is much easier to defeat the gridlock at the UN Security Council by adequately disclosing intelligence and having more inclusive approaches, e.g., intel on Fordo that led to multilateral sanctions. Future developments in technology would similarly better serve the nonproliferation regime if they are brought forward via international programs (e.g., as is being attempted with the CTBTO), rather than secretive national ones.

While the conclusions reached here were based specifically on a review of nuclear programs, similar inferences can be made for chemical and biological ones as well. Transparency and international collaboration through relevant institutions (such as the OPCW) can be vital, as seen with Syria's chemical program in 2013. Intelligence gathering efforts can be equally distorted by biases and organizational problems. Technology plays an as crucial role in the gathering of intelligence. MASINT is similarly vital to such activities and can benefit from further development.

## References

Albright, D. (1994). How much plutonium does North Korea have?, *Bulletin of the Atomic Scientist*, September/October issue, pp. 46–53, 50.

Albright, D. and Brannan, P. (2008). ISIS Report: The Al-Kibar Reactor: Extraordinary Camouflage, Troubling Inspections, *Institute for Science and International Security*, May 12, Available at: www.isis-online.org.

Albright, D. and Hibbs, M. (1992). North Korea's plutonium puzzle, *Bulletin of the Atomic Scientist*, November issue, pp. 36–40, 48.

Albright, D., & Sticker, A. (2015). Iran's nuclear program. In *United States Institute for Peace*. www.usip.org [Accessed 19 April 2016

Australian Broadcasting Corporation. (2003). Spinning the tubes, transcript from *Four Corners* broadcast, October 27.

Bermudez, J. S., Jr. (1991). North Korea's nuclear Programme. *Jane's Intelligence Review, 3*(9).

Black, I. and Morris, B. (1991). Israel's secret wars: A history of Israel's intelligence services, Grove Weidenfeld, pp. 333–334.

Braut-Hegghammer, M. (2011). Revisiting attacks and nuclear proliferation risks. *International Security, 36*(1), 101–132.

Brown, S., Toosi, N. and Dallison P. (2015). World powers hail nuclear deal as 'new era', *Politico*, July 14.

Burrows, W. (1994). Critical mass: The dangerous path for Superweapons in a fragmented world, Simon and Schuster.

Charbonneau, L. (2004). Iran keeps tabs on nuclear officials, *Washington Times*, April 29.

CIA. (1986). North Korea: Potential for nuclear weapons development, In: National Security Archives (NSA) electronic briefing book (EBB) 87, North Korea and Nuclear Weapons: The Declassified US Record.

CIA. (1989). NORTH KOREA: Nuclear program of proliferation concern, *Special Analysis*, March 22, p. 1.

Cockburn, A., & Cockburn, P. (1999). *Out of the ashes: The resurrection of Saddam Hussein*. HarperCollins.

Cordesman, A. H. (2000). *Iran and Nuclear Weapon: A Working Draft, center for strategic and international*. Washington, DC: Studies.

Corera, G. (2006). Shopping for bombs: Nuclear4 proliferation, global insecurity, and the rise and fall of the a. Q. Khan network, Oxford University Press.

Cortright, D. and Lopez, G. A. (2005). Bombs, carrots and sticks: The use of incentives and sanctions, *Arms Control Today*, March 1.

CTBTO. (2016). Overview of the Verification Regime, Available at:. www.ctbto.org [Accessed 10 April 2016].

Davis, J. C., & Kay, D. A. (1992). Iraq's secretive nuclear weapons program. *Physics Today, 45*(7), 21.

Deutsch, J. (1996). *Testimony of the Director of Central Intelligence Agency*, senate governmental affairs committee, permanent subcommittee on investigations, weapons proliferation, March 20.

DeYoung, Karen and Shear, M. D. (2009). US, Allies say Iran has secrete nuclear facility, *Washington Post*, September 26.

Director of Central Intelligence. (1981). Intelligence assessment, *Implications of Israeli Attack on Iraq*, July 1.

Duefler, C. A. Jr. (2004). *Comprehensive Report of the Special Advisor to the DCI for Iraqi Weapons of Mass Destruction*, central intelligence agency, September 30.

Follath, E. (2015). Assad's secret: Evidence points to Syrian push for nuclear weapons, *Spiegel Online*, January 9, Available at: www.spiegel.de.

Ford, P. S. (2004). Israel's attack on Osiraq: A model for future preventive strikes?. PhD. Naval Postgraduate School.

Gellmand, B. and Pomfret, J. (1998). US action stymied China sale to Iran, *Washington Post*, March 13.

Gertz, B. (1994). US defuses effort by Iran to get nukes, *Washington Times*, November 24.

Gertz, B. (1997). Hwang says N. Korea has atomic weapons, *Washington Times*, June 5.

Gertz, B. (2003). 2nd N. Korean nuclear site not likely, *Washington Times*, July 22.

Gordon, M. R. (2002). In Bush's 'axis of evil', why Iraq stands out, *New York Times*, September 2.

Government of Israel. (1981). Ministry of Foreign Affairs and atomic energy commission, The Iraqi Threat – Why Israel Had to Act, Jerusalem.

Hedges, S. J. and Cary, P. (1991). Saddam's secret bomb, *US News & World Report*, November 25.

Hersh, S. (2003). The stovepipe, *New Yorker*, October 27.

Hertz, B. (1990). Saddam close to nuclear weapon, *Washington Times*, November 28. Iraqi atomic energy commission (1992), preface, In: *Iraqi Nuclear Programme 1956–1991* (Baghdad: Al-Adib).

Hymans, J. E. C. (2010). When does a state become a 'nuclear weapons state'?, In: *Forecasting Nuclear Proliferation in the 21st Century*, 1, Stanford University Press.

IAEA. (1995). Seventh report of the director general of the International Atomic Energy Agency on the implementation of the Agency's plan for future ongoing monitoring and verification of Iraq's compliance with paragraph 12 of resolution 687 (1991), April 11.

IAEA. (2004). Communication dated 26 November 2004 received from the permanent representatives of France, Germany, the Islamic Republic of Iran, and the United Kingdom, concerning the agreement signed in Paris on 15 November 2004, Available at: www.iaea.org.

IAEA. (2008a). Syrian Arab Republic: Research Reactor Details – SRR-1, *International Atomic Energy Agency*, Available at:. www.iaea.org [Accessed 20 April 2016].

IAEA. (2008b). Implementation of the NPT safeguards agreement in the Syrian Arab Republic, *Report by The Director General to the Board of Governors*, November 19,. Available at: www.iaea.org.

IAEA. (2009a). Implementation of the NPT safeguards agreement in the Syrian Arab Republic, *Report by The Director General to the Board of Governors*, February 19,. Available at: www.iaea.org.

IAEA. (2009b). Implementation of the NPT safeguards agreement and relevant provisions of security council resolutions 1737 (2006), 1747 (2007), 1803 (2008) and 1805 (2008) in the Islamic Republic of Iran, *Report by the Director General*, November 16, Available at: www.iaea.org.

IHS Global Limited. (2015). Nuclear, Syria: Proliferation, *Jane's CBRN Assessment*,. Available at: www.janes.ihs.com [Accessed 15 April 2016].

Iraqi Atomic Energy Commission. (1992). *Preface, In: Iraqi Nuclear Programme 1956–1991*. Baghdad: Al-Adib.

Kahn, J. and Sanger, D. E. (2005). US-Korean deal on arms leaves key points open, *New York Times*, September 20.

KEDO. (1994). *Agreement Framework between the United States of America and the Democratic People's Republic of Korea*, signed October 21.

Kessler, G. (2004). N. Korea nuclear estimate to rise, *Washington Post*, April 28.

Kessler, R. (1990). Argentina to ink research reactor deal soon with Syria, says CNEA, *Nucleonics Week*, May 31.

Kroenig, M. (2009). Importing the bomb – Sensitive nuclear assistance and nuclear proliferation. *Journal of Conflict Resolution, 53*(2), 161.

Lorieux, C. (2003). Le chah envisageait l'option militaire, *Le Figaro*, September 15.

Miller, J. (2001). Iraqi tells of renovations at sites for chemical and nuclear arms, *New York Times*, December 20.

Montgomery, A. H., & Mount, A. (2014). Misestimations: Explaining US failures to predict nuclear weapons programs. *Intelligence and National Security, 29*(3), 357.

Nuclear Threat Initiative (2016a). Country Profile: North Korea – Nuclear,. Available at: www.nti.org [Accessed 14 April 2016].

Nuclear Threat Initiative (2016b). Country Profile: Iran – Nuclear,. Available at: www.nti.org [Accessed 19 April 2016].

Nuclear Threat Initiative (2016c). Country Profile: Syria – Nuclear,. Available at: www.nti.org [Accessed 4 May 2016].

Raas, W., & Long, A. (2007). Osirak Redux? Assessing Israeli capabilities to destroy Iranian nuclear facilities. *International Security, 31*(4), 7.

Reiter, D. (2005). Preventative attacks against nuclear programs and the 'success' at Osiraq. *Nonproliferation Review, 12*(2), 355.

Reuters. (2009). House passes Iran Gasoline sanctions bill, *Reuters*, December 15.

Richelson, J. T. (2007). Spying on the bomb: American nuclear Intelligence from Nazi Germany to Iran and North Korea, W. W. Norton.

Risen, J. (2003). Russia helped US on nuclear spying inside North Korea, *New York Times*, January 20.

Risen, J. and Miller, J. (2000). CIA tells Clinton an Iranian A-bomb can't be ruled out, *New York Times*, January 17.
Sanger, D. E. (1994). North Koreans say nuclear fuel rods are being removed, *New York Times*, May 15.
Sanger, D. E. (1991). Nuclear Activity by North Korea Worries the U.S., *New York Times*, November 10.
Sangers, D. E. (2003). US sees quick start of North Korea nuclear site, *New York Times*, March 1.
Sanger, D. E. (2012). Obama order sped up wave of cyberattacks against Iran, *New York Times*, June 1.
Sanger, D. E. and French, H. W. (2003). North Korea prompts US to investigate nuclear boasts, *New York Times*, May 1.
Sciolino, E. (1991). Report says Iran seeks atomic arms, *New York Times*, October 31.
Sciolino, E. (1992). US agencies split over North Korea, *New York Times*, March 10.
Sinha, S. and Beachy, S. C. (2015). Timeline on Iran's nuclear program, *New York Times*, April 2.
Smith, J. P. (1978). Iraq's nuclear arms option, *Washington Post*, August 8, p. 14.
Smith, J. P. (1991). Iraq's secret A-arms effort: Grim lesson to the world, *Washington Post*, August 11.
Smith, R. J. (1993). US analysts are pessimistic on Korean nuclear inspections, *Washington Post*, December 13.
Stevenson, R. W. (2002). North Korea begins to reopen plant for processing plutonium, *New York Times*, December 24.
Stone, R. (2005). Profile: Jafar Dhia Jafar. *Science, 309*(5744), 2158–2159.
Tetrais, B. (2004). Kahn's nuclear exports: Was there a state strategy?, In: *Getting MAD: Nuclear Mutual Assured Destruction – It's Origin and Practice*, Carlisle.
Times of Israel. (2015). IDF 'more ready than ever' to strike Iran, security official says, August 25,. Available at: www.timesofisrael.com.
US Congress. (1981). House committee on foreign affairs, how long would it take for Iraq to obtain a nuclear explosive after its research reactor began operation?, *CRS Report for Congress*, hearings: Israeli attack on Iraqi nuclear facility, 97th Congress, 1st sess., June 25.
US Congress. (1992). House committee on energy and commerce, Nuclear Nonproliferation: Failed efforts to Curtail Iraq's Nuclear Weapons Program.
US Congress. (1993). House committee on foreign affairs, Iraq's Nuclear Weapons Capability and IAEA Inspections in Iraq.
US Congress. (2004). Senate select committee on intelligence, *Report on the US Intelligence Community's Prewar Intelligence Assessments on Iraq*, July 7.
Webb, K. (1994). Carter goes into second and crucial round of talks with Kim Il sung, *Agence France-Press*, June 17.
Weisman, S. R. (2003). North Korea said to offer small nuclear steps at a price, *New York Times*, April 29.
Wilson, R. (1983). A visit to the bombed nuclear reactor at Tuwaitha, Iraq. *Nature, 14*(19), 373–376.
Woodward, B. (1986). CIA aiding in gulf war, *Washington Post*, December 15.

# Organizational Legitimacy and Open Source Intelligence

**Lindsey R. Sheppard**

**Abstract** Technological developments of the modern era have brought the Intelligence Community (IC) to a transitional time in the intelligence space. Information and data are becoming increasingly available alongside increased access and ease in computational analysis capabilities. Outside of the IC and in the public realm, Open Source Intelligence (OSINT) products are those that are openly available to the general public and were produced using open source data and sources. OSINT products may include analysis from satellite imagery (GEOINT), ground images (IMINT), and statements or comments by humans (HUMINT), etc. Within the institution of OSINT analysis, non-governmental organizations are increasing their ability to inform and influence the public, including decision makers. While the increase in data and analysis capability is correlated with the increase in OSINT products, a causal relationship could not be detected. Instead, as public OSINT as an institution gains legitimacy, public OSINT organizations are gaining legitimacy. An investigation of organizational legitimacy, or the types of legitimacy organizations hold and how they build and manage legitimacy, reveals that technology is not the driver, but instead it is an enabling and supporting mechanism.

## 1 Introduction

Technological developments of the modern era have brought the Intelligence Community (IC) to a transitional time in the intelligence space. While technology may have initially driven the arrival at such a time, factors outside of the technical realm impact how the community responds and evolves. The evolution of technology was driven by military, inherently government, needs where case officers and handlers had the advantage of time in determining how best to integrate developments into the field. The scientists, engineers, and technologists of the CIA in the Cold War could iterate with the needs of the agents to drive technology development and deployment strategy. The IC in the United States adapted to the emergence of technology while maintaining the human-in-the-loop drivers. The findings and

L. R. Sheppard (✉)
Georgia Institute of Technology, Atlanta, Georgia, USA



M. E. Kosal (ed.), *Technology and the Intelligence Community*, Advanced Sciences and Technologies for Security Applications,
https://doi.org/10.1007/978-3-319-75232-7_9

developments of the IC stayed securely behind closed doors, hidden from the view of the domestic and international general publics (Wallace et al. 2009). The advent of the Information Age, however, is presenting opportunities and challenges for the IC.

Information and data are becoming increasingly available alongside improved access to, and ease in computational analysis of said data. One way this impacts the IC is in the field of Open Source Intelligence (OSINT). Open Source Intelligence analyses are those "produced from publicly available information that is collected, exploited, and disseminated in a timely manner to an appropriate audience for the purpose of addressing a specific intelligence requirement" (Benes 2013). The information used in OSINT is referred to as open source information. Open source information is "publicly available information that anyone can lawfully obtain by request, purchase, or observation" (Benes 2013). Within the IC, OSINT may not be a stand-alone product, but it may contribute insight and information to a larger product. OSINT may become classified or restricted distribution government documents. For example, reports and products produced by the NSA or CIA may use open source information and refer to such findings as OSINT even though the product is not available to the general public. In the context of this paper, OSINT refers to publicly available Open Source Intelligence. That is, finished OSINT products that are openly available to the general public and were produced using open source data and sources. OSINT products may include analysis from satellite imagery (GEOINT), ground images (IMINT), and statements or comments by humans (HUMINT), etc. All of the raw intelligence, however, must be publicly available data (e.g., Google Maps is open source GEOINT) to fall under the rubric of OSINT.

The organizations investigated in this chapter are all OSINT producers: Arms Control Wonk and Center for Nonproliferation Studies, Bellingcat, 38 North, and the Strategic Security Blog and Federation of American Scientists. Each of these organizations produces OSINT analysis for consumption by the general public, ranging from broad news outlet discourse to field-specific professionals and publications. Within the OSINT institution, organizations outside of the government are increasing their ability to inform and influence the public, including decision makers. While the increase in data and analysis capability is correlated with the increase in OSINT products, a causal relationship could not be detected. Instead, as OSINT as an institution gains legitimacy, public OSINT organizations are gaining prominence. An investigation of organizational legitimacy, or the types of legitimacy organizations hold and how they build and manage legitimacy, reveals that technology is not the driver, but instead is a support mechanism.

Organizations are characterized with having three types of legitimacy: normative, pragmatic, and cognitive. Normative, or "moral", legitimacy is gained by an organization that "reflects socially acceptable and desirable norms, standards, and values" (Brinkerhoff 2005). Pragmatic legitimacy is reflected in how the organization fulfills its constituents' self-interest (Brinkerhoff 2005). Cognitive legitimacy is gained by an organization when that organization "pursues objectives and activities that society understands and values as appropriate, proper, and desirable" (Brinkerhoff 2005). OSINT analysis is gaining legitimacy, towards institutionalization, particularly through normative and pragmatic legitimacy linked to improved access to resources and pragmatic legitimacy linked to improved organizational results (Diez-Martin et al. 2013). Technology plays a role in the institutionalization of OSINT by providing a means of managing legitimacy while the resulting performance expectations act as a source of legitimacy. Analysis of open source GEOINT, IMINT, and traditional OSINT sources to produce products that address a specific intelligence requirement (e.g., the success of a DPRK SLBM test (Dill 2016a, b)) allow OSINT as an institution to conform to the IC, inform the public in a manner it expects of the IC, and manipulate the public's perceptions (similar to the way advertising and social media campaigns aim to influence perception) – all strategies for managing legitimacy. These strategies are enabled by technology, but ultimately it is not the technology that institutionalizes OSINT. The performance expectations and standards and best practices, the sources of legitimacy, are what stake-holders and constituents assess as legitimate.

## 2 Emergence of Open Source OSINT

Not even two decades ago, the information that was relevant to both the IC and the general public was limited to well-known or easily monitored sources and media. Open source information, or "publicly available information that anyone can lawfully obtain by request, purchase, or observation" (Benes 2013) is used to produce Open Source Intelligence products and enhance or supplement traditional intelligence products. The first notable OSINT products in the United States came out of the Foreign Broadcast Monitoring Services that were tasked with processing foreign propaganda radio broadcasts. As the IC learned to leverage open source information, print media, such as newspapers, journals, magazines, and audio and visual media made up the information sources. Gathering intelligence from these open sources was not always timely; as an example, CIA officers did not learn of a Soviet agent's fate until two years after the agent's arrest, interrogation, and execution, when the events were published in Soviet propaganda paper *Pravda* (Wallace et al. 2009). Peer review and publishing cycles can decrease the timeliness of print information when OSINT must be "collected, exploited, and disseminated in a timely manner to an appropriate audience for the purpose of addressing a specific intelligence requirement" (Benes 2013).

The future of intelligence gathering and processing is rapidly being shaped by the emergence of new technologies and the vast quantities of information that follow. The amount of open source information now available to an analyst far exceeds that of their antecedent counterparts. The IC faces a challenge in making sense of and managing intelligence data in a time-efficient manner (Noubours et al. 2013). In addition, much of the information now at the disposal of intelligence analysts is also available to the general public. Further, the computational power necessary to leveraging the advantage of the information itself is both more widely available and requires less skill to exploit. While the human-in-the-loop plays an important part of

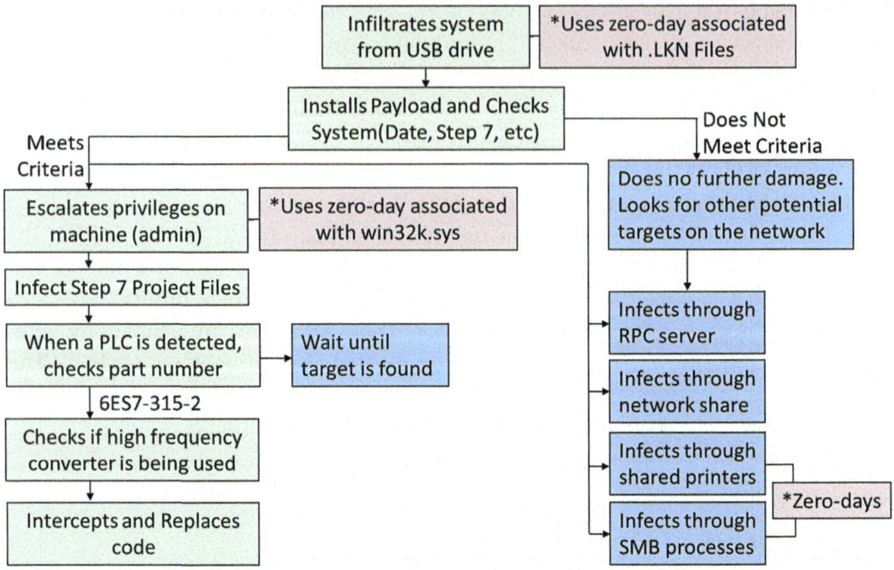


**Fig. 1** Public visibility associated with publication media

the IC's processing, exploitation, and dissemination (PEDs) cycle, the capabilities of technology are more closely intertwined with the analyst.

The availability of information, however, does not mean that such information is verified, validated, and trusted by the IC. OSINT analysis provides an example of technology's role in the intelligence institutions of the future. Open source analysts present a growing challenge to the IC with respect to the legitimacy of the open source OSINT produced, particularly when such products could support action at the national and international levels. Using only publicly available data, OSINT analysts are now producing thorough intelligence analyses that allow the public to peer into the world of opaque regimes such as the Democratic People's Republic of Korea (DPRK) (Lewis et al. 2014). These reports and analyses draw upon satellite imagery, ground imagery, traditional print sources, video broadcasts, and social media in a way that has not been possible before. OSINT seems to be gaining prominence and, as a result, is increasingly informing public discourse. It follows that OSINT may soon be, if not already, a more critical piece of intelligence puzzles within the IC itself.

This evolution, in terms of visibility to a general audience, is depicted in Fig. 1. Visibility of OSINT products are moving from a targeted, specific audience, to a general public audience based on publication media.

It would seem that the emergence of widely available information and technical analysis capabilities has resulted in this increased visibility. That is, technology per se is a supporting and enabling mechanism by which OSINT groups are informing public discourse and understanding.

This chapter identifies four primary groups or entities that provide OSINT analysis on international security topics. Their analyses and products are made available to the general public with high-level goals to predict future developments, help policy and decision makers avoid surprises by shedding light on closed or sensitive topics, make a variety of open data more meaningful and sensible (Fleisher 2008). Each group is briefly detailed in the following sections.

## 2.1 38 North, US-Korea Institute at SAIS

38 North is a program of the US-Korea Institute (USKI) at the Paul H. Nitze School of Advanced International Studies (SAIS) at Johns Hopkins University. Launched in 2006, the US-Korea Institute at SAIS aims to be a center of Korea-relevant studies in the Washington D.C. area and increase the understanding of Korea and Korean affairs (US-Korea Institute at SAIS 2007). The 38 North website provides informed analysis of North Korea to both expert audiences as well as the general public. 38 North focuses not only on coverage of WMDs and security issues, but it also provides analysis on domestic and foreign affairs in the economic, political, and social arenas (38 North 2009).

38 North posts content from a wide range of contributors, not just those professionals closely affiliated with the program and website. Many of the publishing authors on 38 North are primarily connected with other academic, research, or corporate institutions; in fact, analysts from the Center for Nonproliferation Studies (CNS) have cross-published analysis on 38 North. With respect to analysis of events within the DPRK, photographs or ground imagery and satellite imagery provide the bulk of source material (Satellite Imagery 2009). The goal of such publications is to use openly available sources to synthesize meaningful insights into and shed light upon the reclusive nation. Furthering the discussion with respect to DRPK, 38 North will publish content on DPRK-relevant events, such as those at the United Nations (Foreign Affairs 2009).

## 2.2 Bellingcat

Bellingcat is a website that publishes content by investigative journalists using open source data and social media as the basis for their analysis. The youngest of the four organizations, Bellingcat was founded by Eliot Higgins in 2014 but features content by a variety of contributing authors (Bellingcat 2016). At its inception, Higgins' summarized his goal that the website "will unite citizen investigative journalists to use open source information to report on issues that are being ignored" (Higgins 2014). Content topics are dictated more by the author's interests and expertise, provided the work utilizes open source information to draw conclusions and insight on a topic. The site also provides guides as Bellingcat does not produce its own podcast; instead, the site compiles a list of OSINT relevant episodes from other podcasts, often featuring authors that contribute to the site. Aside from Higgins, Bellingcat contributors maintain primary affiliations with other organizations and institutions (Contributors 2016).

## 2.3 Center for Nonproliferation Studies at Monterey and the Arms Control Wonk Blog

Arms Control Wonk (ACW) is a website founded by Dr. Jeffrey Lewis, the Director of the East Asia Nonproliferation Program at the James Martin Center for Nonproliferation Studies (CNS) at the Middlebury Institute of International Studies at Monterey. The site was founded in 2004 and is self-described as the "home for everything that is "too wonky or obscene" for publication about nuclear weapons" featuring the "leading voices on arms control, disarmament, and non-proliferation" (Arms Control Wonk 2016). Specifically with respect to OSINT products, Lewis and his CNS colleagues are the primary content-generators for the site (Authors 2016).

In addition to writing on ACW, Lewis is a regular contributor to the newsmagazine *Foreign Policy*, writing on topics similar to those discussed on ACW. ACW is the primary means by which analysis completed at the East Asia Nonproliferation is released to the public. Content by the authors may appear on other websites as well. However, the website is not directly affiliated with the program aside from the credentials of the contributing authors. When presenting analysis and expertise outside of ACW, the OSINT contributors state their affiliation with CNS (CNS in the Media 2016). For the purposes of this paper, ACW and CNS are referred to as a singular entity, connected through the contributing authors.

## 2.4 Federation of American Scientists and Strategic Security Blog

The Federation of American Science (FAS) is a non-profit organization that provides science-based analysis in the national and international security domains to educate policymakers, the public, and the press. FAS focuses primarily on topics related to WMDs: proliferation of nuclear weapons, nonproliferation efforts, nuclear and radiological terrorism, nuclear energy safety and security, biological and chemical weapons, and government secrecy practices. The organization was founded in 1945 as the Federation of Atomic Scientists and draws members from academic, non-profit, and government communities. FAS staff and fellows are experienced professionals from the engineering and sciences with a Board of Sponsors comprised of influence science and policy professionals (About FAS 2016).

FAS regularly publishes issue briefs, reports, public interest reports, as well as articles and op-eds. With respect to OSINT, the primary publication source for FAS's Nuclear Information Project is on the FAS Strategic Security Blog, the Bulletin of the Atomic Scientists Nuclear Notebook, the Stockholm International Peace Research Institute (SIPRI) Yearbook's World Nuclear Forces overview, and magazine publications. As stated by FAS, "As a primary source for reliable information on nuclear weapons, the project is a frequent advisor to governments, parliamentarians, the news media, institutes, and non-governmental organizations" (Nuclear Weapons 2016). Hans M. Kristensen is the director of the Nuclear Information Project and is the primary cited author on the Strategic Security Blog. While FAS typically cites from the more traditional sources or analysis methods for content, the Strategic Security Blog has become a home for analysis that draws on open source ground images, satellite imagery, YouTube videos, and subject matter expertise (Kristensen 2014).

## 2.5 *Use of Technology in OSINT*

Enabled by the evolution of technology, open source analysts now have more information and analysis techniques available to them. These methods are often computationally intensive or dependent on analytic methods and information sources and include techniques such as satellite imagery analysis, image analysis, 3-D design, and data mining. Of the four groups or entities detailed previously, all utilize technology to some degree in producing their analyses and products. For example, analysis of satellite images combined with analysis from ground images is present in at least one report produced by each organization (Dill 2016a, b; Biggers 2016; Schilling 2016; Kristensen 2014). While shorter products may typically focus on analyzing one information type, larger more intensive products typically fuse insights from many sources. Perusal of archives on all four sites provides examples of analysis on GEOINT, IMINT, video sources, official state releases, news reports, etc. All of the groups release products that result from analysts merging insights from satellite images, ground images, video, official releases, social media, and other open sources. However, not every OSINT content originator is producing analyses that will be picked up by the mainstream media that informs the public's knowledge and understanding of security topics.

FAS maintains its position as a provider of science-based analysis and expertise and makes available an archive of FAS contributions, in both print and video, to the news since 2010. This archive is provided in addition to the regular publications from the organization (Press Center 2016) (News Archive 2016). As an example, Hans Kristensen of FAS provided commentary to CNN in the form of an interview on the B61–12 nuclear bomb as well as written contribution through the Strategy Security Blog throughout the second half of 2015 (Kristensen and McKinzie 2016; CNN Video 2015; Kristensen 2015). CNS maintains a more recent database on its experts' contributions to print and video news media with records from 2015 to the present (CNS in the Media 2016). For example, Jeffrey Lewis, Melissa Hanham, and Catherine Dill provided commentary through print media, Arms Control Wonk, and interviews on DPRK's increased ballistic missile test activity throughout the beginning of 2016 (CBC Player 2016). Recently, media mentions of 38 North products appear in early 2016 as well, with reports from the website used as source material on DPRK reporting (McConnell et al. 2016; McKirdy 2016; Brunnstrom 2016). Media citations of 38 North work however, appear to be less ubiquitous than citations of FAS-Strategic Security Blog and CNS-Arms Control Wonk Blog contributors. Bellingcat's presence in the US media thus far is scarce; the website is primarily referenced by Ukrainian and Russian sources when cited in the media following its report on the Malaysian Airlines Flight 17 (MH17) that was shot down over eastern Ukraine on July 17, 2014 (Google Search 2016).

OSINT products, and the technology that enables the analysis and production, do not yield the same media visibility for the four organizations. The remainder of this chapter explores organizational legitimacy as the causal factor for the increase in media visibility of OSINT analysis.

## 3 Organizational Legitimacy and OSINT

Each of the four organizations utilize a combination of subject matter expertise and computational analysis techniques when producing an OSINT product. However, their outcomes with respect to presence in the media, and as a result presence in public discourse, vary. One possible explanation is the varying degrees of legitimacy held by each of the four organizations.

Legitimacy of an organization in general, including those outside of the OSINT realm, is perception of social acceptance (Vergne 2011). Legitimacy is linked to organizational benefits of alignment with social acceptance the expectations of "purpose, endeavor, and outcomes" (Brinkerhoff 2005). Three types of legitimacy are distinguished in the literature on the subject: normative (or "moral") legitimacy, pragmatic legitimacy, and cognitive legitimacy. Across all three types, a study has documented links from greater legitimacy in general to improved access to resources and improved organizational results. The same study demonstrated the link between pragmatic legitimacy and improved organizational results, and pragmatic and moral legitimacy and improved access to resources (Diez-Martin 2013).

### *3.1 Measures of OSINT Legitimacy*

So far organizations have not turned their lenses towards the United States in a way that the domestic public would view as harmful to national security. Military technologies and capabilities may be assessed, sometimes building on declassified documents, such as analysis of the US Air Force's next-generation nuclear-capable bomber aircraft (Kristenson 2016). This could partly be due to the transparency in US affairs but also a means by which the organization would break a social contract with domestic stakeholders. Intelligence is inherently a national security topic, and OSINT methodology should be used to understand those topics relevant to US national security not undermine it.

### 3.2 Normative Legitimacy

Normative legitimacy, also referred to a moral legitimacy, is held by an organization that "reflects acceptable and desirable norms, standards, and values" and "meets normative judgements about outputs/results, procedures and technologies, structures, leaders, and personnel" (Brinkerhoff 2005).

Normative legitimacy is derived into four sub-types in the literature: "judgments about outputs and consequences, evaluations of procedures and techniques, assessments of categories and structures, and evaluations of leaders and personnel" (Brinkerhoff 2005; Suchman 1995). The first category, judgements about outputs and consequences, results from evaluation of OSINT products against the type of organization from which the product originated. This is, essentially, the organization "doing the right things" and is particularly easy to judge for organizations that produce tangible and measurable outputs (Brinkerhoff 2005). The close association of 38 North, ACW, and SSB with a higher institution helps build these sites' normative legitimacy in this first category because the parent institutions meet these expectations. Of the three, SSB is most closely associated with FAS, being directly located on the home page. 38 North prominently displays that it is associated with USKI on its header, whereas the association to Johns Hopkins University less prominently displayed. Contributing authors to 38 North, however, do not share a common affiliation with USKI and as a result the legitimacy of the academic institution is not leveraged as directly as SSB. ACW is not affiliated directly with CNS, however reading the content quickly conveys the affiliation of its OSINT contributors with CNS. The remaining source, Bellingcat, is always presented as an independent site, a personally owned blog, which does not set any preconceived standards for products among its constituents.

The second type of normative legitimacy, evaluations of procedures and techniques, is gained when constituents view the organization as following accepted norms, standards, techniques, and procedures. While OSINT as an institution does not have accreditation standards, regulatory or legal oversight the way some institutions do, OSINT analysis has developed standards methods for analysis. In this area of normative legitimacy, all four of the organizations follow what constituents expect as practice and procedure to create an OSINT product. The groups publish their source content along with the analysis, such as the satellite imagery, video, press releases, etc. This form of legitimacy is summarized as "doing things right" (Brinkerhoff 2005).

The third type of normative legitimacy is the constituents' assessments of categories and structural characteristics, a value derived from recognition that the organization is "right for the job" (Brinkerhoff 2005). Here the higher institution affiliation of 38 North, ACW, and SSB affords these groups more normative legitimacy than a personally owned website, such as Bellingcat. It makes sense that institutions like USKI, CNS, and FAS would directly sponsor or have affiliations with the content authors on these sites with respect to security topics. 38 North focuses specifically on analysis of the DPRK as an extension of the US-Korea Institute. ACW conducts research and analysis on nuclear and missile topics as an extension of the Center for Nonproliferation Studies. Strategic Securities Blog paints with broader strokes, covering domestic security topics as well as international, which remains aligned with the goals of FAS as an institution.

The fourth type of normative legitimacy sees SSB and ACW break away from 38 North in deriving legitimacy from the "personal status, reputation, and charisma of individual organizational leaders and staff" (Brinkerhoff 2005). This does not mean that 38 North or Bellingcat's contributors are any less legitimate, in fact there is sometimes cross-over between websites. Constituents perceive legitimacy based on the characteristics and reputation of the leaders of the organization. One source of this legitimacy is the manner in which the sites present their authors to their audiences. Authors on SSB are presented with a photo as well as text description of qualifications and affiliation with the higher institution. Bellingcat presents a photo and/or text description of the author, when available. ACW presents a stylized "cartoon" picture of the content author with the article; a statement of qualifications is found on a separate page.

While 38 North presents what constituents "expect" of a news publication in terms of focusing more on the content as opposed to the author, OSINT analysis still depends heavily on the expertise of the analyst producing the work. In this way, SSB leans most heavily on the status of "expert" author to convey credibility to the content presented. ACW, 38 North, and Bellingcat all require familiarity with the website for the credibility of the authors to be conveyed. However, ACW OSINT authors, like FAS, maintain a level of legitimacy based on the personal reputation, with the assumption that a constituent is aware of the personal status and reputation of the authors if he or she is visiting the site.

### 3.3 *Pragmatic Legitimacy*

Pragmatic legitimacy is attained when an organization "fulfills needs and interests of its stakeholders and constituents" when it "exchanges goods and services that constituents want, and receives support and legitimacy" (Brinkerhoff 2005). "Pragmatic legitimacy emerges as a function of exchange relationships between and organization and its immediate stakeholders" (Brinkerhoff 2005). This relationship in OSINT takes the form of published research and analysis and in return, stakeholders support the organizations.

Pragmatic legitimacy may be attained through accordance of legitimacy and influence legitimacy. In accordance legitimacy the relationship between organizations and stakeholders resemble resource dependency models. Organizations provide OSINT research and analysis while stakeholders may confer or withhold legitimacy. In influence legitimacy, the relationship between organization and stakeholder less directly benefits the stakeholder's individual interests, but instead benefits the larger interests of constituents.

All four of the OSINT organizations hold some type of pragmatic legitimacy through one of the two types. In a relatively small OSINT playing field, some accordance legitimacy is present in all four organizations, being the purveyors of a specific type of product. However, websites with a more open policy towards content contributors enter into a more resource dependent relationship where authors may also be stakeholders. Bellingcat's Kickstarter origins and statement of being "for investigative journalists" affords the website accordance legitimacy. In exchange for a venue where analysts may employ new as well as tested OSINT methods, analysts provide the website support and contribute to the long-term survivability and sustainability of the organization. Authors are free to explore topics of their choosing provided the methods and approach meet the over-arching goals of the website. In this way Bellingcat is supported by its authors as well as by its consumers. Bellingcat finds influence legitimacy in the broader goal of furthering OSINT as a practice.

The larger institutional connections of ACW and SSB build legitimacy in the realm of influence legitimacy, where products are targeted at building and sustaining a relationship between the organization and the broader goals of the security community. The research conducted by these organizations strives to achieve policy outcomes in line with national and international security (e.g., nonproliferation), a commitment that aligns with constituents' general interests.

The pragmatic legitimacy afforded to 38 North provides an example of a less clear distinction where the website aims to provide a verified and validated discourse on the DPRK but, through non-USKI authorship, requires the continued contributions of individual stakeholders.

## *3.4 Cognitive Legitimacy*

Cognitive legitimacy is held by an organization when it "pursues goals and activities that fit with broad social understandings of what is appropriate, proper, and desirable" (Brinkerhoff 2005). Organizations with cognitive legitimacy often seem natural or expected when thought of by stakeholders and their presence may be taken for granted. To constituents, what the organization does "makes sense" and perception is formed in two ways (Suchman 1995).

The first means by which organizations "make sense" is through comprehensibility. A cultural framework allows constituents to explain the organization and its behavior as acceptable and resulting in valuable products (Brinkerhoff 2005). Here, the Western culture of openness in government affairs works to the benefit of all the

**Table 1** Summary of organizations and normative legitimacy

| | 38 North & USKI | ACW & CNS | Bellingcat | SSB & FAS |
|---|---|---|---|---|
| *Normative legitimacy* | | | | |
| "Doing the right things" | Focus on security topics and relevant analysis meets expectations of the organization | | Must build expectation that independent and non-institution-aligned source can produce defensible and credible products | Production of reports and publications as well as OSINT products meets expectations of the organization |
| "Doing things right" | OSINT products use sources, follow analytic methods, and present results as expected | | | |
| "Right for the job" | DPRK focus aligned with USKI institutional expertise and research area | Nuclear weapons and missiles focus aligned with CNS institutional expertise and research area | Builds reputation on "being right" for OSINT analysis, specific topic expertise relies on author | Security focus aligned with FAS institutional expertise and research area |
| Leaderships personality and status | Focuses attention on content rather than attributes of the author and does not maintain cohesion with parent institution | Emphasizes personality and reputation of authors, maintains cohesion with parent institution | Relies on reputation of independent authors and contributors | Emphasizes personality and reputation of authors, maintains cohesion with parent institution |

OSINT groups. It is expected that the US government make public its records and documents in support of its decision making and legal authority. The public expects to have a voice in government action at home and abroad. Societal openness and questioning of leadership are legitimate and valued behaviors in Western culture. However, not all states are as open and transparent, particularly those that have been the focus of OSINT analysis. In this way, it makes sense to constituents that expertise in security fields would be turned to answering questions about and shedding light on more closed societies.

The second means by which organizations "make sense" is by being taken for granted ("taken for grantedness"): "Society accepts the organization, its structures, procedures, and activities as so completely understandable and appropriate that no other option is imaginable…" (Brinkerhoff 2005). This type of cognitive legitimacy is not rigid and may evolve as organizations age, but the relative youth of privately owned websites such as Bellingcat works against building cognitive legitimacy in this way. Permanence through time in the eyes of the constituents is one way in which organizations may be taken for granted. The public expects and understands that academic institutions such as CNS and USKI will use their resources in support of educating and informing society.

**Table 2** Summary of organizations, pragmatic and cognitive legitimacy

| | 38 North & USKI | ACW & CNS | Bellingcat | SSB & FAS |
|---|---|---|---|---|
| *Pragmatic legitimacy* | | | | |
| Accordance | Requires contributions and support of non-USKI-affiliated authors for support | Constituents in the form of organizations and institutions afford legitimacy in the form of funding and sponsorship | Kickstarter origins, target at investigative journalists, and more open contribution policy where authors exchange OSINT forum and information in exchange for legitimacy and support | Constituents in the form of organizations and institutions afford legitimacy in the form of funding and sponsorship |
| Influence | Supports broader institutional goals of improved and clear discussion on DRPK | Supports broader institutional goals on nonproliferation | Influence legitimacy in the form of furthering OSINT as a practice, cohesion on specific topics within subsections of the site | Supports broader institutional security goals |
| *Cognitive legitimacy* | | | | |
| Comprehensibility | Western culture (and US culture to a greater extent) that places value on openness and transparency in government creates conducive environment for OSINT | | | |
| "Taken for grantedness" | Leverage scientific background of contributors and expected activities of USKI | Can work to further define policy professionals as technical, as well as policy, analysts | Build upon technical credentials of authors | Leverage scientific background of contributors and expected activities of FAS |

However, one area in which OSINT organizations face an obstacle in building cognitive legitimacy is in the background and expertise of their analysts. While it is expected that scientists and engineers will perform "hard" science-based analysis, many OSINT products originate from authors with policy educations and backgrounds. This breaks with the public's ingrained idea of who performs the analysis and research behind OSINT products.

Table 1 summarizes the findings on normative legitimacy by organization. Table 2 summarizes the findings on pragmatic and cognitive legitimacy by organization.

## 4 Conclusions

Assessing the four OSINT organizations chosen for analysis against types of organizational legitimacy provides a means for explaining the emergence of OSINT products, and their originators, in the public discourse. Within the greater context of OSINT gaining legitimacy as an institution, practitioners of open source analysis are building organizational legitimacy differently, even though all have access to roughly the same technological resources. Websites like Arms Control Wonk and Strategic Security Blog rely heavily on the credentials of their parent institutions and the reputation and status of the leadership, building their normative legitimacy.

All four organizations are defining and sustaining relationships with constituents and stakeholders to build pragmatic legitimacy. The differentiation occurs when the relationship is towards the individual interests of the constituents or the broader goals of the community.

The cognitive legitimacy afforded to each organization will grow as OSINT analysis becomes more established in the public domain. In the meantime, organizations leverage affiliations with institutions the public already expects to perform scientific research and analysis. Within that expectation, organizations whose authors meet the expectations of the public retain more legitimacy in the form of meeting expectations. Professionals who find themselves reshaping the expectations of their constituents must work to be taken more for granted.

The result of the differences in organizational legitimacy can be seen in the frequency with which each organization is cited in the greater public discourse. FAS maintains a longstanding reputation of science-based analysis and research in the realm of policy and security consulting that is conferred to SSB. ACW and 38 North, and CNS and USKI, respectively, are gaining legitimacy and as such are more relied upon sources in their fields. ACW and CNS are able to leverage aspects of their organization to build further legitimacy and establish their authors and contributors as subject matter experts capable of informing the general public. Bellingcat, the newest of the organizations, is establishing itself as supporter of OSINT practices in general and is building pragmatic legitimacy above normative and cognitive. As an independently owned website, it cannot leverage the legitimacy of a higher parent institution and must work to build constituents expectations of independent websites.

Further research could be conducted into the strategies for managing and quantitatively measuring legitimacy. With respect to policy goals, an understanding of when public OSINT analysis will become an established institution capable of informing decision makers at the national and international levels is warranted. Public OSINT as in institution faces a steep challenge with the respect to the legitimacy of the products themselves due to the nature of the data. As OSINT organizations further define their practices, procedures, and standards, OSINT products could shift more towards decision support documents instead of more informative products.

## References


About FAS. (2016). Federation of American Scientists. https://fas.org/about-fas/. Accessed 27 Mar 2016.

Arms Control Wonk. (2016). http://www.armscontrolwonk.com/. Accessed 18 Mar 2016.

Authors. (2016). Arms Control Wonk. http://www.armscontrolwonk.com/authors/. Accessed 18 Mar 2016.

Bellingcat. (2016). https://www.bellingcat.com/. Accessed 19 Mar 2016.

Benes L (2013) OSINT, new technologies, education: Expanding opportunities and threats. Journal of Strategic Security 6(3): Supplement.

Biggers C. (2016). Imagery Confirms Niger's New Cessna 208 Caravan https://www.bellingcat.com/news/africa/2016/01/03/imagery-confirms-nigers-new-cessna-208-caravan/. Accessed 4 Mar 2016.

Brinkerhoff, D (2005) Organizational Legitimacy, Capacity, and Capacity Development. Paper presented at the Public Management Research Association 8th Research Conference, University of South Carolina, 2005.

Brunnstrom D. (2016). Fresh activity seen at North Korea nuclear site: US think tank. http://www.reuters.com/article/us-northkorea-plutonium-report-idUSKCN0XC24Y. Accessed 26 Apr 2016.

CBC Player. (2016). North Korea reportedly tests ballistic missile from a submarine. http://www.cbc.ca/player/play/672422467978. Accessed 24 Apr 2016.

CNN Video. (2015). Vladimir Putin responds to reports of new US nukes. http://www.cnn.com/videos/world/2015/09/28/us-nukes-in-germany-todd-dnt-tsr.cnn. Accessed 2 Apr 2016.

CNS in the Media. (2016). James Martin Center for Nonproliferation Studies, Middlebury Institute of International Studies at Monterey. http://www.nonproliferation.org/experts/cns-in-the-media-2/. Accessed 21 Mar 2016.

Contributors. (2016). Bellingcat. https://www.bellingcat.com/contributors/. Accessed 19 Mar 2016.

Diez-Martin, F., Prado-Roman, C., & Blanco-Gonzalez, A. (2013). Beyond legitimacy: Legitimacy types and organizational success. *Management Decision, 51*(10), 1954–1969.

Dill C. (2016a). Video analysis of DPRK SLBM Footage. Arms Control Wonk. http://www.armscontrolwonk.com/archive/1200759/video-analysis-of-dprk-slbm-footage/. Accessed 15 Jan 2016.

Dill C. (2016b). Convoy Directions from the 816th Launch Brigade. http://www.armscontrolwonk.com/archive/1201074/convoy-directions-from-the-816th-launch-brigade/. Accessed 15 Feb 2016.

Fleisher, C. (2008). Using open source data in developing competitive and marketing intelligence. *European Journal of Marketing, 42*(7/8), 852–866.

Foreign Affairs. (2009). 38 North, US-Korea Institute at SAIS. http://38north.org/category/03-foreign-affairs/. Accessed 17 Feb 2016.

Google Search. (2016). Search for Bellingcat mentions in the news. https://www.google.com/?gws_rd=ssl#q=bellingcat&tbm=nws. Accessed 27 Apr 2016.

Higgins E. (2014). Bellingcat, for and by citizen investigative journalists – Kickstarter. https://www.kickstarter.com/projects/1278239551/bellingcat/updates/. Accessed 19 Mar 2016.

Kristensen H. (2014). Rumors About Nuclear Weapons in Crimea. https://fas.org/blogs/security/2014/12/crimea/. Accessed 15 Mar 2016.

Kristensen H. (2015). General Cartwright Confirms B61–12 Bomb. http://fas.org/blogs/security/2015/11/b61-12_cartwright/. Accessed 3 Apr 2016.

Kristensen H. (2016). New B-21 Bomber or B-2 Mod 1? https://fas.org/blogs/security/2016/03/b-21/

Kristensen H, McKinzie M. (2016). Video Shows Earth-Penetrating Capability of B61–12 Nuclear Bomb. http://fas.org/blogs/security/2016/01/b61-12_earth-penetration/. Accessed 2 Apr 2016.

Lewis J, Hanham M, Lee A (2014) That Ain't my truck: Where North Korea assembled its Chinese transporter-erector-launchers. *38 North*. http://38north.org/2014/02/jlewis020314/. Accessed 9 Feb 2016.

McConnell D, Todd B, Kwon K (2016) North Korea-South Korea tensions: Warning shots fired. http://us.cnn.com/2016/01/13/asia/north-korea-south-korea-dmz-warning-shots/index.html. Accessed 11 Apr 2016.

McKirdy E (2016) 'Suspicious activity' at N. Korea nuke site. http://www.cnn.com/2016/04/05/asia/north-korea-yongbyon-nuclear-complex-activity/. Accessed 7 Apr 2016.

News Archive (2016) Federation of American Scientists. http://fas.org/press-center/news-archive/. Accessed 1 Apr 2016.

North. (2009). US-Korea Institute at SAIS. http://38north.org/about/. Accessed 12 Mar 2016.

Noubours S, Pritzkau A, Schade U (2013) NLP as an essential ingredient of effective OSINT frameworks. Paper presented at the military communications and information systems conference (MCC), Saint-Malo, France, 7–9 Oct 2013.

Nuclear Weapons. (2016). Federation of American Scientists. https://fas.org/issues/nuclear-weapons/. Accessed 27 Mar 2016.

Press Center. (2016). Federation of American Scientists. https://fas.org/press-center/. Accessed 1 Apr 2016.

Satellite Imagery. (2009). 38 North, US-Korea Institute at SAIS. http://38north.org/category/sat-analysis/. Accessed 17 Mar 2016.

Schilling J. (2016). North Korea Tests a Submerged-Launch Ballistic Missile, Take Three. http://38north.org/2016/01/jschilling011215/. Accessed 20 Jan 2016.

Suchman, M. (1995). Managing legitimacy: Strategic and institutional approaches. *Academy of Management Review, 20*(3), 571–610.

US-Korea Institute at SAIS. (2007). Johns Hopkins University. http://uskoreainstitute.org/about/. Accessed 17 Mar 2016.

Vergne, J. (2011). Toward a new measure of organizational legitimacy: Method, validation, and illustration. *Organizational Research Methods, 14*(3), 484–502.

Wallace R, Melton H, Schlesinger H. (2009). Spycraft: The secret history of the CIA's Spytechs, from communism to Al-Qaeda. Plume.

# Strategic Offensive Cyber Operations: Capabilities, Limitations, and Role of the Intelligence Community

**Allison J. Mahvi**

**Abstract** In 2015, the Director of National Intelligence began his discussion of global threats during his testimony to the Senate Armed Services Committee stating that "[Cyber] attacks against us are increasing in frequency, scale, sophistication and severity of impact." In the past, rhetoric around cyber security has focused on defensive measures – strengthening US systems to prevent cyber-attacks from disclosing sensitive information or causing service outages for critical infrastructure. More recently, intelligence, national security, and military leaders have discussed the need for offensive cyber abilities in order to understand and deter the operations of our adversaries.

Advancements in cyber capabilities are outpacing understanding of the risks and implications of cyber conflicts. This chapter explores the technology behind sophisticated offensive cyber-attacks and the role of the intelligence community (IC) in collecting zero-day errors and writing code in order to develop the US cyber-arsenal. This will be investigated using a case-study framework focused on events of covert operations to collect information or damage infrastructure that were allegedly carried out by state actors. Using information about the cyber tools used today, predictions about the future of cyber sabotage and espionage will be presented.

**Keywords** Offensive cyber tools · Cyber-sabotage · Cyber-espionage · Intelligence

## 1 Introduction

Cybersecurity is one of the fastest growing sectors of US military and intelligence communities. In 2016, President Obama proposed to increase the cybersecurity budget to $19 billion to help protect government and private networks and to

A. J. Mahvi (✉)
Georgia Institute of Technology, Atlanta, Georgia, USA
e-mail: amahvi@gatech.edu



M. E. Kosal (ed.), *Technology and the Intelligence Community*, Advanced Sciences and Technologies for Security Applications,
https://doi.org/10.1007/978-3-319-75232-7_10

enhance the ability for the US to respond to an attack (Volz and Hosenball 2016). The United States is susceptible to attacks from both state and non-state actors, which could have lasting effects on critical infrastructure (e.g., electrical grid or water processing), the financial system, and national security. Although non-state actors pose a persistent threat, the malware involved will generally not be for espionage activities and will be limited in its capabilities because of the large fiscal, human, and intelligence resources required to develop highly sophisticated codes. However, many state actors, including Russia, China, and North Korea, have advanced cyber-capabilities that could cause significant damage to civilian and military resources.

Past rhetoric surrounding cybersecurity has focused largely on safeguarding our own networks (White House 2003; CSIS 2008; Herrera-Flanigan 2011), however recently discussions have shifted to developing offensive cyber weapons (Schonberg 2013; Sanger 2016). These types of offensive operations are generally developed and executed through cooperation between the military and the intelligence community, specifically US Cyber Command (USCYBERCOM), the National Security Agency (NSA), and the Central Intelligence Agency (CIA).

This chapter introduces some of the technological aspects of offensive cyber operations for covert or clandestine missions and the role of the intelligence community in the development and deployment of the malicious codes. Malware designed for use in wartime scenarios (primarily developed by USCYBERCOM) is not discussed. The goal of this chapter is to project future trends of the capabilities of cyber-tools used by the intelligence community. This is accomplished by investigating the techniques used in past sophisticated cyber-attacks to understand the potential and limitations of cyber-weapons.

## 2 Technology Overview

Cyber-weapon is a broad term that describes a vast spectrum of malicious code or malware. At one end of the spectrum are low threat viruses that do not damage the underlying system but can cause a denial of service (DOS). For example, a simple low-threat bug might artificially increase the amount of traffic to a website, making it impossible for the intended users to logon or use it. Although such an attack could cause fiscal damage, they are not of greatest concern to government agencies involved in cybersecurity. On the other end of the spectrum are sophisticated autonomous or semi-autonomous codes designed for espionage or sabotage that can cause substantial physical or operational harm. Generally, sophisticated malware is costly to develop and require detailed knowledge about the target system. This chapter will focus on the technological aspects and consequences of sophisticated malware allegedly developed by state actors.

Malware can be used as a tool for espionage, sabotage, or wartime operations. Although the code will be tailored for its specific goal, there are some common methods that are usually exploited. Generally, sophisticated malware leverages zero-day errors, which are errors in a system's software that are not yet know by the software developer. If a security threat has not been seen before, the bug can infiltrate the system and cause damage before a patch is created. Zero-day errors are very valuable and are bought and sold around the world. There are three markets for zero-days: the white, grey, and black markets. Sellers participating in the white market will disclose errors they discovered to major software providers. Although some competitions allow winners to sell their zero-days to these companies for a high price (~$50,000), generally reporting the error to the software provider will conclude in the smallest payday. The black market primarily deals with criminal organizations, and the grey market deals with governments. In the "legitimate" grey market, zero days or zero-days plus the malware that can be used to exploit them can sell for between $5000 and $250,000 (Miller 2007). Although surveillance software is export controlled through the Wassenaar Arrangement (Wolf 2015), it is hard to track the trade of zero-day errors and exploits, so selling them to governments on the grey market is the safest method to trade exploits without major legal ramifications, especially if the seller negotiates with their own government.

Although zero-day exploits are becoming more common because of the high value market that has developed, the majority of malware does not use zero-day errors (Park 2015). Instead the program will use a known error and either attempt to exploit it before it is patched, or intruders will hope their target does not have up-to-date programs and security software on their device. Additionally, there are sometimes clever ways to circumvent security software or infiltrate programs that do not require a zero-day error.

Once a bug infiltrates a device through a program error or some other method, the malware will install a rootkit on the machine. A rootkit contains all of the blocks of code used by the malware to perform a variety of tasks, including masking itself from detection or providing the hacker access through a backdoor (Kim et al. 2012). Rootkits can reside in the area of the computer used for running user applications or in the core of the operating system called the kernel (Zetter 2014). Rootkits installed in the application level are easier to identify by anti-virus software and have less "privileges" (access) on the machine. Kernel rootkits have the highest operating system privileges and are extremely difficult to detect or remove from a computer, however they are also difficult to develop. If there is a bug in a kernel rootkit, there can be serious issues with the computer, warning the user that there is malware on their device.

There are other aspects of malicious code that are similar, especially if programs are created by the same organization or by people who have seen a successful malicious bug. More details of how malicious codes infiltrate a computer and execute their missions will be discussed on a case-by-case basis later in this chapter.

### 2.1 *Role of Intelligence Community (IC)*

Highly sophisticated cyber weapons require time, secrecy, computer science experts, and detailed intelligence on the target to be developed. Because of these limitations, complex cyber-attacks have historically been carried out by state-actors. The structure of cyber operations in the United States is relatively opaque; however, the Intelligence Community is certainly involved in both the development and deployment of cyber-sabotage and cyber-espionage tools because of their covert and clandestine nature (Rosenbach and Peritz 2009; NCRRDP 2013). The ultimate goal of cyber developers is to create a tool that can go undetected for as long as possible, without any activities being directly attributed back to the attackers.

The National Security Agency (NSA) is charged with gathering and interpreting Signals Intelligence (SIGINT) and protecting US networks (Kennedy 2008). They are also purportedly charged with developing cyber-tools to aid in SIGINT intelligence gathering and cyber sabotage efforts (Sanger 2012). As discussed, sophisticated tools require coding expertise but also detailed intelligence about the target. This means that the NSA must work with other IC agencies to leverage information from HUMINT, IMINT, or MASINT to create effective tools. Additionally, NSA must work with the Central Intelligence Agency (CIA) when deploying cyber tools during covert operations and potentially while deciphering the collected information.

Although cyber tools have been a game changing technology for SIGINT, it has not fundamentally changed intelligence gathering in general. Rather sophisticated malicious software is simply a new tool in the IC's arsenal. Traditional forms of intelligence are still needed to get a full picture of the situation, and an informed "human in the loop" is necessary to target the efforts of cyber and interpret the results. As cyber tools become better, the IC must use it wisely and appropriately to avoid unintended consequences.

## 3 Case Studies

Three case studies are examined in this paper to understand the operations and targets of past cyber-attacks. The case studies examine complex malicious code used for espionage and sabotage that allegedly originated from state-actors.

### 3.1 *Attack on DoD Classified Computers (Discovered in 2008)*

In 2008, a flash drive was inserted into a laptop on a military base in the Middle East in the US Central Command Area of Operation. The flash drive was infected with a malicious bug that spread through the classified and unclassified Department of

Defense computer networks. The code was designed to transfer data back to foreign servers (Lynn 2010), however the amount of data extracted is currently unknown. The attack was officially attributed to an unspecified "foreign intelligence agency" (Lynn 2010), but many suspect that the code was specifically of Russian origin (Mick 2010). This attack was unprecedented at the time and is the most severe breach of the US classified network acknowledged to date.

The malicious code started by copying itself onto the servers of the unclassified military network, called the Non-Secure Internet Protocol Router Network (NIPRNET). Although there were security measures at the time, the NIPRNET was susceptible to unauthorized users because it was an extremely large network and was connected to the internet (US-China Commission 2008). Once the virus infected the unclassified network, it spread to removable flash drives connected to a computer accessing the NIPRNET with the hope that a user would transfer the flash drive to a different computer connected to the Secure Internet Protocol Router Network (SIPRNET). The SIPRNET contains classified documents and therefore has several security protocols and is air-gapped from the NIPRNET and internet (Rid 2013). Based on government reports, the virus spread to the SIPRNET, compromising classified information (Lynn 2010).

Little is known about the architecture of the malicious code used in the SIPRNET attack or how much data was successfully exported to foreign servers. Since the virus was never found on civilian computers, it was not analyzed by industry or academic security professionals who typically publish their work. Nonetheless, this case shows that even robust computer networks can be infiltrated, and that cybersecurity is a big concern for all countries including those with the most advanced cyber operations.

## 3.2 *Stuxnet (Discovered in 2010)*

In November 2009, Iran replaced between 900 and 2000 uranium centrifuges in the Natanz Nuclear Facility (Zetter 2014). Although nuclear centrifuges are fragile and prone to failure, this replacement rate is unusually high. About a year later, a distributer of security software in Iran contacted VirusBlokAda about a persistent problem with their computers. After some investigation by the company, the computers were found to have a highly sophisticated bug that was not detectable by any antivirus software (Schneier 2010).

This bug was later named "Stuxnet." The general architecture of the malware is shown in Fig. 1. The malware infiltrated computers using an infected USB drive. When a normal USB drive is inserted in the computer, the computer scans the content to identify the types of files on the device using the. LNK association feature. This is an essential process of the system and cannot easily be disabled. The Stuxnet virus used a zero-day error associated with the. LNK feature to deposit and hide a rootkit on the machine (Falliere et al. 2011). The code was able to do this

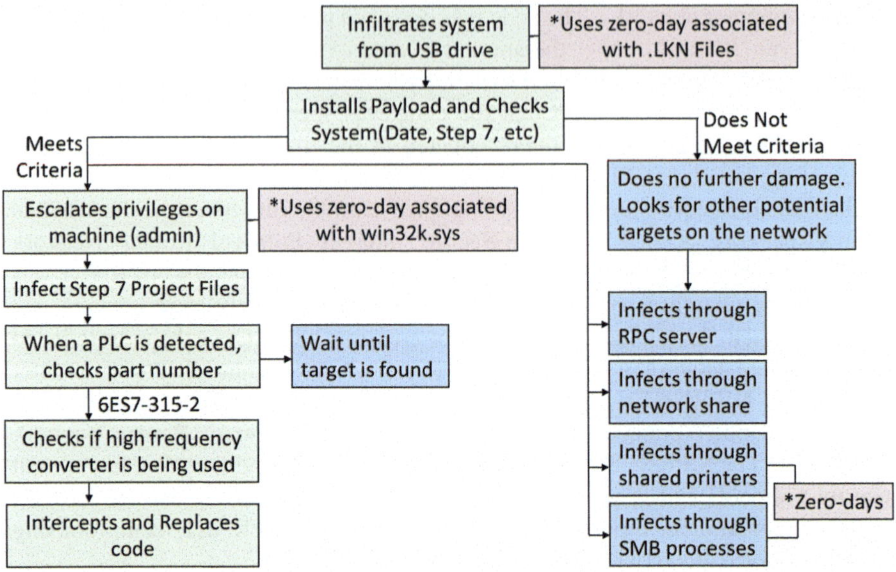


**Fig. 1** Summary of the procedures carried out by the Stuxnet virus

without raising any red flags for the user, which usually occur when software without a certificate (permission) attempts to install itself on a machine.

Stuxnet was specifically designed to operate on computers with Siemens Step 7 industrial software and a specific brand of Siemens programmable logic controllers (part number 6ES7–315-2) (Rid 2013). The combination of software and hardware that were required to trigger Stuxnet primarily occurred in only one location: the Natanz nuclear enrichment plant in Iran (Rid 2013). This prevented the malware from infecting unintended sources, but it required very specific knowledge of the processes used by the target. If the machine did not operate programmable logic controllers (PLCs), Stuxnet would not infect the computer but instead look for other machines to infect on the network using a zero-day errors associated with computers connected to a network or to a single printer. If the correct hardware and software were present on the machine, the virus used an additional zero-day to deposit itself on the PLC and run the virus. The malware would first sit until the Step 7 software was used to program a PLC controller. When the code was uploaded onto the PLC, Stuxnet would intercept it and insert a malicious block of code that could change the inputs into the frequency controller. Again, the malware was picky about its targets and would only unleash an attack on a system with a frequency controller that operated above 800 Hz (Chien 2010). These types of controllers have a limited number of applications and are currently export-controlled in the US by the Nuclear Regulatory Commission (NRC) (Zetter 2014).

Stuxnet operated by first recording normal data for 13 days. It later used that data to send normal reading back to the controller while modifications were made to the operation of the centrifuges. After the initial wait period, the code would increase the frequency of the centrifuge drive to 1410 Hz for 15 minutes, which is close to the maximum frequency that the Iranian centrifuges could handle without resulting in catastrophic failure. After the increase, Stuxnet returned the system back to normal for about one month, and then it set the frequency to 2 Hz for a short period of time (Chien 2010). These changes in frequency continued until the centrifuge failed from fatigue. The long periods between changes in the frequency made it difficult to pinpoint the problem or the source.

There are many aspects of Stuxnet that make it interesting. The code is very sophisticated, and although it spread rapidly to many computers, it zeroed in on a specific target to attack. To infiltrate the Iranian nuclear facility and cause real damage, an immense amount of clandestine information was needed about the plant, and significant testing on PLCs and centrifuges was required to ensure success. Additionally, the code was intricate and used an unprecedented amount of zero-day errors. The complexity and motivations of the malware made it apparent that it was developed by a state-actor.

Although no government has officially acknowledged their role in Stuxnet, a *New York Times* article reported that the code was developed as a joint effort between the US NSA and the Israeli Defense Unit 8200 (Sanger 2012). This information was collected from unnamed government sources that were involved in a larger cyber development project named "Operation Olympic Games." After the code was designed, the program was likely passed off to the CIA, who launched a covert mission to deposit the weapon on the target computer network.

In summary, Stuxnet was a highly targeted malicious code that was aimed at the Natanz Nuclear Enrichment Facility in Iran. The code relied on information about the process and equipment used at the facility, and expert knowledge of how the enrichment process could be compromised. Additionally, Stuxnet was complex, registering as the largest malware program detected at the date of discovery, with a total size of about 500 kilobits and 15,000 lines of code (Zetter 2014). The creation of the Stuxnet virus would have required a lot of time and cooperation between intelligence gathering organizations to develop, therefore it was almost certainly done by a state-actor. Although this code has a specific target, the framework is relatively general. Some experts fear that aspects of it could be copied and used for a similar attack on industrial systems.

### *3.3 Flame (Discovered in 2012)*

In April 2012, Iran's oil ministry reported that a malicious code had deleted all documents and data files from their computers (Zetter 2014). One month later, several malware research laboratories, including a group from the Budapest University of Technology and Economics (Bencsáth et al. 2012) and Kaspersky Labs (Kaspersky Lab 2012), published reports on the code, giving a preliminary description of its operation and goals. The malware has been given several names (SKyWIper, Flamer), however it is now commonly known as *Flame*.

Flame is a sophisticated, multi-purpose espionage software. To date it is the largest malicious software ever discovered, at 20 MB (Lee 2012) and over 650,000 lines of code (Zetter 2014). Flame is designed as a series of twenty modules that can be installed on the host machine depending on the purpose of the mission. These modules include codes that can record keystrokes, monitor spoken language over Skype of in the general vicinity of the device using the microphone, take screenshots when a program of interest is open (email, instant messenger, AutoCAD software, etc.), or collect documents saved on the local device. Flame was primarily found on computers in the Middle East, including Iran, Israel, Sudan, Syria, and Lebanon (Munro 2012; Lee 2012). Because of the size, locations that the software was found, and the similarities to Stuxnet, some researchers and journalists have suggested that the US developed and deployed *Flame* as part of a confidential cyber-program called Operation Olympic Games (Paganini 2012).

The method used to infect the first computer in a network is still unknown. Once on a network, *Flame* used several methods to download the main module onto different computers. Some of these techniques are similar to Stuxnet, including the. LNK exploit and the shared printers exploit described above. Additionally, *Flame* uses Microsoft's Windows Update to infect other computers on the local network (Bencsáth et al. 2012). Windows Update is a piece of software that allows Microsoft to send new versions and security patches of their programs to millions of computers. Windows Update periodically sends beacon signals to Microsoft's servers to see if there are new software versions to download. *Flame* intercepts the beacon signals from the infected machine and sends out the main module of *Flame* to that computer disguised as a ligament update (Zetter 2014). This hack is well designed and very dangerous. In *Flame*, the program only compromises the Windows Update feature on local networks, but the same approach could have been theoretically used on Microsoft's servers, which would quickly spread malicious software around the world. Microsoft quickly created a patch to fix the vulnerability, which ironically was distributed to customers through a Windows Update.

After the main module of *Flame* infected a device, it contacted one of the 80 command and control servers dispersed around the world (Zetter 2012). Once the machine is identified to the attackers, they decide if it is a computer of interest. If it is not the targeted computer, *Flame* had a method to completely erase itself from the hard drive to avoid potential detection. If the computer was a target machine, *Flame* created a back door so that the attackers could install any additional modules that they may need depending on the goals of the mission (Zetter 2014). The modules would then go to work collecting data from the computer. It periodically compressed and encrypted the information and sent it back to the command and control server to be analyzed.

*Flame* is an interesting example of the current capabilities of cyber espionage. The software is very flexible and allowed the attackers to collect a wealth of information from their targets. It is also highly targeted because of its non-autonomous nature, allowing the commanders to choose the machine and the types of information they want to gather. Additionally, *Flame* was a successful attack. The virus is thought to have been in the wild, undetected for as long as five years. The amount of sensitive information that may have been collected over that time is staggering.

## 4 Future

Cyber operations have profoundly affected intelligence gathering and sabotage efforts in the past and will likely be widespread in the future. Advances in coding strategies and artificial intelligence will increase the capabilities of cyber in the coming years and will require accelerated efforts to protect physical and digital infrastructure. These advances raise both opportunities and concerns for the intelligence community. In the words of James R. Clapper; then Director of National Intelligence:

> "The consequences of innovations and increased reliance on information technology… on both our society's way of life in general and how we in the Intelligence Community specifically perform our mission will probably be far greater in scope and impact than ever. Devices, designed and fielded with minimal security requirements and testing, and an ever-increasing complexity of networks could lead to widespread vulnerabilities in civilian infrastructures and US Government systems. These developments will pose challenges to our cyber defenses and operational tradecraft but also create new opportunities for our own intelligence collectors." (Clapper 2016)

Although cyber tools are a threat to our national security, they are also a game changing technology for the intelligence community. An unprecedented amount of information about processes and plans of governments and terrorist groups are stored on computer networks. Although this information is often heavily guarded, past attacks have shown that malware can find creative ways into even air-gapped networks. The ability for the intelligence community to access sensitive information with well-designed malicious code will in some ways change the methods for intelligence and will undoubtedly grow the signal intelligence sector of the community, as evident by the budget for the NSA (Sahadi 2013). However, the extent of cyber espionage and sabotage tools will be different because of the resources required for each.

Instances of cyber espionage will likely increase. Although malicious code for cyber-spying is costly to develop, it is generally flexible and can be used for many purposes. As evident by the *Flame* attack, sophisticated cyber espionage tools will be able to monitor all electronic information created, discussed, or viewed on or near an infected computer. The easiest targets will be terrorist groups or unprotected nation states, however all countries including those currently leading in cyber development (the United States, China, and Russia) will also be vulnerable to sophisticated attacks. The malicious code developed by nation states will be confidential and great effort will be taken to keep the development and methods of cyber tools secret because it will help prevent attribution of detected attacks or information about the architecture from being transferred to adversaries.

Although cyber espionage attacks will increase in frequency, cyber sabotage missions (especially attacks that cause physical damage) will likely remain rare. This is because the development of such code must be single use malware designed for a specific purpose in order to cause substantial damage, which is both time intensive and costly. One illustrative example is the power generation system in the United States. For years, reporters and government officials alike have discussed the vulnerabilities of the grid and the catastrophic failures that could result from a cyber-attack (Mooney 2017; Tehan 2017). It is certainly true that the grid is vulnerable, especially now that the consumer-side of transmission systems is connected to the internet. A well designed cyber-attack may be able to cause a regional blackout for some period of time, which would have temporary social and economic effects. It is theoretically possible to cause substantial, lasting damage if malware were able to destroy a component of a baseload plant or a portion of the power conversion system, but these types of attacks would be extremely difficult to develop and would require intimate knowledge about the process and controls systems. The possibility of causing physical damage to industrial processes was demonstrated by Stuxnet and the 2007 turbine demonstration conducted by Idaho National Labs (Zetter 2007), but any attack would have to avoid physical safety systems and be tailored to the control system present in that particular plant. There are vulnerabilities in these processes, but the fiscal, human, and intelligence resources to develop malware to make a lasting impact on the US system makes it unfeasible for most potential attackers. Additionally, an extremely sophisticated attack may be able to take out a plant or subset of plants, but the malware could not be used to compromise a different type of power plant (even one with slightly different control mechanisms), let alone be useful for a different type of industrial process. Therefore, these tools will be one-off codes, further increasing the costs and usefulness of developing and deploying the cyber-weapon. In summary, cyber sabotage events like Stuxnet will likely be rare in the future. The resources that are required to develop sophisticated sabotage malware will limit them to use state actors, and traditional weapons may be used in lieu of their cyber counterparts.

## 5 Conclusion

Malware is a game-changing tool for the intelligence community. Currently cyber-tools can complete complex espionage and sabotage missions without putting people into potentially dangerous situations. The use of malicious code for intelligence gathering will likely increase in the future as flexible software becomes available and as state actors improve their coding expertise. However, the development and use of cyber-tools cannot be done in a vacuum. It will require the integration of different organizations to develop the codes and analyze the information collected. Additionally, it is crucial that cyber-tools are focused by a "human in the loop" to identify targets and combine intelligence information to see the larger picture. As the IC has learned from the past integration of intelligence gathering technologies, they must not become over-reliant on cyber tools but instead use them in conjunction with traditional intelligence gathering methods.

## References

Bencsáth, B., Buttyán, L. and Félegyházi M. (2012). Pék, G. sKyWIper (aka Flame aka Flamer): A complex malware for targeted attacks. *CrySyS Lab*, [online]. Version 1.05. Available at: https://www.crysys.hu/skywiper/skywiper.pdf. Accessed 23 Apr 2016.

Center for Strategic and International Studies (CSIS). (2008). *Securing Cyberspace for the 44th President*, [online]. Washington DC: Center for Strategic and International Studies. Available at: https://www.nitrd.gov/cybersecurity/documents/081208_securingcyberspace_44.ppd. Accessed 31 March 2017.

Chien, E. (2010). *Stuxnet: A Breakthrough* [Blog]. Symantec security response. Available at: https://www.symantec.com/connect/blogs/stuxnet-breakthrough. Accessed 18 Apr 2016.

Clapper, J. (2016). *Worldwide treat assessment of the US intelligence community*. Washington, DC: Office of the Director of National Intelligence.

Falliere, N., Murchu, L. and Chien, E. (2011). W32.Stuxnet Dossier. *Symantec Security Response*. Version 1.4. p 1–68.

Herrera-Flanigan, J. (2011). Mission 4: Safeguarding and Securing Cyberspace. *Nextgov*, [online]. Available at: http://www.nextgov.com/cybersecurity/cybersecurity-report/2011/02/mission-4-safeguarding-and-securing-cyberspace/54283/ Accessed 31 March 2017.

Kaspersky Lab. (2012). *Kaspersky Lab and ITU Research Reveals New Advanced Cyber Threat*, [online]. Woburn, MA: Kaspersky Lab. Available at: http://usa.kaspersky.com/about-us/press-center/press-releases/2012/kaspersky-lab-and-itu-research-reveals-new-advanced-cyber-threa. Accessed 23 Apr 2016.

Kennedy, R. (2008). *Of knowledge and power: The complexities of National Intelligence*. Connecticut: Greenwood Publishing Group.

Kim, S., Park, J., Lee, K., You, I., & Yim, K. (2012). A brief survey of rootkit techniques in malicious codes. *Journal of Internet Services and Information Security, 3*(4), 134–147.

Lee, D. (2012). *Flame: Massive cyber-attack discovered, researchers say*. BBC News, [online]. Avaliable at: http://www.bbc.com/news/technology-18238326. Accessed 10 Mar 2016.

Lynn, W. (2010). Defending a new domain: The Pentagon's Cyberstrategy. *Foreign Affairs*, [online] 89(5) Available at: https://www.foreignaffairs.com/articles/united-states/2010-09-01/defending-new-domain?gp=66687%3A31ac65264-c9a4440. Accessed 26 Feb 2016.

Mick, J. (2010). USB stick led to worst cyber attack on US military; Russia suspected. *Daily Tech*, [online]. Available at: http://www.dailytech.com/USB+Stick+Led+to+Worst+Cyber+Attack+on+US+Military+Russia+Suspected/article19458.htm. Accessed 18 Apr 2016.

Miller, C. (2007). The legitimate vulnerability market: Inside the secretive world of 0-day exploit sales. *Independent Security Evaluators*, [online]. Available at: http://www.econinfosec.org/archive/weis2007/papers/29.pdf. Accessed 12 Mar 2016.

Mooney, C. (2017). *New Obama report warns of changing 'Threat environment' for the electricity grid*. The Washington Post, [online]. Avaliable at: https://www.washingtonpost.com/news/energy-environment/wp/2017/01/06/new-obama-report-warns-of-changing-threat-environment-for-the-electricity-grid/?utm_term=.52382d1dc2c1. Accessed 09 Oct 2017.

Munro, K. (2012). Deconstructing flame: The limitations of traditional defences. *Computer Fraud & Security, 2012*(10), 8–11.

National Commission for the Review of the Research and Development Programs of the United States Intelligence Community (NCRRDP). (2013). *Special Topic White Paper: The IC's Role within US Cyber R&D* [online]. Washington D.C. Available at: https://fas.org/irp/eprint/ncrdic-cyber.pdf . Accessed 31 March 2017.

Paganini, P. (2012). *Flame and stuxnet, the union is strength. security affairs.* [online]. Avaliable at: http://securityaffairs.co/wordpress/6373/intelligence/flame-and-stuxnet-the-union-is-strength. html. Accessed 15 Apr 2016.

Park, R. (2015). *Guide to Zero-Day Exploits* [Blog]. Symantec security response. Available at: https://www.symantec.com/connect/blogs/guide-zero-day-exploits. Accessed 01 Apr 2017.

Rid, T. (2013). *Cyber war will not take place.* New York: Oxford University Press Inc.

Rosenbach, E. and Peritz, A. (2009). Cyber security and the intelligence community. Memorandum. *Confrontation or Collaboration? Congress and the Intelligence Community.* Belfer Center for Science and International Affairs. Harvard Kennedy school.

Sahadi, J. (2013). What the NSA costs Taxpayers. *CNNMoney*, [online]. Available at: http://money.cnn.com/2013/06/07/news/economy/nsa-surveillance-cost/. Accessed 01 May 2016.

Sanger, D. (2012). Obama order Sped up wave of cyberattacks against Iran. *The New York Times*, [online]. Available at: http://www.nytimes.com/2012/06/01/world/middleeast/obama-ordered-wave-of-cyberattacks-against-iran.html. Accessed 18 Apr 2016.

Sanger, D. (2016). US cyberattacks target ISIS in a new line of combat. *The New York Times*, [online]. Available at: https://www.nytimes.com/2016/04/25/us/politics/us-directs-cyberweapons-at-isis-for-first-time.html?_r=0. Accessed 31 March 2017.

Schneier, B. (2010). The story behind the Stuxnet virus. *Forbes*, [online]. Available at: https://www.forbes.com/2010/10/06/iran-nuclear-computer-technology-security-stuxnet-worm.html. Accessed 15 Apr 2016.

Schonberg, M. (2013). Defining the DOD Rode in National Cybersecuity. MSS. United States Army War College.

Tehan, R. (2017). *Cybersecurity: Critical infrastructure authoritative reports and resources.* Congressional Research Service [online]. Report No. R44410. Available at: https://fas.org/sgp/crs/misc/R44410.pdf. Accessed 31 May 2017.

US China Economic and Security Review Commission. (2008). *Report to Congress of the US – China Economic and Security Review Commission*, [online]. 110th Congress, 2nd session. Washington D.C.: USCC.. Available at: https://www.uscc.gov/sites/default/files/annual_reports/2008-Report-to-Congress-_0.pdf. Accessed 31 March 2017.

Volz, D. and Hosenball, M. (2016). Concerned with cyber threat, Obama seeks big increase in funding. *Reuters*, [online]. Available at: http://www.reuters.com/article/us-obama-budget-cyber-idUSKCN0VI0R1 Accessed 05 Apr 2016.

White House. (2003). *The National Strategy to Secure Cyberspace* [online]. Washington, DC. Available at: https://www.us-cert.gov/sites/default/files/publications/cyberspace_strategy.pdf. Accessed 31 March 2017.

Wolf, K. (2015). Wassenaar arrangement 2013 plenary agreements implementation: Intrusion and surveillance items. *Federal Register*, [online]. Available at: https://www.federalregister.gov/documents/2015/05/20/2015-11642/wassenaar-arrangement-2013-plenary-agreements-implementation-intrusion-and-surveillance-items. Accessed 15 Apr 2016.

Zetter, K. (2007). Simulated cyberattacks shows hackers blasting away at the power grid. *Wired*, [online]. Available at: http://www.wired.come/2007/09/simulated-cyber/. Accessed 25 Apr 2016.

Zetter, K. (2012). *Meet 'Flame,' The Massive Spy Malware Infiltrating Iranian Computers.* Wired, [online]. Avaliable at: https://www.wired.com/2012/05/flame. Accessed 02 Feb 2018.

Zetter, K. (2014). *Countdown to zero day: Stuxnet and the launch of the world's first digital weapon.* New York: Crown Publishing Group.

# Promise and Perils of Big Data Science for Intelligence Community

**Karan P. Jani and Anmol Soni**

**Abstract** Collecting, processing, and analyzing digital data in bulk holds critical importance today more than it has at any period in past, or, arguably, in future. Big data science is influencing the global financial, industrial, academic as well as defense sectors. With the exponential rise of open source data from social media and increasing government monitoring, big data science is now closely aligned with national security policies, and the intelligence community. This chapter reviews the role that big data sciences can play in supporting functions of the intelligence community. A major part of the chapter focuses on the inherent limitations of big data, which can affect and even disrupt the chain of operations of intelligence agencies from gathering information to anticipating surprises. The limiting factors range from technical to ethical issues. The chapter concludes that there is a continuing need for experts with domain knowledge from intelligence community to efficiently guide big data analysis to fill any gaps in knowledge. As a case study on limitations of using big data, work in nuclear intelligence using simple analytics is examined to show why big data analysis in certain cases may lead to unnecessary complications.

## 1 Introduction

> *"When a person monitors a database, it's small data, & when a database monitors a person, it's called big data." –Anonymous*

We live in a knowledge economy, where data is the new currency. The amount of digital data generated in just the two years from 2011–13 surpassed the total data collected in all of recorded history (SINTEF 2013). By 2020, the estimated number of devices connected to the internet will be three times the total human population on earth (Chandrasekaran 2015). This exponentially increasing data has and will continue to profoundly impact the socio-political and economic structure of society.

K. P. Jani (✉) · A. Soni
Georgia Institute of Technology, Atlanta, Georgia, USA
e-mail: karan.jani@gatech.edu; anmolsoni@gatech.edu

 
M. E. Kosal (ed.), *Technology and the Intelligence Community*, Advanced Sciences and Technologies for Security Applications,
https://doi.org/10.1007/978-3-319-75232-7_11

The grand challenge in the coming decade will be to shape global strategic policies that adapt, account for, and comprehend the information from such large data sets.

Collecting and analyzing big data (multiple rows and columns) that represent a physical system has been the driving force of modern sciences. Many of the standard data analytics tools used today have their roots in solving a complicated numerical problem in basic sciences, viz. Monte Carlo (Metropolis 1987; Andrieu et al. 2003). This analysis of large, complex data sets has led to the discovery of fundamental particles of universe like Higgs Boson and understanding the origins of life by decoding the genome. On the industrial front, dependence on data substantially increased post the 'dot com' revolution in early 90s. The major asset of technology's "Frightful Five"- Google, Facebook, Amazon, Microsoft, Apple - lies in their ability to synergize user databases with third party commerce (MIT Sloan 2017; Manjoo 2017). The analysis tools adopted by industries are the ones first developed in academia.

For a nation's Intelligence Community (IC), data and its analytics are central to its mission to anticipate surprise. Over the last century, sensitive communications have scaled from Morse code to telephones to digital data, and at each phase, the IC has evolved itself and its technical infrastructure to sustain its mission and serving national security priorities (Richelson 2002). However, the last decade has been particularly distinct and challenging for IC as the tactics of collecting, processing, and analyzing information have undergone a radical change. Several factors combine to add new dimensions to the role of data in IC. These include the scale at which data is produced, the rate at which it is shared on open source platforms, the formats in which it is circulated and, most significantly, the direct and indirect ways in which this data is interlinked with the nation's critical infrastructure. It is therefore argued that more than during any era in the past, the IC today needs to broaden its reach to include domain expertise in the entire spectrum of data analytics - from the excessive unstructured data collection schemes of industries to the sophisticated analysis tools being developed in academia. As represented by the Venn diagram in Fig. 1, within the overlap of big data, whether open or restricted, from big businesses and the science of data mining lies the interest sphere of IC in enhancing national security strategies. The collaborations between academia and industry towards big data analytics have increased rapidly, primarily due to the dependence on one group towards the other (academia develops new techniques, industry funds research and provides data) (Jain et al. 2014; Mitroff and Sharpe 2017). In the process, breakthrough developments in artificial intelligence and machine learning have occurred, setting the trajectory for future disruptive technologies that will rely on Big Data. In the context of IC, the partnerships with industry since the 9/11 have added a critical workforce; however, tensions have existed in terms of budgeting and accountability of managing databases (Rosenbach and Peritz 2009). Therefore, long term strategic planning in the IC has been aimed at strengthening research, development and human resources in the field of big data sciences.

The motivation behind this chapter is to investigate the scope and limitations of data sciences in intelligence gathering. Section 2 defines and lists the properties of what constitutes as a big dataset. The key sectors in which big data interfaces with

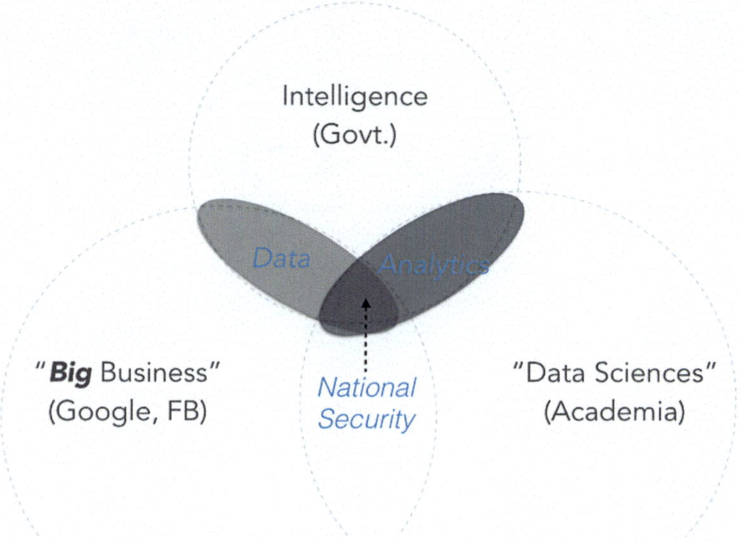


**Fig. 1** A Venn diagram overlap of the areas of focus in big data between academia, industry, and government

IC is reviewed in Sect. 3. In doing so, the role of agencies such as Intelligence Advanced Research Projects Activity (IARPA) in shaping national security from big data is highlighted (Sect. 3.1). Set of such programs and agencies from other countries is also listed. Section 4 dissects the core limitations of big data analytics, especially in the context of its dependence in IC. For these, several examples from recent events to demonstrate the technical and ethical limitations are provided. The case study, Sect. 5, on the role of data analytics in nuclear intelligence resonates the need of tacit knowledge instead of simply relying on big data sciences. Section 6 provides a conclusion to this study and lists scenarios for future investigation on the need to address knowledge gap in big data sciences for IC.

## 2 Defining a "Big Data"

Even though the term is used widely, it is difficult to find a unified definition of the term big data in its prolific usage and application across different fields and over time (Ward and Barker 2013). The first instance of the "big data problem" being defined in literature was in 1997 when NASA defined it as follows. "Visualization provides an interesting challenge for computer systems: data sets are generally quite large, taxing the capacities of main memory, local disk, and even remote disk. We call this the problem of big data" (Cox and Ellsworth 1997). The issue being discussed by the NASA researchers in this paper was focused less on the complexity of the dataset itself and more on visual interpretation of the underlying data, which

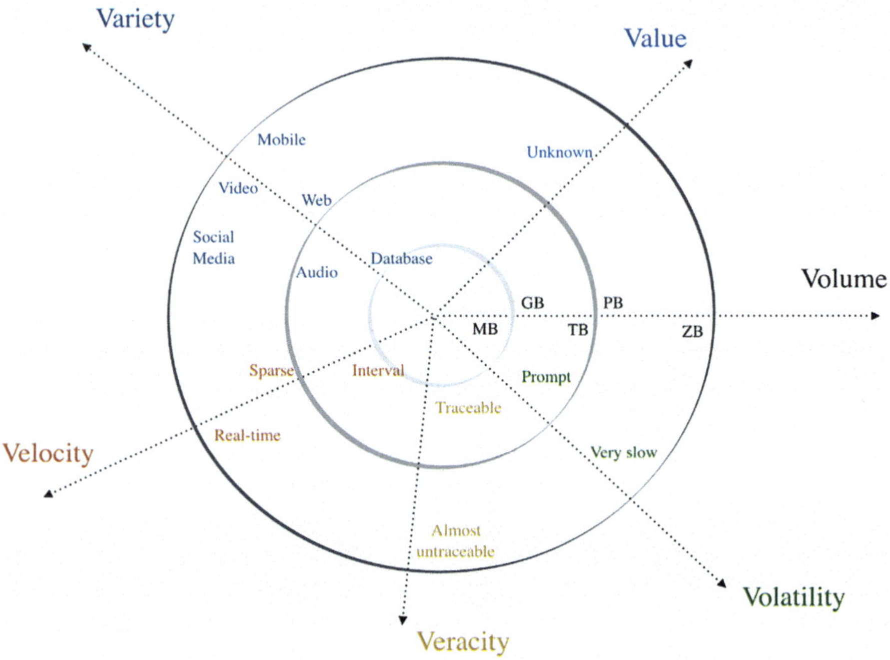


**Fig 2** The **V**s of big data. Each concentric circle with increasing radius refers to increasing complexity in analyzing the data. The volume scales for 15 orders of magnitude (from megabytes to zettabytes). Such diagrammatic representation of Vs of big data is commonly found in the literature (one such example: Soubra (2012))

demanded a better technological solution. They were dealing with data on fluid mechanics, which had a known solution under good approximation, but the visualization demanded interaction between each fluid element to form the clumps and the turbulence.

The understanding of "big data" has evolved significantly since then, but the primary concern regarding the storage of datasets resonates the precedent set by NASA's definition. Presumably, Cox and Ellsworth (1997) referred to the analysis of data as an identifying characteristic of big data. This aligns well with the two aspects of big data that Ward and Barker (2013) identify – storage and analysis. The most standard definition of "big data" that can be found in contemporary literature is "Datasets whose size is beyond the ability of typical database software tools to capture, store, manage, and analyze" (McKinsey Global Institute 2012). This definition is robust and invariant to any complexity of data set there may be. In this sense, it captures the universal "big data problem."

The four fundamental properties of "big data" highlighted in any such definition are (i) *capture*: acquiring the database, (ii) *store:* memory of the database, (iii) *manage*: transferring the database, and (iv) *analyzing* the database, interrelated to creating a practical challenge for solving big data problem. The complexities arising from these four properties are characterized in the literature as the "Vs" of big data

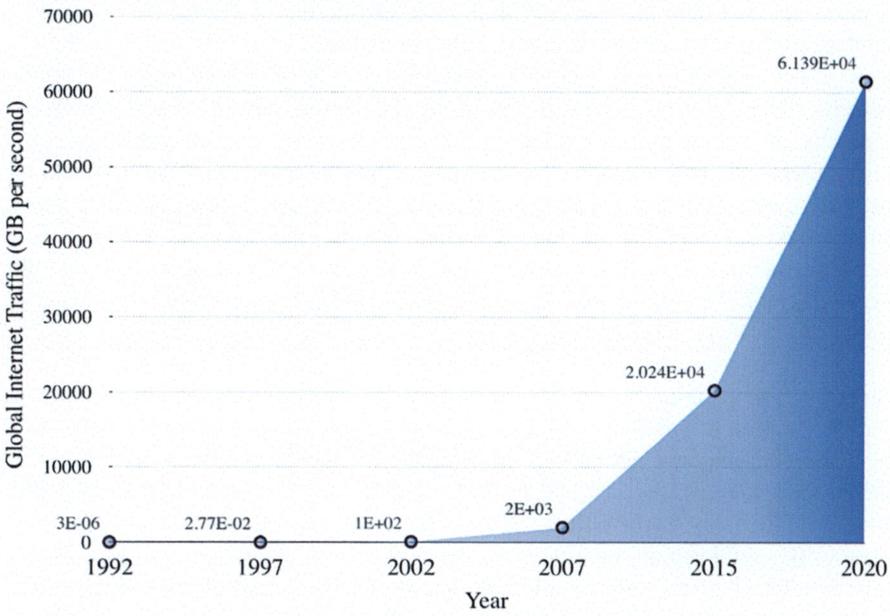


**Fig. 3** The exponential rise of global internet data in three decades (above 10 orders of magnitude). Data points have been obtained from Cisco 2015. The gap between stated years is not up-to-scale

(Roberto 2013, Gandomi and Haider 2015). In the context of IC, there are six such "Vs" that play a crucial role. Fig. 2. represents these "Vs" and their behavior at each circle of complexity.

**Volume** This fundamentally defines the size of "big data." The total data generated on the internet in 1992 was ~1000 bytes per second. Every 5 years since 2002, the data has increased by a factor of 10 (Cisco 2015). By 2020, the internet data is expected to reach ~$10^{21}$ bytes per year. This exponential trend of global internet traffic is demonstrated in Fig. 3. Every single day, Facebook alone generates ~$10^{15}$ bytes (1 PB) of new data (Wiener 2016). From the analytics perspective, our ability to find meaningful trends with this increasing volume gets subsequently challenged at each circle shown in Fig. 2.

**Variety** This describes the complexity of data types which can be in the form of a simple tabular data set, text in webpages and even things which are heavier and complex to dissect such as audio, videos, photos and live feed on social media. This also includes data that are being stored by service providers, as well as in some case government monitoring. Over 80% of all the data collected from the internet is unstructured, i.e. it simply cannot be expressed in some rows and columns or described using known models (Lant 2017). The degree of unstructured data grows as the volume of data increases. With the rise of Internet of Things (IoT), most of big data analysis is likely to be unstructured in nature.

**Velocity** This describes the rate of acquiring data before one can start the analysis. Typically, the data with lower volume comes structured and at predictable and regular time intervals (for example, weather data, flight information). The text data on internet emerging in web pages typically release information to a source sparsely. But heavier data and more unstructured data sets emerging from mobile phones and social media come in real time. Ideally, the newly available dataset is merged with an existing or older dataset, which can at times prove highly non-trivial (for instance, in the case of social media feed).

**Veracity** The accuracy and authenticity of a digital dataset remains a big unknown. For information of smaller volumes, one may be able to systematically trace back the origin, but once we get bulk of information that has been passed over by multiple internet sources, the original host becomes almost untraceable (for example, YouTube videos by terror outfits). Also, a large fraction of raw datasets of importance to IC could remain unstructured, hence making it challenging to find its origins. In a lot of cases, one cannot guarantee if what is available is a full dataset or just partly acquired in a thread. All these factors may lead to significant error bars for the predictions by IC.

**Volatility** This property refers to the latency (processing time) by which one can analyze a given dataset and provide a meaningful result. The larger in volume and non-linear the data set (such as videos, images and text on Twitter for specific keywords), the more computationally intensive it is to process and get any useful information. For time-sensitive scenarios involving big data, machine learning algorithms (supervised on prior samples) promise a much higher probability of generating meaningful analysis than the domain experts.

**Value** Whether there is an immediate value in utilizing computational (and human) resources for processing and analyzing big data is usually the hardest to predict initially. In cases of large collection of unstructured data (for example, scanned documents of incoming refugees in language other than English), the language analytics tools using brute force methods may labor significant computing time and not reveal any new insight about the applicant than learnt from conventional intelligence expertise.

## 3 Big Data and Intelligence Community

> *"The IC's routine work of collection, processing, exploitation, and dissemination, and analysis is still largely organized on the Cold War model of seeking out sparse and secret information."* Symon and Tarapore (2015)

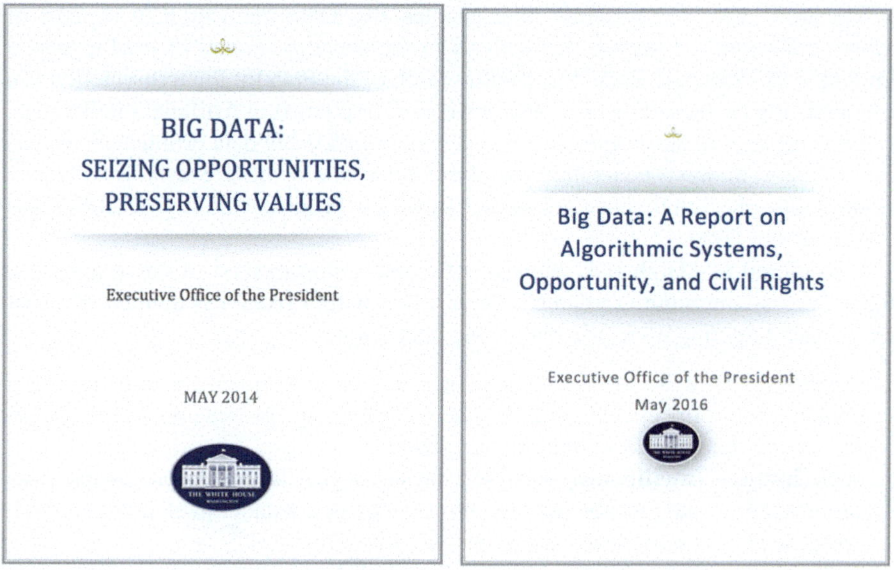


**Fig. 4** Covers of the White House policy reports on Big Data

## *3.1 Policy on Adoption of Big Data*

In recent years, the United States government has placed high importance on developing a national strategy to address the big data problem and to incorporate big data in decision making. The White House released two reports addressing national priorities aligned with big data (Fig. 4) (White House 2014, 2016). Though the main focus is on the ethical issues and policy formation related to using big data across the spectrum, these reports list some of the highlights of its usage in the context of defense and intelligence community:

"During the most violent years of the war in Afghanistan, DARPA deployed teams of data scientists and visualizers to the battlefield... these teams ... fused satellite and surveillance data to visualize how traffic flowed through road networks, making it easier to locate and destroy improvised explosive devices" (White House Report 2014).

In 2011, the Big Data Interagency Working Group (BD IWG) was formed with the aim of focusing on developing techniques of managing and analyzing large data. The IWG released its report titled "The Federal Big Data Research and Development Strategic Plan) in 2016. The IWG consists of members from different agencies including the Department of Defense. The key DoD agencies included are DARPA, NSA, OSD, and the Service Research Organizations from the Army, Airforce and Navy. The report provides strategies to help build the collection, analysis and storage framework for all government agencies. Perhaps the most important strategies in the context of Defense is Strategy 3 which talks about a building and enhancing a "research cyber infrastructure that enables Big Data innovation in support of agency missions." However, the analysis in the specific context of intelligence is limited in the Strategic Plan. A keyword search of the Plan for the term intelligence revealed only two results both of which were in the context of Artificial Intelligence.

In an attempt to develop in-house expertise to tackle big data problem as well as in investing in next generation breakthroughs, the United States Intelligence Community, the Intelligence Advanced Research Projects Activity (IARPA) was established in 2006.

Considered as "High risk, high payoff" and a startup type of environment to foster "cross-community research", the research within IARPA is divided into four areas, each targeting a specific part of big data science:

1. **Analysis**, where the focus is to develop tools that maximizes insights into a big data. Current projects under this theme include image processing, video analytics, natural language learning, among others.
2. **Anticipatory Intelligence**, in which the main goal is to get forecasting from more accurate and prompt. Specific projects are being undertaken in threat modeling, social network analysis and analytic tradecraft.
3. **Collection**, where innovative ways of quality data collection from unconventional sources are being developed. Examples of such include bioinformatics, optical interferometry and chromatography.
4. **Computing**, where in research spans from fundamental developments in quantum information to refining superconducting electronics.

IARPA promotes collaboration with industries, academia. IARPA also deals dominantly with open source data and engages general public in data sciences puzzles. Two successful past programs of IARPA in the open source sector include:

**FOREST (Forecasting Science and Technology)** which initiated crowdsourced, forecasting tournaments to predict disruptive technologies. Thousands of scientists and academician worldwide participated in this, making it "world's largest S&T forecasting tournament" (www.SciCast.org).

**OSI (Open Source Indicators)** in which tools were developed to analyze social media trends and other publicly available for predicting warning about real world events such as natural disasters, protests and disease outbreaks. This program has been credited to be the first to realize the Ebola outbreak and notified to the US health officials (Harbert 2015).

## *3.2 Role of Big Data*

The amount of data collected just from surveillance has increased by 1600% since September 11, 2001 (Young 2012). There are over 7 million computing devices currently within the United States armed services, which is expected to increase to 15

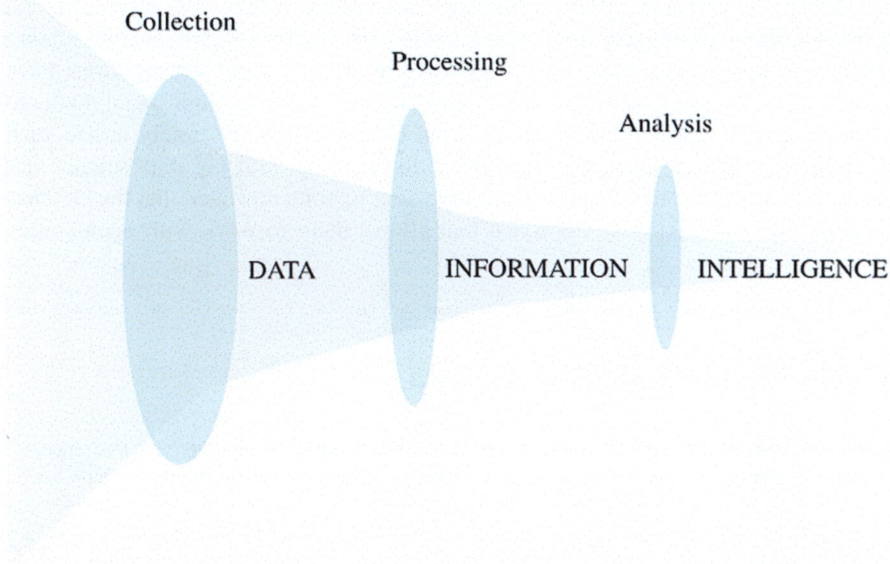


**Fig. 5** Schematic diagram describing the relation between big data and intelligence community. The three stages referring - collection, processing, analysis - increasingly cleans the data to make it more **V**aluable. Such diagrammatic representation of data and intelligence is commonly found in the literature (one such example: Joint Intelligence (2013))

million by 2020. Some of the primary areas of national security where big data is being generated include Maritime Security, Cyber Security, Money Laundering, Multi-INT Analysis, and Space Situational Awareness (Fahey 2012).

The information analysis of IC, for any given form of data set is shown in Fig. 5. It is a three-stage structure and with each step the **V**alue of dataset increases (i.e. noise decreases). With the rise of big data, each conventional approach of IC to these three steps gets too time consuming.

One hopes that by developing the tools that are already available in the standard big data analysis in industry and academia, the information processing in IC can benefit in three major ways: (i) the collection and characterizing of data becomes automated, (ii) the processing time for analyzing complex data structure almost reduces to being real time and (iii) the presentation of result can be refined to only key features that led to effective decision making.

The application of big data sciences to Intelligence Community (IC) and subsequently to National Security is evident. However, big data is one of those rare disruptive technologies that escalated in industry and academia at a much higher rate compared to standard defense research. The level of access to user data and sophisticated mining tools that are available with IT giants like Facebook, Google and IBM, exceeds the scale of the current generation of infrastructure deployed by the IC. At the same instance, open source information available through social media has developed into to a new subtype of intelligence, social media intelligence or *SOCMINT* (Omand et al. 2012). Unlike the conventional forms of intelligence (such as human, signal or imagery intelligence), SOCMINT allows to gain insights among specific groups on real time events "Twitter trends." Recent investigations have found that the Russian agencies relied on intelligence built through social media to influence the 2016 US Presidential Election (Shane 2017). To timely utilize such massive open source data, the need for IC to catch up with big data science has never been more urgent than now. It is interesting to note, however, that the Defense agencies are now taking an approach that allows them to work with open-source tools.

# 4 The Big Limitations of Big Data

> *"If 50 years of research in artificial intelligence has taught us anything, it's that every problem is different, that there are no universally applicable solutions. An algorithm that is good at chess isn't going to be much help parsing sentences, and one that parses sentences isn't going to be much help playing chess... solving problems will often require a fair amount of... "domain knowledge"—specific information about particular problems, often gathered painstakingly by experts. So-called machine learning can sometimes help, but nobody has ever, for example, built a world-class chess program by taking a generally smart machine, endowing it with enormous data, and letting it learn for itself... Big Data is a powerful tool for inferring correlations, not a magic wand for inferring causality"*
> –Gary Marcus, Professor of Cognitive Science, New York University (2013)

By their very construct, data analytics algorithms (primarily involving machine learning) are designed to solve problems where we have a prior understanding of the complexities in the data sets. Those complexities can be common across spectrums and multiple disciplines, hence one might be tempted to apply, for example, an algorithm used for facial recognition on Flickr to identify black holes from data of Hubble Space Telescope. However, using brute force approach in big data typically leads to a degenerate solution set, i.e., there are too many options for the question posed and thus the likelihood of any meaningful or coherent judgement is extremely low. In this chapter, such primary limitations of big data in the context of IC were explored and highlighted. We further argue that to break that degeneracy and find meaningful correlation in the complex datasets, one must rely fundamentally on domain knowledge.

## *4.1 Technical Limitations*

Just the like the "Vs" of big data, experts in the field of analytics have their own list and ranking of the limitations of big data sciences. But by and large, the five major issues of big data sciences, which are relevant for the discussion in this paper, can be stated as following (March and Davis 2014):

**Big Data Gives Too Many Correlations** If one looks at the trends in the criminal and financial data sets from 2006–2011, there exists a correlation between market share of Internet explorer and number of murders in the United States (Diaz 2013). This can be eliminated as a fluke only because we have a prior knowledge to analyze that such correlation cannot exist (and is of no use). However, in territories where one does not have the domain expertise or a prior result to base our judgement on, it may seem the correlations found in an unstructured data set demonstrate a fundamental feature. Also, the number of correlations one finds in an unstructured data approximately scales with the volume of the data, as there is a larger parameter space being spanned in search of "best fit."

**Big Data Can be Gamed** Every correlation that we see in big data is ultimately just a number. In most cases, this number refers to the peak of some probability distribution. But as soon as one understands the patterns in the dataset, it is trivial to fake and hack the data set such that every time you get the same correlation. Thus, big data analysis could purposefully be misguided.

**Big Data Changes with Time** .This relates to one of the key characteristics of big data identified in Sect. 2 – Volatility. Most of the interesting data sets (political support on social media, stock market, weather forecast), which rely on big data analysis are dynamic, and their trends fluctuate with time. The correlations obtained on the set using such analytics are valid only for a short period of time. Hence, all judgments from big data come with a finite deadline.

**Big Data Is Less Robust** The hype on the success of big data mostly overlooks the fact that the problems it solves are rather common data sets. For more complex data sets, like whether can you predict translation of a word in another language, there have been no breakthroughs, though it does fall under the same category as predicting behavior from social media response. Previously hailed as the big success of big data, the Google Flu Trends failed recently by "predicting more than double the proportion of doctor visits for influenza-like illness than the Centers for Disease Control and Prevention's surveillance" (Lazer et al. 2014).

**Big Data is only a Technique** Like the many technologies of the past and present, big data is no more than just an analysis tool that gives little further heads up than conventional approaches. At no instance can we write super codes that will predict every trend and come back with solutions to most complex issues.

**Big Data Is No Longer Safe** The rise of big data was once termed as an asset ("data is the new oil"). However, keeping this data secure has become big a liability. The current trend in industry of collecting big amount of data to finding its monetary gain is being done with the backdrop of very weak regulation in terms of data protection. The 2017 security breach of credit rating agency Equifax (Norrie 2017), and more recent revelations of the 2016 data breach at Uber (Lee 2017) among many other such incidences at various organizations in last few years, has demonstrated that more efforts, resources, stricter regulations and possibly even oversight are needed for securely storing big data.

## *4.2 Limitations from the Perspective of Intelligence Community*

*"We simply cannot be the slaves to the data"* –Anonymous

### 4.2.1 Big Data Needs Experts

When faced with petabytes of unstructured data sets, scattered across multiple mediums and highly uncertain source of authenticity, one simply cannot use a brute force method and apply machine learning algorithms. To expedite the decision-making process, experts in the field would have to down-sample datasets and prioritize feasible targets. This requires familiarity with the context of the situation and the questions one wants to answer using data analysis.

The primary need for correlations within complex, unstructured data sets is to guide IC in terms of planning and strategy. However, as noted in the previous section, big data analysis can give multiple correlations, some of which are absurd if the domain knowledge is not available *a priori*. Following the outputs from machine learning algorithms will ultimately lead to low-latency in decision making and wastage of key resources. Expert leadership, therefore, is the final step in the intelligence - big data cycle.

**Take for Example, the Case of** the Islamic State (IS) which has been very active in using social media (Twitter) and YouTube to carry forward their agenda. Using such open source intelligence, along with surveillance information and intelligence from ground, in principle, big data can correlate their possible location and resources. But it still does not provide any strategy to the IC for the next course of action. In fact, IS has been wise in using twitter so that it is more populated with similar keywords, leading to correlations. It is in such places the IC needs human intelligence and experts.

### 4.2.2 Big Data cannot Address the Knowledge Gap

One of the prime usages of big data analysis in IC is to bridge the information gaps that exists in unstructured data sets. For example, say there are blurred surveillance pictures of a potential activity of national security concern, and, at first, completely unrelated posts on social media and internet from a particular territory. Machine learning algorithms may pick up trends between such divergent data sets and provide a correlation which otherwise will go unnoticed or have high latency in the conventional approaches of IC. This will be a classic example of filling the information gap. However, what more does it say? If this was the case for Syria, does this imply non-state actors are going to attack a western country in some finite time? One cannot answer such questions without having experts in the field in those regions to join the dots to predict potential actions by non-state actors. This inability of big data to predict causality is where the *knowledge gap* lies.

Much of the knowledge gap is acquired by analyzing "long data" rather than big data (Arbesman 2013). *Long data* in this context refers to understanding the trends over a long period of time. The focus of this book is to demonstrate the transformation of IC over last half century with changing technologies. Using this domain knowledge, we can identify if technologies like augmented/virtual reality will lead to an effect similar to the one observed in the Cold War era due to satellite communication. Big data by itself cannot meaningfully answer such questions. The information bridged by big data is only valid within the set of knowledge at a given instance of time.

### 4.2.3 Example Scenarios

(i) During the Arab Spring, social media played a very important role in spreading the movement across the region (Gerbaudo 2012; Howard et al. 2011). However, with all the information at hand, none of the big data analyses available to IC at the time could predict the trajectory of the protest from Tunisia to Egypt, neither could they have predicted the unrest following a 'successful' revolution of taking down authoritarian regimes. Instead, an expert with the knowledge about the politics and the history of the region could have possibly predicted the fate of these nations.

(ii) The joint India-US remote sensing satellite, NISAR, will be monitoring the Indian Ocean with high resolution for predicting weather. Suppose these heavy databases identify a new Chinese vessel with weapons that directly correlates with previously known information. This surely bridges the intelligence gap. But what are the intentions of Beijing with such new weapons? Big data may not answer that but an expert in the field may point China's past strategies to intimate its neighbor for the claim of oil basins in Indian Ocean (Detsch 2015).

### 4.2.4 Big Data Analysis Has a Finite Lifetime

The current computational infrastructure and algorithms that enable us efficient deployment of big data analytics have a lifetime which is proportional to the size of the data set. The data analysis tools we had available in mid 2000s have already been quashed with the exponential rise of open source data from social media. Hence it should be acknowledged that the current investments that the IC is making in big data sciences will be dusted in less than 5 years from now with the rise of Internet of Things and cloud computing which would require new infrastructure and techniques of analysis. While developments on this front are already underway, this does speak to the large spending and effort needed for a rather short period of time.

For a long-term plan, what the IC can at best do is develop a framework (along with interactive software) within which experts from varying intelligence fields can efficiently conduct:

(a) Processing, exploitation, and dissemination of complex data sets,
(b) The final stage of implementing strategies from the results of analysis.

This needs to be examined using a robust process just like any defense strategy, which largely remains invariant with changing nature of warfare (in this case size of data).

### 4.2.5 Big Data Comes with Ethical Issues

The ethical standards of the IC vis-a-vis those of the public have always remained a point of contention (e.g., see Smith-Colin and Kleinhenz, "Analyzing the Public and State Reactions to Global Surveillance Disclosures: Using Ethical Frameworks to Gain Understanding" in this volume). For example, the leaks by Edward Snowden of large scale NSA surveillance of US citizens came under strong criticism and reinforced the debates on privacy laws. Big data practices in IC cannot be viewed independently of the ethical standards that has been warranted in these debates. Eventually the IC will be compelled to uphold what is being mentioned in literature as *Big Data Ethics* "—a social understanding of the times and contexts when big data analytics are appropriate, and of the times and contexts when they are not" (Richards and King 2013).

The popularly held belief of government overreach also manifests in a recent survey on ethical use of big data (White House Report 2014). The survey conducted on WhiteHouse.Gov, which included 24,092 participants, revealed that a majority of respondents had a strong lack of trust against the usage of big data particularly geared towards national security agencies (see figure on page 79 from the White House Report 2014). The report noted that over 80% participants have serious concerns about transparency in the practices related to big data analysis, as well as the legal boundaries within which such data is being collected. Industries, including social media and technology giants, were surprisingly found less untrustworthy by the participants compared to IC and law enforcement agencies. In fact, 67% of all participants had serious concerns about their data being breached in the name of intelligence. Though a survey has a small sample size and is not statistically representative of the varying demographics of United States, it does highlight a belief commonly portrayed in media on the overreach of government surveillance.

The three major ethical issues (or rather paradoxes) that limit role of big data sciences in IC and have been discussed at length in literature (Richards and King 2013) include:

**Transparency** Even if the big data being used by IC is obtained mostly from open sources and hence hypothetically publicly available, (i.e., transparent) the analysis being conducted, however, is opaque. Eventually, the public, who are the source of this open data, may want to know what decisions are being affected by such open data and may also question the ethical use of monitoring their life.

**Identity** When faced with unverifiable sources of information, the IC may rely on big data analysis to uncover the identity of the source, but this very much threatens the identity, as the analysis can lead to absurd correlations of an identity with criminal intentions. This will end up forcing people to remain anonymous.

**Power** The information from big data is prone to be misused for manipulating public opinions. Governments and dictatorial regimes could allow open source data to flourish so it gains insights on individuals or groups that tend to be anti-establishment. Scenarios for abuse of big data by authority that are being discussed in literature and media include:

(i) During the Arab Spring, the Syrian government lifted the ban on social media sites, since it was easier to locate individuals and movements using the open source data and analytics (Preston 2011).
(ii) (ii) The Supreme Court of India had to reinforce the Privacy Act which could restrict government from collecting mandatory biometrics data (which is linked with bank accounts, mobile sim cards, centralized tests), as it possesses a huge risk of being misused by private contractors in the government as well in spying on individuals in the name of national security (Supreme Court of India 2017; The Wire 2017).
(iii) The *Global Database of Events, Languages, and Tones* (Makuch 2014) has a record of 250 million entries of worldwide conflicts since 1979. It performs real time big data analysis on open source data and highlights regions of future conflicts. Such information can be used by authoritarian governments to tackle any possible conflicts.

The inherent issues of big data sciences thus affect the process of analysis conducted by the intelligence community. It is therefore important to reaffirm the role of intelligence community's leaders, analysts, and experts with domain knowledge for efficient usage of big data. Also, it is important to have a policy and framework in place so that future issues of big data can be sufficiently addressed.

## 5 Case Study: Big Data & Nuclear Intelligence

*"We can't let our treaties get ahead of our monitoring and verification headlights."* Defense Science Board Task Force Report (NextGov 2014)

The role of big data has gained importance recently in the purview of nuclear intelligence. By analyzing real time data from multiple sensors, big data analysis can lead to sophisticated intelligence towards monitoring nuclear deterrence, treaties and power plants. In 2014, a Task Force appointed by the Defense Science Board presented a report to the Office of The Secretary of Defense on "*Assessment of Nuclear Monitoring and Verification Technologies*" (Defense Science Board Task Force Report 2014). In one of the "paradigm shift" conclusions, the Report stated: "Revamping the monitoring framework to identify proliferating early or well before the fact. The framework should: - Adopt / adapt new tools for monitoring (e.g., open and commercial sources, persistent surveillance from conventional war-fighting, "big data" analysis) across the IC, DOD, and DOE."

The monitoring framework described in the report is a three-stage process:

(i) **Access:** Mutual agreements for access to data from the nuclear sites. Any difficulty in access could be identified as potential proliferation.
(ii) **Sensing:** Having sensors to measure radiation from special nuclear material, satellite imagery, the "INTs." This will require technologies beyond those agreed upon in existing treaties.
(iii) **Assessment:** Analyzing the vast amount of data sets coming from stage (ii) using big data science and open source imagery and assess if any threat.

This cyclic monitoring framework, "Access-Sensing-Assess-Iterate" is diagrammatically represented and discussed in further detail in the Defense Science Board Task Force Report 2014 (see Fig. 5–1 on page 53).

The stage (iii) requires dedicated supercomputing infrastructure and software to analyze the information to be presented for quick follow up. The United States has strongly considered the investment in this area of supercomputing, and recently IBM and NVIDIA established a $425 Million facility at Oak Ridge National Laboratory in Tennessee "to advance key research initiatives for national nuclear deterrence" (Energy.gov 2014).

However, in a January 2015 report by the World Institute for Nuclear Security, the experts group presented a warning for such investments:

> *"Application of Big Data technology to nuclear security was probably premature and not warranted financially; the costs of significant Big Data solutions can run to $millions/ annum and* ***the perceived corporate benefits may not justify such investment at present.****"; "Should you rely on external analysts working for a contractor? – The general feeling was that this was not an ideal solution and that* ***in-house expertise was desirable but would take time to develop****"* (World Institute for Nuclear Security 2015).

In light of such warnings, it is important to reexamine the question of what else can be learned about nuclear intelligence from basic data analytics instead of getting carried away with buzzwords (such as "Big Data" and "Supercomputers"). In an attempt to understand the role of tacit knowledge in nations with nuclear powers, scholars in the Sam Nunn School of International Affairs at the Georgia Institute of Technology are examining the correlation between academic journals and the authors in the field of nuclear physics and engineering in Pakistan and Iran (Baxter 2015). This involved searching academic databases for set of keywords that are linked with nuclear weapons (e.g., plutonium, uranium). Each keyword is assigned a ranking and accordingly a score is assigned to the authors.

The correlations, then examined using visual analytics such as network diagrams, helped identify key individuals (as outliers) who played a significant role in passing tacit knowledge. In addition, the study also shows organizational patterns and the decision-making structure in those countries. This provides insights of the inner workings of the government and back channels. Such studies are important for two reasons:

1. They utilize open source data (academic journals), which is structured and is smaller (only few GBs) in size.
2. The simple data analytics lead to very specific questions that can be directly addressed through experts with domain knowledge (why does Pakistan have nuclear intelligence structured around A. Q. Khan, while the name that comes for Iran is a rather lesser known individual?)

As a proof of concept, if the same study is carried using big data tools applied to social media, would it help in gaining further insight in the field of nuclear intelligence? The answer is *probably no*. Rather, it may lead to a completely different conclusion than the one obtained through minimal computational resource and utilizing conventional intelligence and monitoring. If one searches social media, say Twitter (petabytes of data), for the same keywords, it will at first order create a cross correlations with the worldwide trends, and those will likely be bizarre (currently "nuclear" search tag shows tweets about Donald Trump). Clearly, the noise exceeds useful signal in a database like Twitter. This is a classic case on when not to use big data sciences.

As reported earlier by the Defense Science Board Task Force, the social media indicators can be used to monitor treaties and proliferations. This information can then be overlaid with other forms of intelligence. For example, if a tweet originates from a region where other independent sources of intelligence have also indicated incidences of violations, then such information is indeed useful. But as the social media data is dynamic, latency between analyzing relatively older data set and the new incoming information will make it difficult to draw quick conclusions and take swift actions.

## 6 Conclusion and Future Work

This paper reviews the scope of big data in the context of the intelligence community. There is no doubt that big data holds critical importance for a nation, in particular to preserve national security. The US government places critical importance on the growth of technology, policies and ethical frameworks surrounding big data. Over time, policies and mechanisms have been put in place to support and encourage the growth of big data and its usage in government decision making. In the intelligence community as well, a concerted effort is being made to incorporate big data analytics to support on-ground decision making. This, however, comes with caveats and limitations. In scrutinizing the inherent issues in big data, we identify four main limitations where it particularly affects the intelligence community:

**Reliance on Experts for Efficient Use** Even with the proliferation of large datasets and complex mechanisms of analyzing them, experts and humans in the loop will continue to remain engaged with the process in order to efficiently and effectively utilize the data for making critical intelligence analysis.

**Addressing Knowledge Gaps** The idea of long data versus big data is of importance here. Trends over time and expert knowledge are both necessary to draw useful conclusions in this context.

**Temporal Use of Current Infrastructure** Big data is dynamic and constantly evolving. Therefore, decisions and conclusions made at one point in time may not have long shelf-lives thereby limiting their validity.

**Ethical Issues that Will Affect Scope of Analysis** Collection, storage and analysis of big data, especially in surveillance has strong ethical implications. This data makes citizens vulnerable to leaks, thefts, and even wrongful culpability. Having strong political and regulatory frameworks is therefore essential in supporting the growth of big data.

In terms of the hype around big data, we can conclude there is definite help in bridging information gaps but that doesn't make the domain knowledge irrelevant. In fact, for efficient use of big data in intelligence community, the issues stated above reaffirm the need for human input and experts to guidance.

As a case study on potential misuse of big data for intelligence community, we showcase the nuclear intelligence. We conclude that expensive infrastructure and over analyzing of heavy datasets would rather deviate us from conclusions that can be achieved by analytics on moderate sized open source academic data.

The analysis can be further strengthened by scrutinizing the effect of big data and its use in intelligence community towards:

- Arab Spring and using social media as an indicator for predicting real world events (Temple-Retson 2012)
- Crowdsource forecasting tournaments to predict disruptive technologies
- Reliance of intelligence agencies on the private sectors for developing tools on big data sciences

A fruitful exercise for policymakers would be to conduct more ambitious surveys like the White House Report of 2014 to address concerns among different states, age groups and industries regarding Big Data. This can then set the course for outreach activities the government and industry could undertake with the aim of addressing public concerns.

## References

Andrieu, C., De Freitas, N., Doucet, A., & Jordan, M. (2003). An introduction to MCMC for machine learning. *Machine Learning, 50*(1), 5–43.

Arbesman, S. (2013). Stop Hyping Big Data and Pay Attention to Long Data, WIRED, 2013 http://www.wired.com/2013/01/forget-big-data-think-long-data/

Baxter, P. (2015). The False Hope of Nuclear Forensics? Assessing the Timeliness of Forensics Intelligence. *FAS* http://fas.org/pir-pubs/the-false-hope-of-nuclear-forensics-assessing-the-timeliness-of-forensics-intelligence/

Chandrasekaran, N. (2015). Is data the new currency? *World Economic* Forum Retrieved April 29, 2017 from https://www.weforum.org/agenda/2015/08/is-data-the-new-currency/

Cisco VNI Global IP Traffic Forecast. (2015–2020). http://webobjects.cdw.com/webobjects/media/pdf/Solutions/Networking/White-Paper-Cisco-The-Zettabyte-Era-Trends-and-Analysis.pdf

Cox, M. & Ellsworth, D. (1997). *Application-controlled demand paging for out-of-core visualization.* Retrieved April 29, 2017 from https://www.nas.nasa.gov/assets/pdf/techreports/1997/nas-97-010.pdf

Defense Science Board Task Force Report. (2014, January). *Assessment of Nuclear Monitoring and Verification Technologies,* https://www.acq.osd.mil/DSB/reports/2010s/NuclearMonitoringAndVerificationTechnologies.pdf

Detsch, J. (2015). The mixed consequences of Sino-Indian competition in the Indian Ocean, *The Diplomat* https://thediplomat.com/2015/01/the-mixed-consequences-of-sino-indian-competition-in-the-indian-ocean/

Diaz, J. (2013). Internet Explorer vs Murder Rate Will Be Your Favorite Chart Today, *Gizmodo,* http://gizmodo.com/5977989/internet-explorer-vs-murder-rate-will-be-your-favorite-chart-today

Energy.gov (2014, November). Department of Energy Awards $425 Million for Next Generation Supercomputing Technologies. http://energy.gov/articles/department-energy-awards-425-million-next-generation-supercomputing-technologies

Fahey, S. (2012). Big Data Analytics and National Security http://web.stanford.edu/group/mmds/slides2012/s-fahey.pdf

Gandomi, A. Haider, M. (2015). Beyond the hype: Big data concepts, methods, and analytics. International Journal of Information Management. 35 2 2015137–144. ISSN 0268-4012, https://doi.org/10.1016/j.ijinfomgt.2014.10.007.

Gerbaudo, P. (2012). *Tweets and the streets: Social media and contemporary activism.* London: Pluto Press.

Harbert, T. (2015). IARPA's New Director Wants You to Surprise Him, *IEESpectrum,* https://spectrum.ieee.org/computing/networks/iarpas-new-director-wants-you-to-surprise-him#TopPageAnchor

Howard, P. N., Duffy, A., Freelon, D., Hussain, M. M., Mari, W., & Maziad, M. (2011). *Opening closed regimes: What was the role of social media during the Arab spring?*, SSRN, https://doi.org/10.2139/ssrn.2595096

Jain, S., Rosenblatt, M., Duke, J. (2014). Is big data the new frontier for academic-industry collaboration?. Journal of the American Medical Association. 2014;311(21):2171–2172. doi:https://doi.org/10.1001/jama.2014.1845, https://jamanetwork.com/journals/jama/fullarticle/1854520

Lant, K. (2017). Apple Is Using AI to Unlock Previously Unusable "Dark Data", *Futurism* https://futurism.com/apple-is-using-ai-to-unlock-previously-unusable-dark-data/

Lazer, D., Kennedy, R., King, G., & Vespignani, A. (2014). The Parable of Google Flu: Traps in Big Data Analysis. *Science, 343*(6176, March 14), 1203–1205.

Lee, D. (2017, November 22). Uber concealed huge data breach. BBC News. *http://www.bbc.com/news/technology-42075306*

Makuch, B. (2014). The Database Tracking the World's News Could Predict Future Conflicts, *Motherboard,* http://motherboard.vice.com/read/the-database-tracking-the-worlds-news-could-predict-future-conflicts

Manjoo, F. (2017). Tech's frightful five: They've got us. *New York Times*. Retrieved September 16, 2017, from https://www.nytimes.com/2017/05/10/technology/techs-frightful-five-theyvegotus.html?mcubz=1

March, G. & Davis, E. (2014). Eight (No, Nine!) Problems With Big Data. *New York Times*, http://www.nytimes.com/2014/04/07/opinion/eight-no-nine-problems-with-big-data.html?_r=0

Marcus, G. (2013). Steamrolled by Big Data. *New Yorker* http://www.newyorker.com/tech/elements/steamrolled-by-big-data?utm_source=datafloq&utm_medium=ref&utm_campaign=datafloq

McKinsey (2012). McKinsey Global Institute, Big Data: The next frontier for innovation, competition, and productivity Report, June, 2012.

Metropolis, N. (1987). *The Beginning of the Monte Carlo Method*. Los Alamos Science, No. 15, p. 125, http://library.lanl.gov/cgi-bin/getfile?00326866.pdf

MIT Sloan (2017). What's Your Data Worth? *MIT Sloan Management Review*. Retrieved September 16, 2017, from http://sloanreview.mit.edu/article/whats-your-data-worth/

Mitroff, S., Sharpe, B. (2017). Using big data to solve real problems through academic and industry partnerships, Current Opinion in Behavioral Sciences 1891–96, ISSN 2352-1546, https://doi.org/10.1016/j.cobeha.2017.09.013. http://www.sciencedirect.com/science/article/pii/S2352154617301900

NextGov (2014) *Stop Loose Nukes With Big Data And Crowdsourcing, Experts Urge* http://www.nextgov.com/big-data/2014/01/stop-loose-nukes-big-data-and-crowdsourcing-experts-urge/77324/

Norrie J. (2017). Big Data Needs Bigger Security, U.S. News, https://www.usnews.com/opinion/economic-intelligence/articles/2017-09-11/equifax-hack-shows-why-big-data-needs-bigger-security

Omand, D., et al. (2012). Introducing social media intelligence (SOCMINT). *Intelligence and National Security, 27*, 801–823. https://doi.org/10.1080/02684527.2012.716965.

Pomerleau, M. (2015). IC embraces open source intel, even if it is double-edged. *Defense Systems*, https://defensesystems.com/articles/2015/09/25/iarpa-open-source-intelligence.aspx?admgarea=DS

Preston, J. (2011). Syria restores access to Facebook and YouTube, *New York Times*, http://www.nytimes.com/2011/02/10/world/middleeast/10syria.html

Richards, N. M. & King, J.H. (2013). *Three Paradoxes of Big Data. Stanford Law Review.* http://www.stanfordlawreview.org/sites/default/files/online/topics/66_StanLRevOnline_41_RichardsKing.pdf

Richelson, J. T. (2002). *The Wizards of Langley: Inside The Cia's Directorate Of Science And Technology* (p. 386 pp., apps., bibl., index. Boulder, Colo). Westview Press.

Roberto, V. Z. (2013). *Big Data: Challenges and Opportunities*. http://www.odbms.org/wp-content/uploads/2013/07/Big-Data.Zicari.pdf

Rosenbach, E. and Peritz, A. (2009). "The role of private corporations in the intelligence community." Memorandum, "Confrontation or Collaboration? Congress and the intelligence community," Belfer Center for Science and International Affairs, Harvard Kennedy School, July 2009.

Shane, S. (2017). These are the Ads Russia Bought on Facebook in 2016, *New York Times*, https://www.nytimes.com/2017/11/01/us/politics/russia-2016-election-facebook.html

SINTEF. (2013). Big Data, for better or worse: 90% of world's data generated over last two years. *ScienceDaily*. Retrieved April 29, 2017 from www.sciencedaily.com/releases/2013/05/130522085217.htm

Soubra D. (2012). The 3Vs that define Big Data, Data Science Central, http://www.datasciencecentral.com/forum/topics/the-3vs-that-define-big-data

Symon, P. and Tarapore, A. (2015). *Defense intelligence analysis in the age of big data*. Forum of Defense Intelligence and Big Data, 2015.

Temple-Retson, D (2012). Predicting The Future: Fantasy Or A Good Algorithm? *NPR* http://www.npr.org/2012/10/08/162397787/predicting-the-future-fantasy-or-a-good-algorithm

The Wire (2017). FAQ: What the Right to Privacy Judgment Means for Aadhaar and Mass Surveillance, *The Wire* https://thewire.in/170700/right-to-privacy-aadhaar-supreme-court/

U.S. Joint Chiefs of Staff (2013). Joint Intelligence Report http://www.dtic.mil/doctrine/new_pubs/jp2_0.pdf

Ward, J. S., & Barker, A. (2013). Undefined by data: A survey of big data definitions. arXiv preprint arXiv:1309.5821.

White House (2014). Big Data: Seizing opportunities, preserving values. Executive Office of the President. https://obamawhitehouse.archives.gov/sites/default/files/docs/big_data_privacy_report_5.1.14_final_print.pdf

White House (2016). Big Data. A Report on Algorithmic Systems, Opportunity, and Civil Rights. Executive Office of the President. https://obamawhitehouse.archives.gov/sites/default/files/microsites/ostp/2016_0504_data_discrimination.pdf

Wiener J., Facebook (2016). Facebook's Top Open Data Problems, https://research.fb.com/facebook-s-top-open-data-problems

World Institute for Nuclear Security (2015). *Data Analytics for Nuclear Security*, https://www.wins.org/files/30.01.2015_data_analytics_for_nuclear_security_rev_1.0_en_final_1.pdf

Young C. (2012). Military intelligence Redefined: Big Data in the Battlefield, *Forbes*, March 2012 http://www.forbes.com/sites/techonomy/2012/03/12/military-intelligence-redefined-big-data-in-the-battlefield/#367ab43f718f

# Situational Awareness in Megacities

**Margaret L. Loper**

**Abstract** Each day an estimated 180,000 people across the globe migrate to cities. By 2030 cities will account for 60% of the world's population. Cities with populations of ten million or more are called megacities. The problems found in megacities and other urban environments – explosive growth rates, vast and growing income disparity, and a security environment that is increasingly attractive to the politically dispossessed – present great challenges to national and international security. Therefore, monitoring these cities will become increasingly important to provide decision makers with effective predictors of looming instability. The urban environment is becoming increasingly more connected and complex. In the coming decades, we will be surrounded by billions of sensors, devices and machines, the Internet of Things (IoT). Cities and urban areas that benefit from the IoT are commonly referred to as Smart Cities. Based on the rise of IoT adoption around the world, future intelligence techniques for megacities will rely, in part, on Smart City technologies. To this end, there is a need for a common framework that captures military operations, urban operations, emergency response operations, and city behavior. By creating such a framework, the sharing of information and common understanding of instabilities in megacities can be recognized and communicated across different operational needs. With a common intelligence framework, a variety of technologies - data analytics, sensor fusion, augmented reality, cyber security, 3D tracking, and predictive modeling - can be developed to provide situational awareness on the many dimensions of megacities.

## 1 Introduction

A historic transition is underway; each day an estimated 180,000 people across the globe migrate to cities (USAID 2013). By 2030 cities will account for 60% of the world's population and 70% of the worlds GDP (Heilig 2012). Cities with populations of ten million or more are called megacities. There are currently over twenty megacities in the world, and by 2025 there will be close to forty (Harris et al. 2014).

M. L. Loper (✉)
Georgia Tech Research Institute, Atlanta, Georgia, USA
e-mail: margaret.loper@gtri.gatech.edu



M. E. Kosal (ed.), *Technology and the Intelligence Community*, Advanced Sciences and Technologies for Security Applications,
https://doi.org/10.1007/978-3-319-75232-7_12

In the next century, the problems found in megacities - explosive growth rates, vast and growing income disparity, and a security environment that is increasingly attractive to the politically dispossessed - will present great challenges to national and international security. Therefore, monitoring megacities will become increasingly important so to provide decision makers with effective predictors of looming instability that can have global impact.

Traditional military intelligence approaches often focus on the concept of Areas of Operation, which emphasize discrete problem sets and well-defined regions. The problem is that megacities are neither. They are multidimensional, interconnected through globalization and information, and uncontrollable due to issues such as connectivity and informal economies. Therefore, new ideas for collecting intelligence are needed which capture the dynamic nature of the modern urban environment. The Army, NATO, the World Bank, and others are looking at megacities, trying to define indicators for managing or describing these urban environments. There is some overlap, but each group defines their indicators from a different perspective.

The future of megacities indicate that instability could come from a variety of areas, including rapid population growth, income disparity, environmental vulnerabilities, racial or ethnic separation, infrastructure capacity, or hostile actors. In other words, the triggers for instability cross many different operational interests – military, emergency response, health and human affairs, and city management. Thus, intelligence collection must capture all these dimensions. Based on the rise of the Internet of Things and Smart City technologies, cities will be populated with millions of sensors, devices, and machines. As a result, future intelligence techniques for megacities will be data focused. To this end, there is a need for a common framework that captures military operations, urban operations, emergency response operations, and city behavior. By creating such a framework, the sharing of information and common understanding of instabilities can be recognized and communicated across different operational needs.

The rise of smart cities has spurred new work in defining the anatomy of a city. This work looks at the city as a system of systems, capturing behavior from an urban design and city operations perspective. While still emerging, this framework may provide the necessary ontology to capture the dynamic nature of the modern urban environment. With a common intelligence framework, a variety of technologies - data analytics, sensor fusion, augmented reality, cyber security, 3D tracking, and predictive modeling - can be developed to provide situational awareness on the many dimensions of megacities.

## 2 Megacities

World population growth is increasing at a staggering pace. At the start of the Industrial Revolution in 1750, the world's population was 700 million. A little over 200 years later it had reached 3 billion people. Population growth accelerated

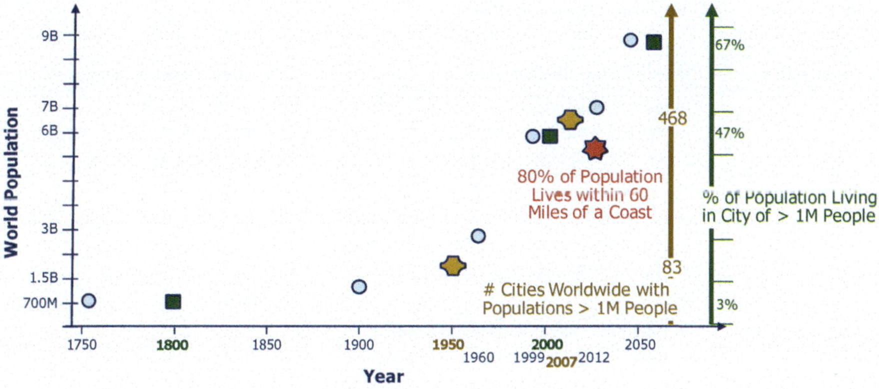


**Fig. 1** World population growth and urbanization

between 1960 and 1999, doubling to 6 billion people (Killcullen 2012). Today, the world population is 7.5 billion, and the United Nations predicts that there will be 9.7 billion humans on the planet by 2050 (United Nations 2017).

As the world's population has grown, the pace of urbanization has been accelerating. In 1800, only three percent of people lived in a city of one million or more; by the year 2000, it was forty-seven percent. In 1950, there were only eighty-three cities worldwide with populations over one million; by 2007 there were 468 (Killcullen 2012). The trajectory is predicted to continue such that "by 2050 two-thirds of the world's population will live in cities…the rise of supercities is the defining megatrend of the 21st century" (Kazan 2009). The population growth and urbanization statistics are illustrated in Fig. 1.

Roughly 1.4 million people across the world migrate to a city every week. In the year 2015, the total population of megacities worldwide was about 359 million and the future growth is increasing. For example, the populations of Jakarta, Delhi, Dhaka, and Karachi tripled between 1975 and 2003 (Kötter and Friesecke 2009). The population, urbanization, and resource trends contributing to the rise of megacities show no signs of disappearing. Urban areas are expected to grow by 1.4 billion in the next two decades, almost entirely in the developing world.

The nature of each megacity's complexity makes it unique. No two cities are the same - their geography, internal or external connections, scale, and global connectedness - defy efforts to map or fully understand them. This means that an attempt to monitor a megacity requires a novel approach to understand each city as a unique context.

There are some observable characteristics in all megacities, to varying degrees. Two sets of characteristics have been identified by very different sources. The first set comes from a report presented at a World Bank conference. Their interest in megacities stems from a land use and urban planning perspective (Kötter and Friesecke 2009). The second set of characteristics comes from a report by the US Army. Their interest in megacities is from the perspective of a future operational environment (Harris et al. 2014). Table 1 shows the characteristics from both

**Table 1** Summary of megacity characteristics

| Combined characteristics | World bank characteristics | US Army characteristics |
|---|---|---|
| Density | Density: High density of inhabitants, industrial assets and production, social and technical infrastructure. | Density: Population, infrastructure, and signals all pose significant challenges. |
| Scale | Dynamism of growth: High spatial and demographic growth, change of land use and consumption of land for settlement purposes often in absence of urban planning. | Scale: Relative size differentiates them from other urban environments and presents fundamental challenges. |
| Infrastructure & Land Use | Settlement, infrastructure and land tenure: Urban planning and public infrastructure can only partially guide the urban development to achieve a proper sustainable structure. | |
| Socio-cultural and Socio-economic Diversity | Socio-economic disparities: Wide range of social standards and social fragmentation as well as social-cultural conflicts. | Context: Unique historical, cultural, local, regional and international context. |
| Threats & Vulnerabilities | Risks and vulnerability: Highly vulnerable to natural and man-made disasters, e.g. floods, earthquakes, landslides etc. are most likely to happen | Threats: Constantly challenged by threats to their stability. |
| Governance | Urban Governance: State, private sector and civil society. | |
| Connectedness | | Connectedness: Cities don't exist in isolation. |
| Flow | | Flow: The movement of people, resources or things into or out of a megacity. |

sources, as well as a combined set of characteristics that will be used for this chapter. Note that the characteristics are not exact comparisons, so the combined characteristics are a best attempt to create a merged set of attributes. In the following sections, the characteristics from each source are described in more detail.

## 2.1 World Bank Characteristics

Megacities come into the international focus of policy and science because of their serious impacts on the global environment, such as enormous land consumption, air pollution, water scarcity, poverty, social segregation, and vulnerability. The following characteristics are typical features that bring megacities into the focus of science, policy, and economy; they are summarized from Kötter and Friesecke (2009):

Density: Megacities have the highest density of inhabitants, industrial assets and production, and social and technical infrastructure. They are the centers of political and economic decisions, they generate a lot of income, and their local economies are important for their rural surroundings.

Dynamism of growth: Megacities are highly dynamic in spatial and demographic growth, change of land use, consumption of land for settlement purpose, and formal and informal urban economic sectors. The local, regional and global markets and the connection with international economies stimulate economic activities.

Settlement, infrastructure and land tenure: Informal and illegal housing areas (squatters) developed by migrants can lead to an extensive settlement structure. In many cases the infrastructure (e.g., public and private transportation, garbage removal, and sewage systems) are not efficient, which has serious consequences on the environment and public health.

Socio-economic disparities: A wide range of social standards, social fragmentation and social-cultural conflicts exist in megacities. The growing socio-economic disparity within megacities, including urban poverty, and the lack of social cohesion is a serious risk.

Risks and vulnerability: Megacities are highly vulnerable to natural and man-made disasters (floods, earthquakes, landslides) due to the high density and large number of inhabitants. Damage will have a dramatic impact on poor people and ethnic minorities.

Urban Governance: Urban Governance includes the state, the private sector and civil society. One of the greatest challenges of megacities is their governability. Centralized top-down strategies are not good models, because of the highly dynamic and highly complex interactions within the megacities and also with their surroundings.

## 2.2 *US Army Characteristics*

Megacities come into the focus of US military operations because a gap exists in the Army's doctrinal understanding of large cities. As a result, the US Army conducted a study of megacities to identify characteristics that would enable them to understand or prepare for these environments. These characteristics are summarized from Harris et al. (2014):

Context: Every megacity must be understood within its own historical, cultural, local, regional, and international context. Knowledge of the rate and characteristics of a megacities' growth may enrich contextual understanding.

Scale: The relative size of megacities differentiates them from other urban environments. Density, connectedness, flow, and context can be studied in any environment, but understanding these elements at scale is what makes megacities a different problem set.

Density: Population, infrastructure, and signals all pose significant challenges. Population density can, intentionally or unintentionally, disrupt flows on fixed capacity lines of transportation and communication in and around the urban environment.

Connectedness: Cities don't exist in isolation. Instantaneous information transfer, robust international surface and air shipping, and mass migration (legal and illegal) connect cities around the world in ways unknown a decade ago.

Flow: Flow is the movement of people, resources or things into or out of a megacity. Vast amounts of energy and other goods flow into the megacity, circulate throughout the urban space, and then waste flows out if the megacity is to remain healthy.

Threats: Megacities are constantly challenged by threats to their stability. The nature of these environments manifest multiple dynamics of observable friction which operate against the city or emanate from within it. These manmade and natural threats contribute significantly to the complexity of the megacity.

## 3 Challenges in Megacities

Rapid urbanization will create economic, social, and governance challenges while simultaneously straining city infrastructure. Thus, the most vulnerable cities will be less able to meet these challenges. The implications for future conflict are profound - more people will fight over scarcer resources in crowded, under-serviced, and under-governed urban areas (Kilcullen 2012).

### *3.1 Resilience and Fragility*

Cities differ widely on their ability to adapt to volatility and stress. Monitoring megacities to predict instability, requires identifying tipping points or triggers which describe the cumulative effect of stressors on the city. Triggers will indicate a dramatic shift in some system (or systems) in the city, from a state of equilibrium to a state of imbalance. Since each city is unique in its composition (people, systems, geography, etc.), there is no universal measure for these tipping points. Each city has an equilibrium which must be understood in its own unique context.

Equilibrium can be conceptualized as a range, from fragile to resilient. Fragile is characterized by weak state capacity and/or weak state legitimacy leaving citizens vulnerable to a range of shocks (Fragile State 2017). These cities respond poorly to adversity, which can make bad situations worse. Resilience is the capability to prepare for, respond to, and recover from significant multi-hazard threats with minimum damage to public safety and health, the economy, and security (Wilbanks 2007). These cities return quickly to a normal state, expending resources to minimize the impact of the adversity.

A city's stable functioning is dependent on systems of finite capacity. When these systems experience demand that exceed their capacity, the support mechanisms erode and become more fragile. There is a strong correlation between highly integrated systems and antifragility. A highly-integrated city, like New York, which can learn from setbacks and design systems that prevent future disruptions from similar events, is said to exhibit anti-fragile characteristics (Zolli and Healy 2013). In comparison, loosely integrated cities show little or no improvement in the aftermath of adverse events, making it more fragile and incapable of meeting the needs of its population (Turchin 2013). If a city does not learn and evolve from adverse events, future events will overwhelm it and make it more vulnerable to triggers.

Although equilibrium is often described as a fixed state, it is more aptly thought of as a process. Megacities will be continuously changing - physically, socially, and politically. This means that the profile of risk in a city will change and evolve. Therefore, the means for monitoring and assessing the state of cities must account for the ever-changing face of "normal."

## 3.2 *The Dynamics of Instability*

Drivers of instability in megacities range from wealth disparity to environmental risk factors. In 2011, the Asian Development Bank estimated that drought, desertification, and soil salinity will be exacerbated by climate change, thus prompting millions of rural people across the Asia-Pacific region to migrate to cities over the coming decades. Rural migration to coastal cities will put more people in low-lying regions where the slightest rise in sea level can cause major disruption. Climate-induced displacement will undermine economic growth, enhance the risk of conflict, and lead to deterioration of social indicators (Asian Development Bank 2011).

Some megacities, particularly highly integrated cities, are capable of coping relatively well when instability arises, while others will have their service and security capabilities quickly overwhelmed. Entirely new concepts are needed to understand the megacity environment. The US Army developed a framework based on context, scale, density, connectedness, and flow as a mechanism to encourage new thinking on the subject. It is not meant to provide a model for understanding all megacities, nor is it meant to take the place of existing analytical tools that help tactical planners focus their efforts. Instead the framework can be used for planning, to provide an urban-centric strategic appreciation of the megacity system (Bailey et al. 2014).

A set of triggers or tipping points, which can be used for monitoring a city are summarized below (Harris et al. 2014).

Population Growth and Migration: Rapid hetero and homogeneous population growth will outstrip a city's governance capability. Many emerging megacities are ill-prepared to accommodate the kind of explosive growth they are experiencing (National Intelligence Council 2008).

Separation and Gentrification: Radical income disparity along with racial, ethnic, and sub-cultural separation are major drivers of instability. They can create subtle tensions, which if allowed to fester, can build over time leading to instability.

Environmental Vulnerability and Resource Competition: Unanticipated weather events and natural disasters can devastate city systems, interrupting governance and service delivery. These events will affect larger populations that are densely packed into urban centers on a scale never before seen. Environmental disasters and resource scarcity (real or perceived) can produce competition and instability, which can rapidly exceed the capability of local authorities (Ogbonnaya 2013).

Hostile Actors: If internal or foreign actors conduct offensive operations which exceeded a city's capacity to contain or defend against them, external intervention could be required (Dias and Salla 2013). This would be especially true if the city is in an allied country or the threat extends its hostilities to the US or its citizens abroad. If these frictions become disruptive to the international system, outside intervention to restore order may be required. The ability of a city to reconstitute its systems or adapt to a new normal is a measure of its resilience.

Capacity: Understanding the systems that keep the city functioning is an essential component of understanding the city itself. Similarly, understanding the surge capacity of the city (emergency response capability, planning and exercising emergency procedures, material resources, etc.) is essential to forecasting the city's ability to return to steady-state and whether external assistance will be necessary.

### *3.3 Case Study: Mumbai Attack*

For connected, developing-world coastal cities under stress, threats to their wellbeing can be summed up in three broad concepts: irregular actors and methods, hybrid threats, and nested networks. An example of this new pattern of conflict in megacities is the November 26, 2008 attack on Mumbai, India. Summarized from Kilcullen (2012), the attack was as follows.

> Members of a non-state armed group (the Pakistani terrorist organization Lashkar-e-Taiba) carried out a seaborne raid on India's largest city, killing 164 people and wounding 308 over a three-day period. The attackers inserted themselves into a coastal fishing fleet to cover their approach to the city. They landed in the urban core of Mumbai, allegedly receiving assistance from a reception party operating in the city's urban areas. The terrorists had support from retired members of the Pakistani Inter-Services Intelligence, who trained them prior to the attack and continued to act as mentors, supporters, and coaches during the raid. The Pakistani handlers walked individual terrorists through specific actions step by step over satellite phones, monitored Indian news channels and the Internet, and used Google Earth for targeting. The terrorists and their sponsors monitored the progress of the attack in real-time via Twitter. The overstretched Mumbai police and counterterrorism units took a long time to react to the size and complexity of a seaborne terrorist attack, which targeted several locations simultaneously in a cluttered downtown coastal area. They were also hampered by the need to seek bureaucratic approval from distant officials in the capital city, New Delhi.

The Mumbai attacks involved a hybrid organization of state and non-state actors, who attacked national and international targets, and leveraged both local and remote networks for support. The terrorists exploited difficulties of an overstretched public safety and policing infrastructure and a heavily populated, highly complex, networked urban littoral environment. Mumbai represents one future face of conflict in a megacity, as this attack has been cited as the template by which the 2015 Paris attacks were modeled (Riedel 2015; and Mehta 2015). Additional sources that describe the Mumbai attack can be found in (CNN 2016; Henderson 2013; Azad and Gupta 2011; Oh et al. 2011).

### 3.4 Understand, Operate and Defend

As stated earlier, there are a number of factors unique to megacities: geographic sprawl, population volume and density, sociocultural and socioeconomic diversity, governance challenges, varying qualities of infrastructure, and regional and global (inter)connectedness. These factors represent risks and challenges that apply to peacekeeping, humanitarian assistance, counterinsurgency, stabilization and reconstruction operations. Over the past century, the US has conducted one major peacekeeping, counterinsurgency or stabilization operation every 25 years, and a small or medium-sized one every five to 10 years (Kilcullen 2012). This pattern is currently independent of policymakers' partisan affiliation or personal preferences. Thus, we can assume this trend will continue. So, what does it mean for these operations to be conducted in megacities?

Humanitarian assistance and disaster relief operations may be impacted by varying qualities of infrastructure. Cities depend on extensive infrastructural support systems to provide food, water, energy, transportation, lodging, information transfer, and other necessities. The sudden loss of a city's infrastructure, due to a natural or man-made disaster, would result in human suffering at a magnitude that far exceeds a disaster in a less centralized, rural setting. Responding to the disaster could be compounded by issues like geography, density or substandard infrastructure, thereby magnifying the environmental and health risks.

Peacekeeping and stabilization operations maybe be impacted by sociocultural and socioeconomic diversity. The inequality between rich and poor, and historically antagonistic religions and ethnicities brought into close proximity could increase tension and conflict. The existence of slums and shanty towns alongside rapidly expanding modern high-rises will create socio-economic divides.

Counterinsurgency and reconstruction operations can be impacted by constrained resources. The sheer size of megacities and the challenges associated with their governance make them ripe targets for violent non-state actor recruitment and funding activities. Also, the emergence of illicit networks which create informal governance structures can replace over-extended and under-capitalized governments. In such situations, irregular actors can create and/or take advantage of black markets, shadow governance, illicit economies, and dark networks to generate revenue and recruit new members over long periods (Serena and Clark 2016). Further, the density of buildings and people in megacities can provide cover and concealment to hostile actors and their operations. Dense urban environments reduce the effectiveness of traditional intelligence, surveillance, and reconnaissance (ISR) platforms the military needs to successfully conduct operations.

The trends are clear - megacities are growing in size and complexity, and they will present an increased security risk. In the next century, the urban environment will be the place where drivers of instability converge. The cities that grow the fastest will be the most challenged, and ignoring these cities can create strategic vulnerabilities. From a military and intelligence perspective, three pillars emerge as crucial for future interactions with megacities; all are drivers for improved situational awareness techniques:

- **Understand**: How do we understand the native infrastructure and exploit the inherent connections and complexity for improved operational effectiveness?
- **Operate**: How does improved situation awareness combined with an understanding of a megacity's attributes drive the development of new tactics, techniques, and procedures for the military, first responder and city operators?
- **Defend**: How do we develop distributed ISR capabilities and situational awareness to recognize the tipping points and triggers of a megacity, so we can defend its vulnerabilities, especially emerging and aggressive cyber-attacks?

To accomplish this, multi-modal data on cities is needed.

## 4 Multimodal City Information

The City Environment (CE), analogous to the military's concept of operational environment, needs to be made accessible to many different users. This includes not only individual military services, but also other governmental agencies, allies, coalition partners, non-governmental partners, first responders, city operation professionals, etc. This access needs to be understood and leveraged by decision makers at the point of need, in order to shape favorable mission outcomes. The data that makes up the CE includes all possible sources (military, civilian, urban, social, and so forth) that may be required for use by the decision maker. The data needs to be integrated and presented in a sense-making way.

In order to present the right information to decision makers, multimodal data is needed that describes and defines the CE. This consists of a wide variety of data, some of it from military sources, some civilian; some of it is structured, some unstructured; some of it is available on demand, and some is unanticipated. Military data includes things such as order of battle information, geospatial information, and information about operations. Civilian data includes information about population demographics, infrastructure, economic factors, political situation, and non-state actor information.

Structured data is data that comes in a well-defined format and that can be ingested and manipulated with standard tools and programs. Unstructured data needs to be located, identified, mined, processed and then analyzed. Finally, there is the distinction between anticipated data and unanticipated data. The former

**Table 2** Sources of data

| Examples of data sources | | |
|---|---|---|
| Departments, Agencies, Allies | Geography, Geospatial, Location | City and Urban Environments |
| INTs | Historical | Health, Medical |
| Military (training, installations, doctrine, health, service, etc.) | Human, Social, Cultural, Behavior | Economic, Financial, eCommerce |
| Sensor (military environments and urban environments) | Social Media present and "to be" (Chat, FB, Instagram, Snapchat, LinkedIn, etc.) | Cyber-attacks, data breeches, security |

might exist within a sharing organization, or within an archive or catalog. The latter, unanticipated data, comes about from intelligence origination and from sources such as social media. It must be scanned for, and techniques should be available to recognize when there is something worth retrieving, within a steady stream of potential data.

Enormous amounts of data are collected daily for a variety of reasons; data sets collected from multiple modalities. Examples of data sources are shown in Table 2. Identifying and integrating these multimodal data sets is critical for achieving situational awareness that provides insight and understanding of multidimensional, interconnected CE. In the following sections, multimodal data that is emerging from Smart Cities, urban environments, and traditional military intelligence will be discussed. All this data will be needed for understanding megacities.

## 4.1 Smart Cities

People move to urban areas with the hope of finding better job opportunities as well as a better standard of living. However, the increasing number of people migrating to urban areas leads to complex issues such as congestion and an increased demand for a limited pool of resources, including energy, water, sanitation, education and healthcare services (Kondepudi 2014). With the growing economic and environmental problems in urban areas, information and communication technologies are being used to help solve these problems.

The urban environment is becoming increasingly more connected and complex. In the coming decades, we will be surrounded by billions of sensors, devices, and machines, commonly called the Internet of Things (IoT). Cities and urban areas that benefit from the IoT are commonly referred to as Smart or Sustainable Cities (SC). Based on the rise of IoT adoption around the world, future intelligence techniques for megacities will rely, in part, on SC technologies.

There are many benefits of using IoT technologies in a city. A smart electrical grid will make cities more efficient by optimizing how energy is used and distributed. Device data will help inform and protect city residents by improving city service monitoring capabilities. Consumers will have better insights on the consumption of personal resources (e.g., energy, water, and gas) and granular neighborhood data (Patterson 2015). From smart energy meters to context aware security devices to more efficient sanitation systems; city infrastructures and services will change with new interconnected systems for monitoring, control, and automation.

To develop a common understanding of SC, the International Telecommunication Union (ITU) analyzed 116 definitions obtained from a variety of sources including: academia and research communities, government initiatives, international organizations (United Nations, ITU, etc.), corporate/company profiles, user centric definitions, trade associations, and standards development organizations. Based on their analysis, a definition was approved by the ITU Focus Group on Smart Sustainable Cities (Kondepudi 2014):

> A smart sustainable city is an innovative city that uses information and communication technologies and other means to improve quality of life, efficiency of urban operation and services, and competitiveness, while ensuring that it meets the needs of present and future generations with respect to economic, social and environmental aspects.

Cities are the second or third largest target area for IoT, with a projected economic impact totaling somewhere between $1 trillion and $1.6 trillion by 2025 (Manyika 2015). Increasingly, megacities are becoming smart, deploying instrumentation and creating open data sources. Table 3 shows a mapping between the projected megacities of 2030 (United Nations 2017) and the world's smartest cities, as identified in 2014 (IESE 2014). Note that there are numerous rankings of the world's smartest cities, but there is no uniform criteria by which cities are evaluated for being "smart." As such, many emerging initiatives, like Singapore's investments toward becoming a "Smart Nation", Smart Cities India (Government of India 2017), and the 200 smart city projects in China (Li et al. 2015) don't often make any of the published rankings. Further, Moscow is investing in numerous smart city projects, and in South Africa a $7.4 billion smart city project is already underway. Table 3 gives a notional mapping, with the cities in dark gray appearing in both published lists (megacity and smart city). Megacities that are shown in light gray have emerging smart city initiatives not included in all rankings. One thing is clear, the relationship between megacities and smart cities is growing.

Regardless of whether it's a megacity or a dense urban environment, cities are becoming smarter – and the trend will continue into the future. These cities will create rich information environments using commercial technology – not specialized government or defense instrumentation. These cities will provide growth opportunities for civilians (health, sustainability, mobility, etc.,) and they also represent challenges for defense and intelligence organizations.

**Table 3** Megacities that are becoming smart

| Rank | Megacities in 2030 | Rank | Smart Cities 2014 |
|---|---|---|---|
| 1 | Tokyo Japan | 1 | Japan-Tokyo |
| 2 | Delhi India | 2 | United Kingdom-London |
| 3 | Shanghai China | 3 | USA-New York |
| 4 | Mumbai (Bombay) India | 4 | Switzerland-Zurich |
| 5 | Beijing China | 5 | France-Paris |
| 6 | Dhaka Bangladesh | 6 | Switzerland-Geneva |
| 7 | Karachi Pakistan | 7 | Switzerland-Basel |
| 8 | Al-Qahirah (Cairo) Egypt | 8 | Japan-Osaka |
| 9 | Lagos Nigeria | 9 | South Korea-Seoul |
| 10 | Ciudad de México (Mexico City) Mexico | 10 | Norway-Oslo |
| 11 | São Paulo Brazil | 11 | USA-Philadelphia |
| 12 | Kinshasa Democratic Republic of the Congo | 12 | USA-Los Angeles |
| 13 | Kinki M.M.A. (Osaka) Japan | 13 | USA-Dallas |
| 14 | New York-Newark USA | 14 | Denmark-Copenhagen |
| 15 | Kolkata (Calcutta) India | 15 | Netherlands-Eindhoven |
| 16 | Guangzhou, Guangdong China | 16 | Netherlands-Amsterdam |
| 17 | Chongqing China | 17 | Australia-Sydney |
| 18 | Buenos Aires Argentina | 18 | Sweden-Stockholm |
| 19 | Manila Philippines | 19 | USA-Chicago |
| 20 | Istanbul Turkey | 20 | USA-Baltimore |
| 21 | Bangalore India | 21 | Australia-Melbourne |
| 22 | Tianjin China | 22 | USA-Minneapolis-Saint Paul |
| 23 | Rio de Janeiro Brazil | 23 | Austria-Linz |
| 24 | Chennai (Madras) India | 24 | Israel-Haifa |
| 25 | Jakarta Indonesia | 25 | USA-Houston |
| 26 | Los Angeles-Long Beach-Santa Ana USA | 26 | Germany-Munich |
| 27 | Lahore Pakistan | 27 | Austria-Vienna |
| 28 | Hyderabad India | 28 | Germany-Berlin |
| 29 | Shenzhen China | 29 | Canada-Toronto |
| 30 | Lima Peru | 30 | Canada-Ottawa - Gatineau |
| 31 | Moskva (Moscow) Russian Federation | 31 | Finland-Helsinki |
| 32 | Bogotá Colombia | 32 | United Kingdom-Nottingham |
| 33 | Paris France | 33 | Germany-Cologne |
| 34 | Johannesburg South Africa | 34 | Israel-Tel Aviv |
| 35 | Krung Thep (Bangkok) Thailand | 35 | Germany-Stuttgart |
| 36 | London United Kingdom | 36 | United Kingdom-Liverpool |
| 37 | Ahmadabad India | 37 | Sweden-Gothenburg |
| 38 | Luanda Angola | 38 | France-Lille |
| 39 | Thành Pho Ho Chí Minh (Ho Chi Minh City) Viet Nam | 62 | China-Beijing |
| 40 | Chengdu China | 66 | Thailand-Bangkok |

## 4.2 Urban Data in Modern Cities

Cities are complex systems. This complexity, combined with ubiquitous IoT sensors in the environment and a proliferation of communications channels, is creating unprecedented amounts of human intelligence and machine data (Dignan et al. 2013). Local governments are looking to data and analytics technologies to improve their services. Large US cities are using data analytics to solve specific problems in areas such as health, transportation, sanitation, public safety, economic development, sustainability, street maintenance, and resilience. For example:

- The city of Los Angeles shares road closure, safety, and other data with app providers to improve driving, reduce congestion, and promote safety. In return, the app providers, such as Waze, share real-time crowd-sourced reports of issues encountered on the streets from more than 1.5 million users to the city's emergency management, police, fire, transportation, street services, sanitation, and other departments (PCAST 2016).
- The New York Fire Department started using data mining and predictive analytics to determine which of New York City's one million buildings are most likely to erupt in a major fire. They now examine 7500 factors across 17 city-agency data streams and use artificial intelligence to track trends city-wide (PCAST 2016).
- In Wake County, North Carolina, cardiac arrest victims have a better chance of survival due to new, analytics-driven recommendations from the county's Emergency Medical Services (EMS). Based on an analysis of 20 years of data about cardiac arrest patients, Wake County EMS changed its recommendations for how long to conduct CPR from 25 minutes (the industry standard) to 60 minutes or more. A study found that using the new CPR guidelines saved 100 people in the first year (Henderson 2015).

The US is not alone - governments in the United Kingdom, Germany, China, India, Brazil, and Singapore have stepped up with considerable organization and resources to become leaders in urban innovation, thereby positioning their countries for the emerging multi-trillion-dollar SC economy.

- One of the UK's largest police forces deployed an intelligence analytics platform across the entire enterprise. The mission-critical system contains 12 million documents, 9 million structured records and provides real-time intelligence 24 hours a day. It is used by more than 40,000 officers and police personnel daily and can be accessed securely by other government agencies. The platform provides integration with confidential, highly specialized and highly secure protected units. With the intelligence management system in place, information can be acted upon in real time to protect the public around the clock (Henderson 2015).
- The Øresundsbron between Denmark and Sweden is a five-mile expanse of bridge and tunnel that connects the two countries and two major metropolitan areas: Copenhagen in Denmark and Malmö in Sweden. With the help of analytics, it also connects its "customers" with destination ideas based on their unique

likes and dislikes. Using SAS Customer Intelligence, the bridge's marketing and customer-service teams share meaningful, relevant offers unique to each of the 180,000 travelers who own toll passes (Henderson 2015).

- Anticipating another 1.1 billion people moving into Asian cities in the next 20 years, the Asian Development Bank is allocating $18 billion per year in grants and loans to help transform cities. Other countries are also generating showcase innovations and, in so doing, improving the quality of life for some of their poorest people (PCAST 2016).

There are challenges to the new modern city, however. Dignan et al. (2013) addresses the "data rich, information poor" nature of today's cities and identifies some data-related challenges that face modern cities. These challenges are summarized below.

Data challenges in developing versus developed world cities are quite different. Cities such as New York City and London face the need to retrofit infrastructure to become smarter. Their requirement to drive efficiency, improve operations, improve service levels for citizens, and lower costs comes from the need to compete more effectively with other cities for human capital, investment, and job growth.

Cities such as Mumbai, Nairobi, Lagos, and Karachi, which struggle to provide basic services to large segments of their populations, are unable to handle the incoming data, given their reliance on manual processes and their lack of automated records digitization. Infrastructure tends to be more basic, with less instrumentation and therefore less incoming sensor-based data. On the other hand, they are often unburdened by legacy technology systems and have an expanding tax base due to a rapidly growing middle class. This results in an opportunity to invest in high-return improvements to city operations, infrastructure, data-driven governance, and the use of "citizens as sensors" to inform urban planning and crisis intervention.

The nature of data in an always-on society is a result of urbanization and the spread of connectivity through society. In the Asia-Pacific region, smartphones are heading toward ubiquity, and in developing countries they are rapidly picking up market share. These always connected societies are leading to new opportunities, threats, and governance challenges for governments. Cities' constituents are becoming accustomed to the ease and speed with which they can access data to make online purchases, connect with friends, or reserve a table at a restaurant. In the future, they will have the same capabilities to access data for community needs.

The open data movement has increased citizen trust in government, enabled information sharing between agencies, and provided the basis for co-creation by government and citizens. However, a number of obstacles arise around open data, such as the need for data management and analytical tools that can test for data validity and address privacy issues. Without this validity and privacy, automatically making incoming data universally accessible creates an opportunity to "game" the system and create the potential for decisions to be based on inaccurate data.

There is pressure to collect and utilize more types of data as the basis for better service delivery and decision-making. There are four main types of data sources: factual, mathematical, observational, and emotional. Connecting across multiple data sources will be necessary to create a single picture of the city. City leaders must consider how each of these sources will be used and what the positive and negative impacts may be.

There are many examples of analytics serving the greater good in cities around the world and many challenges cities face to operationalize data for decision making. The move toward data and analytics hold the key to improving situational awareness on cities.

## 4.3 *Military Intelligence*

Cities are deploying IoT technology and using data to improve efficiency and deliver better services to their citizens. Can that technology benefit the military, who will operate in cities to provide services such as stabilization, reconstruction, and human assistance? The key is how to better integrate the data that is emerging about cities with the military intelligence process. Military intelligence is a discipline that uses information collection and analysis approaches to provide guidance and direction to commanders in support of their decisions (Military Intelligence 2017). In order to provide an analysis, the commander's information requirements are identified, and then incorporated into intelligence collection, analysis, and dissemination. Information requirements may be related to terrain and impact on vehicle or personnel movement, disposition of hostile forces, sentiments of the local population, and capabilities of the hostile order of battle. Intelligence activities are conducted at all levels, from tactical to strategic:

- Strategic intelligence is focused on broad issues such as economics, political assessments, military capabilities, and intentions of foreign nations and non-state actors. Such intelligence is analyzed in combination with known facts about the area in question, such as geography, demographics, and industrial capacities.
- Operational intelligence is focused on support or denial of intelligence at a tier below strategic leadership, which is dedicated to the design of practical demonstration.
- Tactical intelligence is focused on support to operations attached to the battlegroup. At this level, briefings are delivered to patrols on current threats and collection priorities. Once patrols are completed, they are debriefed to obtain information which can be included in further analysis and communication.

Military intelligence concepts have been used for decades to provide a snapshot of the operating environment. One of these concepts is Intelligence Preparation of the Battlefield (IPB), which is the systematic process of analyzing the mission variables of enemy, terrain, weather, and civil considerations in an area of interest to determine their effect on operations (Army 2014). The IPB is a continuous process of analyzing the threat and environment in a specific geographic area, using four steps: define the operational environment, describe (environmental) effects on operations, evaluate the threat/adversary, and determine threat/adversary courses of action (COAs).

To be effective the IPB must "accurately define the commander's area of interest (AO) in order to focus collection and analysis on the relevant aspects of the mission variables. Relevant is defined as having significant effect on friendly operations and threat/adversary operations, and population in a unit's AO" (Army 2014). An AO is an area defined by the commander that is large enough to accomplish the mission and protect the force (Joint Pub 2017). There is a tendency to treat it as a discrete region that can be circled on a map and discussed in a vacuum and not to take into account how the variables explaining dense urban areas are increasingly interconnected. The problem in the modern dense urban environment - the impact of connections and linkages, facilitated by globalization, advances in communication technology, and media access - challenge the traditional idea of a unique operational environment that can be isolated for analysis by an intelligence team (Wolfel et al. 2016). These connections that exist in modern cities, especially megacities, must be addressed as part of the analysis.

One of the largest issues with the basic definition of IPB is that it does not take into account how the variables explaining dense urban areas are increasingly interconnected (Wolfel et al. 2016). Change in urban areas is not a unidirectional process in which the individual agent influences change in the area of interest. Instead, as a result of interacting with the area of interest, the agent is also influenced as a result of social and environmental change. This process is called the Duality of Structure (Giddens 1979). This duality is important in understanding dense urban areas. We often look directly at the influence of the mission variables, take them as well-defined and unchanging, and do not address the recursive nature of society in which the enemy, the terrain, and civil considerations often change rapidly based on the actions of agents in the area (Wolfel et al. 2016). For example, the enemy might try to shift dimensions by moving from surface to subterranean or use gray zone tactics, such as developing a disinformation campaign that can fundamentally shift the societal characteristics of the area of operations (Mazarr 2015).

We are no longer able to extract terrain from the society and analyze it separate of civil considerations. The terrorist attacks on Mumbai in 2008 used the interaction of terrain (littoral situation of Mumbai) and civil considerations (the unregulated nature of the fishing fleet) to explain how the terrorists were able to gain access to the region virtually undetected (Kilcullen 2012). Neither the terrain, nor the civil considerations alone were adequate to explain the situation in Mumbai – the complex connection between the two were needed to have clear situational awareness.

In addition to IPB, other foundational analytical tools for intelligence operators to provide information to promote situational awareness about the operating environment include (Hedges 2016):

- Areas, Structures, Capabilities, Organizations, People, and Events (ASCOPE)
- Sewage, Water, Electricity, Academics, Trash, Medical, Safety, and Other Considerations (SWEAT-MSO)
- Political, Military, Economic, Social, Information, Infrastructure, Physical Environment, and Time (PMESII-PT)
- Mission, Enemy, Terrain and Weather, Troops and Support Available, Time Available, Civil Considerations (METT-TC)

- Diplomatic, Information, Military, Economic, Financial, Intelligence, and Law Enforcement (DIMEFIL)

IPB products are not helpful to complex operational environment. These products fail to adequately unify the different elements found within the environment while seeking alignment to a threat or system of opposition (Hedges 2016). The construct does not afford a clear step for tying together "like" elements into a single operational data layer and modeling the relationship of that system on other systems.

## 5 City Analytic Framework

The next step in determining a strategy for situational awareness in megacities is defining a construct that represents multimodal city information. Future DoD missions in megacities and gray zone conflicts can include kinetic military operations but also stability, reconstruction, and humanitarian assistance and disaster response (Kilcullen 2012). The key is to define a city analytic framework that is not specific to military, urban, or emergency response operations, rather a framework that can work across these uses.

One approach is to look at the city as an organism – a system of systems. This concept isn't novel in itself, for example IBM's Smarter Cities initiative categories a city as the interaction of three types of systems (Dignan et al. 2013):

1. City planning/management systems (e.g., agency administration)
2. Infrastructure systems (e.g., transportation)
3. Human services systems (e.g., social programs)

City systems often overlap and interconnect, so they cannot be considered in isolation; systems addressing the environment, for example, can be both infrastructure-oriented and planning/management-oriented. The challenge, then, is to design solutions that address specific needs in any one system and also interconnect with data coming from other systems, in order to move toward a comprehensive view of a city's operations. This leads to key ideas that help define a city analytic framework (Wolfel 2016):

- Context is needed via an ecosystem definition.
- A standardize lexicon, taxonomy, and ontology is needed, as a starting point.
- The ontology needs to grow into a set of interconnected networks, i.e., the network of networks.

### *5.1 Anatomy of a City*

An ontology that represents a city as a system of systems is the Anatomy of a City, developed by the City Protocol Society (City Protocol Society 2015). The City Protocol Society (CPS), formed in 2012, includes city leaders, technologists, urban

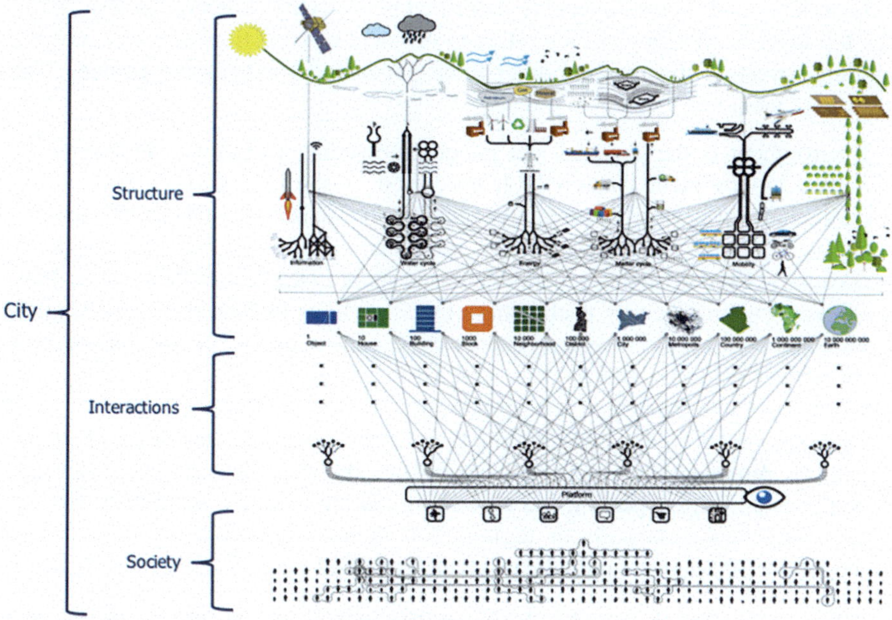


**Fig. 2** Anatomy of a city (City Protocol Society 2015)

designers, architects, and citizen-leaders from 33 cities, 20 major businesses, 14 universities, and 20 other organizations. Membership is internationally diverse, including participants from Amsterdam, Netherlands; Dubai, United Arab Emirates; Charlotte, NC; Dublin, Ireland; Barcelona, Spain; Bandung, Indonesia; and Montevideo, Uruguay. Therefore, the products and projects pursued by CPS include a multi-national perspective about cities, their organization and operations.

The goal of the CPS is to define an interoperable city platform, which will allow cities to communicate and operate across silos and across communities.

> "A city is a system of systems and interactions that fosters emergent human behavior. It can be regarded as an arrangement of, and set of relationships between, multiple layers of a relatively large and permanent human settlement, with an administrative and legal status supported by local laws, and one that is recognized as such worldwide. The world of cities has ambiguous and porous political, economic and social borders, and governance is carried out locally, pragmatically and collaboratively. Since people's needs and the chronic global societal challenges are mainly fulfilled and dealt with in Cities, a common platform for generating solutions together and sharing knowledge within and between cities is needed" (City Protocol Society 2015).

A product developed by the CPS is called the *Anatomy of a City*. This document defines a common language describing the city ecosystem as three key system elements. As shown in Fig. 2 this includes: *Structure* - a set of physical structures; *Society* - the living entities that make up a city's society; and *Interactions* - the flow of interactions between structure and society. The Anatomy of a City helps understand and map interconnections between different city systems.

Each layer and its sub-layers are described below (City Protocol Society 2015). The top layer is *Structure*, which refers to physical constructions in a city, i.e., the building, streets, subways, and other three-dimensional macroscale edifices. The sub-layers include:

- Environment is the physical and geographic setting of the city, including the natural environment ("nature"). It is formed by the three basic elements: air, earth, and water, which interact dynamically in a seasonally dependable way.
- Infrastructure is the connective structures that enable people to get resources, especially from the environment, and bring them to the city or that enable flows or cycles within the city itself. These infrastructures include those that support communications, the water and energy cycles, the matter cycle that supports the movement of goods and food as well as the resultant waste, the mobility networks, and nature or green infrastructure.
- Built Domain is organized according to the approximate number of people that it can accommodate on a physical basis. Within the Built Domain, an object corresponds to a single person, and a dwelling, building, block, neighborhood, district, city and metropolis or region each increase the scale by an order of magnitude. Private and public spaces are contained within each level of scale.

The bottom layer is *Society*, which comprises the living entities of the city and ecosystem. The society sub-layers include:

- Citizens can be broken down into: person, family, visitors, organizations, and businesses.
- Governance refers to the process of running a government and focuses on the effectiveness of the executive branch of government. In city government, the head is normally the mayor.

The middle layer is *Interactions*, which includes the flow of interactions between the structure (top layer) and society (bottom layer). The Interaction sub-layers include:

- Functions including living, working, education, shopping, transport, caring for health, education, the performing arts, and more.
- Economy influences urban innovation and the everyday city operation and the life cycles of services provided by cities.
- Culture is the language, traditions, beliefs, values, and the way that people organize their concepts of the world etc., i.e., the non-material assets in the city anatomy.
- Information includes the City Operating System; performance indicators and indexes; tools and applications; city ontology; and an information portal for open data and specific learning protocols and related resources.

This ontology provides a useful construct for thinking about a wide range of cities and smart city communities, and enables mapping a wide variety of multimodal data from all sources. In Table 4, the characteristics of a megacity, as described in

**Table 4** Mapping city analytics framework and megacity characteristics

| Anatomy of a City Layer | | Megacity Characteristic |
|---|---|---|
| Structure | Environment | Infrastructure & Land Use |
| | Infrastructure | Density; Scale; Threats & Vulnerabilities; Flow; Connectedness |
| | Built Domain | Density; Scale; Infrastructure & Land Use |
| Society | Citizens | Density; Socio-cultural and Socio-economic Diversity; Threats & Vulnerabilities |
| | Government | Governance |
| Interactions | Functions | Socio-cultural and Socio-economic Diversity; Threats & Vulnerabilities; Flow |
| | Economy | Socio-cultural and Socio-economic Diversity; Threats & Vulnerabilities |
| | Culture | Socio-cultural and Socio-economic Diversity; Threats & Vulnerabilities |
| | Information | Density; Connectedness |

Table 1, are mapped to the city analytic framework. This shows that we can capture the unique aspects of a megacity and its tipping points with this ontology.

## 5.2 City Environment

Intelligence about the city environment is useable knowledge about the external environment that can be used for tactical and strategic decision making. This data or knowledge can come from many sources and in many formats. One source of data is from a smart city perspective. As mentioned earlier, the ITU identified eight categories of terms related to the definition of smart sustainable city (Kondepudi 2014). These include:

1. Quality of life and lifestyle;
2. Infrastructure and services;
3. Information & communications technologies, communications, intelligence and information;
4. People, citizen and society;
5. Environment and sustainability;
6. Governance, management and administration;
7. Economy and finance; and
8. Mobility.

Another source of data is from a military perspective. The six basic traditional intelligence sources identified by the Director of National Intelligence (DNI) (What is Intelligence 2017) include: Human Intelligence (HUMINT), Geospatial Intelligence (GEOINT), Imagery Intelligence (IMINT), Measurement and Signature Intelligence (MASINT), Signals Intelligence (SIGINT), and Open Source

**Table 5** Traditional military intelligence mapped to smart city dimensions

| Smart city dimensions | Military intelligence |
|---|---|
| Quality of life and lifestyle | HUMINT, MASINT, OSINT,MEDINT, IoTINT |
| Infrastructure and services | GEOINT, MASINT, OSINT, IoTINT |
| Information & communications technologies, communications, intelligence | OSINT, SIGINT, CYBIN/DNINT, IoTINT |
| People, citizen and society | HUMINT, OSINT, MEDINT, IoTINT |
| Environment and sustainability | GEOINT, MASINT, MEDINT, IoTINT |
| Governance, management and administration | HUMINT, OSINT, SIGINT, CYBINT/ DNINT, FININT, IoTINT |
| Economy and finance | OSINT, SIGINT, CYBIN/DNINT, FININT |
| Mobility | HUMINT, OSINT, IoTINT |

Intelligence (OSINT). A few additional types of intelligence-gathering disciplines have also been defined (List of Intelligence Gathering Techniques 2016).

In 2016, the Army Science Board (ASB) chartered a working group to look at Internet of Things. In that study, they identified a new type of emerging intelligence IoT data ("IoTINT"), when fused with other intelligence data, leads to better situational awareness and understanding for soldiers operating in megacities (ASB 2016). As described in their final report, IoT data can provide information such as vehicles, through position and location, traffic control, and vehicle occupancy; buildings via occupancy, occupant location, appliance status, stock (food, etc.) inventory, and sewer contents; and medical records, such as disease trends, vaccination status, and medical stock inventory.

Such information can provide detailed knowledge of the inner workings of a city, or at a much finer grain level, the routine of a single individual. In the case of a military action requiring precise interdiction or targeting, IoT surveillance can provide the warfighter with information on ingress and egress points, target locations, patterns of behavior, and differentiation among enemy "red" forces, unknown "grey" forces, and non-combatants.

The city analytic framework provides a construct for mapping this data, regardless of source, in a uniform way. Table 5 shows a notional mapping for how military intelligence categories can be used to collect data about a smart sustainable city. The traditional INTs defined by DNI do not provide enough fidelity to capture all the types of data in a smart city. Therefore, the broader set of intelligence categories defined in (List of Intelligence Gathering Techniques 2016) and (Army Science Board 2016) were used for the analysis.

The dimensions of a smart city (and indirectly military intelligence data) can then be mapped to the city analytic framework to show that, indeed, this framework is able to capture and represent the wide variety of multimodal data that will be important for improved situational awareness in megacities. This mapping is shown in Table 6.

**Table 6** Smart city dimensions mapped to city analytic framework

| City analytic framework | | Smart city dimensions |
|---|---|---|
| Structure | Environment | Environment and sustainability |
| | Infrastructure | Infrastructure and services; Mobility |
| | Built Domain | Infrastructure and services |
| Society | Citizens | People, citizen and society |
| | Government | Governance, management and administration |
| Interactions | Functions | Quality of life and lifestyle; Mobility |
| | Economy | Economy and finance |
| | Culture | People, citizen and society; Quality of life and lifestyle |
| | Information | Information & communications technologies, communications, intelligence and information |

## 6 Situational Awareness in Megacities

In the previous sections, we defined megacities and their attributes, the challenges that can result from instabilities, the multimodal data that can be known about megacities, and an analytic framework for organizing and understanding that data. The next step is how to use the analytic framework and data to provide better situational awareness and understanding before instabilities occur. This problem falls under the goals of the Department of Defense (DoD) Third Offset Strategy - a long-term competitive strategy to strengthen our military's competitive edge. The offset strategy is about finding the right combination of technologies, operational and organizational constructs to enable the US to maintain global supremacy and bolster conventional deterrence. The DoD's Third Offset strategy focuses on five key areas (Hicks et al. 2017):

- Autonomous learning systems. This includes using machines to make decisions in places where human reaction times aren't fast enough, e.g., indications and warnings in cyber defense. It also deep learning systems that are able to analyze big data sets (e.g., Facebook posts) and find patterns from it.
- Human-machine collaborative decision-making. This pairs a human and their ability to think on the fly with a computer's quick problem-solving methods using artificial intelligence, so as a team better, faster decisions can be made.
- Assisted human operations. This area includes wearable electronics, apps, heads-up displays, and even exoskeletons to help warfighters in all possible situations.
- Advanced manned-unmanned systems operations. Instead of a machine and human collaborating to make better decisions, this area is focused on humans and unmanned systems working together as combat teams to perform missions.
- Network-enabled autonomous weapons and high-speed projectiles. Everything online today must be prepared for cyber-attacks. In the event that an attack occurs, military systems must still be able to operate without connectivity, e.g., without GPS.

Megacities represent new challenges to situational awareness, in that traditional stove-piped approaches to visualize and understand systems, people, and their interactions cannot be relied on. Therefore, autonomous learning systems, human-machine collaboration, and assisted human operations are all part of improving situational awareness. They are key for helping the warfighter understand the multidimensional, interconnected, and uncontrollable elements of the complex environments they will be operating in. The following sections will cover a few of the technologies that can be used to assist the warfighter in this challenge.

## *6.1 Intelligent Operations Using Data*

To effectively operate in megacities, the warfighter will need access to data from multiple sources and have tools to correlate it accurately, analyze it rapidly, and visualize it to enable fact-based decision-making. In today's world, that leads to the discipline of data science and analytics. Social, mobile and cloud, analytics and data technologies are the core disruptors of the digital age. The DoD has been slow to adopt emerging technologies, instead waiting years for them to mature. These disruptors, however, have caused the rate of change of innovative solutions to increase; as people iterate on these technologies they drive new solutions at a much faster pace. Now is the time to leverage these new data technologies to make the right information, available to the right people, at the right time, to solve the warfighters need. Below are a number of current data trends that are relevant for transforming the military's understanding and use of data.

- Artificial Intelligence and Machine Learning: Rich data representations of the world that can be used to better learn new classification and prediction tasks over time from many related tasks.
- Augmented Reality: A live direct or indirect view of a physical, real-world environment whose elements are augmented by computer-generated sensory input such as sound, video, or GPS data.
- Automated Composition Engines: Solutions that combine finding, analyzing, and assembling data into natural language formats, improving the speed and quality of content creation and delivery.
- Big Data Analytics: Process of examining large data sets containing a variety of data types to uncover hidden patterns, unknown correlations, market trends, user preferences, and other information.
- Blockchain: Public ledger/record of digital events shared between different parties or computing nodes that are geographically and computationally isolated; a record can only be updated if there is a consensus of the majority of the participants in the system.
- Crowd Sourcing: Obtaining information or input about a particular task by enlisting the services of a number of people (paid or unpaid) via the Internet or some closed set of users.

- Data Discovery & Preparation: User-generated data discovery tools reduce time and complexity of preparing data for analysis, existing tools include Tableau, Alteryx, Trifacta, Paxata, and Lavastorm.
- Democratization of Data: More easily collect data from around the world; user-friendly tools enable people to parse the information to secure meaningful value.
- FOG Computing: Enable smart machines to perform part of the analytics locally and only send the prepared data to the cloud.
- Identity Management & Enterprise Standards: Security that provides a system for enforcing fine-grained, role-based authorization to data and metadata stored on machines.
- Internet of Things: Pervasive presence around us of sensors, devices, and machines that communicate and collaborate to provide services
- Predictive Analytics: Using techniques from data mining, statistics, modeling, machine learning, and artificial intelligence to analyze current data to make predictions. This can include predicting uncollected data, latent variables, or future events.
- Reputational Feedback Mechanisms: Buyers' and sellers' rate one another and share information about their interactions. Examples of this include Uber, eBay, and eCommerce.
- Smart Machines/Devices: Fully automating tasks and removing human control; augmenting the cognitive and physical performance of individuals that feels like an extension of their own abilities.

The military intelligence community will need to piece together a comprehensive and actionable intelligence picture from millions of bits of data, such as cellphone communications, social media postings, and financial transactions. They will need to develop urban analytics that help them understand the connectivity of the modern urban center, and leverage the power of graph analytics to enable a city's complexity to be thought of as a network of networks. They will need tools that enable multidimensionality and visualization, so they can track people in three-dimensional space (e.g., subterranean through high rise tower). The challenge is the ability to manage and interpret the unending stream of data and represent it in a way that is meaningful and timely. The failure to do so will "exacerbate the difficulties associated with operating in megacities, prolong conflicts therein, and create circumstances where hostile groups can exploit physical and virtual sanctuaries largely unobserved by US forces" (Serena and Clark 2016).

## 6.2 *Models, Simulations and Wargames*

In a February 2015 memo, Deputy Secretary of Defense (DepSecDef), Robert Work said he expects wargames to "pursue an innovative third offset strategy, avoid operational and technological surprise, and make the best use of our limited resources." The DepSecDef believes wargames can be used to "better address threats that would benefit from a more global strategic perspective" (CSIAC 2016). So how do wargames apply to megacities and situational awareness?

A wargame is a strategy game that deals with military operations of various types, real or fictional. When used by the military to study warfare, a wargame could refer to a simple theoretical study or a full-scale military exercise. Wargames explore and represent some feature or aspect of human behavior directly bearing on the conduct of war, even if the game subject itself does not concern organized violent conflict or warfare (Wargaming 2017).

Wargames can be non-computational, such as board games, or they can use computer-based models and simulations for their execution. There are several definitions of models, simulations and the Modeling and Simulation (M&S) discipline. The US DoD defines these terms as follows in their online glossary (MSCO 2011):

- Model: a physical, mathematical, or otherwise logical representation of a system, entity, phenomenon, or process.
- Simulation: a method for implementing a model and behaviors in executable software.
- M&S: the discipline that comprises the development and/or use of models and simulations.

Simulation is one of the fundamental intellectual tools, alongside mathematics and experimentation, for understanding the behavior of both natural and artificial systems. M&S applications are rapidly increasing in scale and complexity as systems become more complex and interconnected. For example, consider the use of M&S to understand megacities. As discussed previously, one must view cities as a system, which means modeling the interdependencies among critical infrastructures such as transportation, water, and energy, as well as interactions with social processes and policy. Each of these systems and infrastructures is a large, complex adaptive system in its own right. Creating simulation models that capture the behaviors and interactions among these infrastructures and social-economic processes is even more challenging (Fujimoto 2017).

M&S is complementary to the exploitation of data analytics: "While models derived purely from data analytics offer much benefit, they do not include behavioral descriptions of the system under investigation that are necessary for prediction of dynamic system behaviors that are necessary for what-if experimentation, or analysis of situations where sufficient data, or the right data are not available, e.g., due to privacy or other concerns" (Fujimoto 2017). Further, cloud computing and high performance computing capabilities now make M&S technologies more broadly accessible, enabling the embedding of simulations into operational environments.

So how can simulations and wargames be used to improve situational awareness. The sheer volume of people, vehicles and buildings present in megacities can complicate an already challenging mission (Serena and Clark 2016). The electronic and cyber fog formed by the pervasive presence of cheap and encrypted mobile communications devices can overwhelm the military's optical-electrical surveillance platforms and limit its capacity for effectively targeting and tracking the activities of hostile actors. Simulation can be used to evaluate courses of action for specific missions, enabling warfighters to determine the most effective alternatives for accomplishing their goals.

Simulation is also effective for dealing with the issue of scale found in a megacity. When deploying to a city, it is impractical for warfighters to physically control a city or urban population of tens of millions of people spread over hundreds of square miles. Instead, they will be focused on a subset of the city - a mission corridor – that relates to the task at hand. For example, if the goal is to provide humanitarian assistance, the mission corridor would be the part of the city where food and supplies are stationed, and how those resources are then transported and delivered to civilians is part of the simulation task. Similarly, if the mission is reconstruction after a natural disaster, the mission corridor would be the part of the city where the damaged infrastructure is located. With simulation, a low-fidelity view of the city as a whole can be provided, adding detail and resolution only to the mission corridor, as required, to enable a better understanding of behavior and courses of action.

Modeling and simulation has many uses for understanding a megacity, including:

- Wargaming the complexity of the city for the warfighters advantage
- Assessing the impact of combinatoric complexity on courses of action
- Determining location and types of sensors needed for mission specific actions
- Forecasting or predicting city triggers and tipping points using real world sensor data (e.g., IoT)
- Real-time course of action analysis and re-planning (e.g., during a mission)
- Training, testing, and mission rehearsal

Computer simulation has been around since the mid-1940's, and modern distributed simulation has been around since the 1990's. However, the underlying computing platforms and technologies exploited by M&S have undergone dramatic changes in the last decade. These advances create new opportunities, and challenges, for modeling and simulation to achieve even greater levels of impact. For megacities, the challenge remains for how to model massive scale connected cities and validate those simulations so we can ascribe some credibility to the answers they produce. Other challenges include how to integrate autonomous learning into a simulation, enabling human-simulation collaboration. For example, humans and simulated agents collaborating in a wargame, so as a team they can make better, faster decisions.

## *6.3 Examples of Future Capability*

Three use cases are described that illustrate how the emerging trends in data and simulation can be used to provide the warfighter with the right data at the right time, in order to shape favorable mission outcomes in megacities.

### 6.3.1 Case 1: Situational Awareness

A warfighter is patrolling a location in a city. They hold up a device on which they see overlays of many types of data: restaurants, businesses, hotels, areas of interest, recent crime stats, social activity, frequent visitors, staff, previous patrol history, crowd sourced data, etc. Some of this data may have been provided by previous warfighters or security personnel on past patrols; other of the data could have been crowd sourced or collected from local or state governments. Based on the information portrayed on the device, the warfighter can request additional data (e.g., search a repository) based on the specific mission they are executing. The warfighter may also request data from nearby sensors that are deployed in the city (e.g., IoT, FOG). Once the data has been integrated into the device and decision making process, the warfighter annotates the overlays with current information from their patrol, using a voice or written interface. They also rate the quality of the data and overlays based on its relevance to their mission. The data is stored in a repository, and shared with appropriate people at appropriate level of trust.

### 6.3.2 Case 2: Courses of Action

A warfighter encounters a situation that requires some decision for which they don't understand the best course of action (e.g., take route x vs y, enter shop a vs b). They hold up their device to acquire situational awareness (as described in Case 1). Based on the data presented, the warfighter states a question and launches predictive analytics tool. The data returned is quickly turned into a visualization of options, providing the warfighter with several trade-offs. If they are convinced which action to take, they can crowd source additional data from warfighters in the area, or those that have performed this mission recently. Once the warfighter makes a decision, they execute the mission and rate the options based on their context. The mission and outcome can be entered into a blockchain, such that the information is digitally signed by the warfighter and available for others to use in the future.

### 6.3.3 Case 3: Training

A warfighter approaches a situation (or is assigned a mission) that requires a particular skill set or knowledge. They use their device to gain situational awareness (as described in Case 1). Using cognitive reasoning techniques, the device determines the skills needed for the situation / mission; matches these skills against the personnel records, context and situational awareness data; and then creates training set for the warfighter. The device then searches the simulation repository for appropriate training simulations, and searches the data repository for an appropriate data set. The device launches a composition engine to automatically create/author scenarios based on current data. The training is then launched from cloud. The device can use augmented reality, local sensors, and devices as needed to supplement the training.

## 7 Conclusion

Much work in megacities is from an US Army operations perspective or from the World Bank related to sustainability. The rise of smart cities has spurred work in defining the anatomy of a city; this is from an urban design and city operations perspective, but it can be applied to megacities. Military intelligence approaches often focus on areas of operation which emphasize discrete problem sets and well-defined regions. The megacity city is neither. There is a need for a common intelligence framework that captures military operations, urban operations, emergency response operations, and city behavior. From this framework, multimodal data from a variety of sources, using new or existing technologies can be captured and used for improved situational awareness. Future situational awareness techniques will be data focused; non-traditional data sources, fusion, visualization, and prediction will all be critical in the future.

**Acknowledgements** Thank you to Georgia Tech's Sam Nunn Security Fellows program for encouraging me to think about this problem. Thanks to Ben Riley for numerous conversations on this topic, as well as the idea of City as a Sensor. This work was motivated by conversations with Dr. Tom Burns of DARPA, looking at the intersection of smart cities, megacities and gray zone operations. The author gratefully acknowledges the Georgia Tech Research Institute, who supported this work as part of the Distributed Urban Sensing Technologies (DUST) strategic initiative.

## References

Army, US. ( 2014). Intelligence Preparation of the Battlefield/Battlespace, ATP 2-01.3 MCRP 2-3A.

Army Science Board, (2016). The Military Benefits and Risks of the Internet of Things. Army Science Board Fiscal Year 2015 Study, Department of the Army Office of the Deputy Under Secretary of the Army Washington, D.C. 20310–0103.

Asian Development Bank. (2011). Climate-Induced Migration in Asia and the Pacific. Available at:. https://www.adb.org/features/climate-induced-migration asia and pacific [Accessed 20 June 2017].

Azad, S., & Gupta, A. (2011). A quantitative assessment on 26/11 Mumbai attack using social network analysis. *Journal of Terrorism Research, 2*(2).

Bailey, M. M., Dixon, L. R., Harris, C. M., Melin, C. S. M. D. H. M. A. J., & Russo, S. R. (2014). A Proposed Framework for Appreciating Megacities: A US Army Perspective. *Journal Article April, 21*(9), 12.

CNN. (2016). Mumbai Terror Attacks Fast Facts. *CNN website*. Available at:. http://www.cnn.com/2013/09/18/world/asia/mumbai-terror-attacks/index.html [Accessed 20 June 2017].

City Protocol Society. (2015). City Anatomy: A Framework to support City Governance, Evaluation and Transformation. City Protocol Agreement (CPA-I_001-v2). Available at: http://cityprotocol.org/publications/ [Accessed 20 June 2017].

CSIAC. (2016). Wargaming Introduction. Cyber Security and Information Systems Information Analysis Center (CSIAC) Modeling and Simulation Journal, Special Edition on Wargaming, Vol 4, no 3. Available at:. https://www.csiac.org/journal-issue/modeling-and-simulation-special-edition-wargaming/ [Accessed 4 July 2017].

Dias, C. N., & Salla, F. (2013). Organized crime in Brazilian prisons: the example of the PCC. *International Journal of Criminology and Sociology, 2*, 397–408.

Dignan, J., Shah, N., & Tunvall, F. (2013). *Deriving Insight from Data for Smarter Urban Operations*. Europe: Ovum. http://www-01.ibm.com/software/city-operations/offers/Intelligent%20Urban%20Operations%20Whitepaper%20-%20IBM_Ovum%20branded%20FINAL.PDF. [Accessed 20 June 2017].

Fragile state. (2017). In Wikipedia, The Free Encyclopedia. Retrieved July 4, 2017, from https://en.wikipedia.org/w/index.php?title=Fragile_state&oldid=786573098.

Fujimoto, R., et al. (Eds.). (2017). *Research challenges in modeling and simulation for engineering complex systems*. Cham: Springer.

Giddens, A. (1979). *Central problems in social theory: Action, structure, and contradiction in social analysis* (Vol. 241). Univ of California Press.

Government of India. (2017). Smart Cities Mission. Available at:. http://smartcities.gov.in/ [Accessed 4 July 2017].

Harris, M., Dixon, R., Melin, N., Hendrex, D., Russo, R. and Bailey, M. (2014). Megacities and the United States Army: preparing for a complex and uncertain future. Chief of staff of the Army strategic studies group Arlington VA.

Hedges, W. (2016). An Analytic Framework for Operations in Dense Urban Areas. *Journal Article Mar*, *11*(6), p.29 am.

Heilig, G.K. (2012). World urbanization prospects: the 2011 revision. United Nations, Department of Economic and Social Affairs (DESA), Population Division, Population Estimates and Projections Section, New York.

Henderson, B. (2013). Mumbai terror attacks: the making of a monster. *The Telegraph [online], 12*. Available at: http://www.telegraph.co.uk/news/worldnews/asia/india/9985109/Mumbai-terror-attacks-the-making-of-a-monster.html. Accessed 20 June 2017.

Henderson, P. (2015). 10 Ways analytics can make your city smarter. *SAS Insights* [online]. Available at:. http://www.sas.com/en_us/insights/articles/big-data/10-ways-analytics-can-make-your-city-smarter.html [Accessed 27 June 2017].

Hicks, K., Hunter, A., Ellman, J., Samp, L., & Coll, G., (2017). *Assessing the third offset strategy*. A Report of the CSIS International Security Program, Center for Strategic and International Studies. Available at: https://www.csis.org/analysis/assessing-third-offset-strategy. Accessed 4 Jul 2017.

IESE. (2014). *IESE Cities in Motion Index 2014. Center for Globalization and Strategy*. IESE Business School: University of Navarra. http://www.iese.edu/en/multimedia/ST-0333-E_tcm41-159595.pdf [Accessed 20 June 2017.

Joint Pub. (2017). Joint operations, Joint Publication 3–0.. Available at: http://www.dtic.mil/doctrine/new_pubs/jp3_0.pdf [Accessed 20 June 2017].

Kazan, C.. (2009). Sprawl! Is Earth Becoming a Planet of Supercities? *The Daily Galaxy* [online].. Available at: http://www.dailygalaxy.com/my_weblog/2009/06/is-earth-becoing-a-planet-of-supercities-leading-expert-says-yes.html [Accessed 20 June 2017].

Kilcullen, D. J. (2012). The city as a system: Future conflict and urban resilience. *Fletcher F. World Aff., 36*, 19.

Kondepudi, S.N. (2014). Smart sustainable cities analysis of definitions. *The ITU-T Focus Group for Smart Sustainable Cities*.

Kötter, T., & Friesecke, F. (2009). *Developing urban indicators for managing mega cities. Land Governance in Support of the MDGs: Responding to New Challenges*. USA: Washington DC.

Li, Y., Lin, Y. and Geertman, S. (2015, July). The development of smart cities in China. In *Proceeding. of the 14th International Conference on Computers in Urban Planning and Urban Management* (pp. 7–10).. Available at: http://web.mit.edu/cron/project/CUPUM2015/proceedings/Content/pss/291_li_h.pdf [Accessed on 20 June 2017].

List of intelligence gathering disciplines. (2016). In Wikipedia, The Free Encyclopedia.. Retrieved July 4, 2017, from https://en.wikipedia.org/w/index.php?title=List_of_intelligence_gathering_disciplines&oldid=741041290.

Mazarr, M.J., 2015. Mastering the Gray Zone: Understanding a Changing Era of Conflict. Strategic Studies Institute and US Army War College Press.

Manyika, J. (2015). The Internet of Things: Mapping the value beyond the hype. McKinsey Global Institute. Available at:. http://www.mckinsey.com/insights/business_technology/the_internet_of_things_the_value_of_digitizing_the_physical_world [Accessed 17 June 2017].

Mehta, N. (2015). What we can learn from Paris. *The Times of India* [online]. http://blogs.timesofindia.indiatimes.com/academic-interest/what-we-can-learn-from-paris/ Retrieved June 20, 2017.

Military intelligence. (2017). In Wikipedia, The Free Encyclopedia. Retrieved July 4, 2017, from https://en.wikipedia.org/w/index.php?title=Military_intelligence&oldid=787298595.

MSCO. (2011). DoD Modeling and Simulation (M&S) Glossary. Available at:. https://www.msco.mil/MSReferences/Glossary/MSGlossary.aspx [Accessed 4 July 2017].

National Intelligence Council (US). (2008). Global trends 2025: a transformed world. National Intelligence Council.

Ogbonnaya, U. M. (2013). Arab Spring in Tunisia, Egypt and Libya: A comparative analysis of causes and determinants. *Alternatives: Turkish Journal of International Relations, 12*(3).

Oh, O., Agrawal, M., & Rao, H. R. (2011). Information control and terrorism: Tracking the Mumbai terrorist attack through twitter. *Information Systems Frontiers, 13*(1), 33–43.

Patterson, D. (2015). Cities first to benefit from Internet of Things, if we can write better software. *Tech Republic* [online].. Available at: http://www.techrepublic.com/article/vint-cerf-cities-first-to-benefit-from-internet-of-things-if-we-can-write-better-software/ [Accessed 27 June 2017].

PCAST. (2016). Technology and The Future of Cities. Executive Office of the President, Presidential Council of Advisors on Science and Technology (PCAST). Available at:. https://www.whitehouse.gov/sites/whitehouse.gov/files/images/Blog/PCAST%20Cities%20Report%20_%20FINAL.pdf [Accessed 4 July 2017].

Riedel, B. (2015). Modeled on Mumbai?: Why the 2008 India Attack Is the Best Way to Understand Paris. *Brookings Blog, November, 14*.. Available at: https://www.brookings.edu/blog/markaz/2015/11/14/modeled-on-mumbai-why-the-2008-india-attack-is-the-best-way-to-understand-paris/ [Accessed 20 June 2017].

Serena, C. and Clark, C., (2016). A New Kind of Battlefield Awaits the US Military — Megacities. *Reuters, The Great Debate blog* [online].. Available at: http://blogs.reuters.com/great-debate/2016/04/06/is-the-u-s-military-prepared-to-fight-in-megacities/ [Accessed 27 June 2017].

Turchin, P. (2013). Modeling social pressures toward political instability. *Cliodynamics: The Journal of Theoretical and Mathematical History, 4*(2).

United Nations, Department of Economic and Social Affairs. (2017). *World Population Prospects: the 2010 Revision*. Available at. http://esa.un.org/wpp/Documentation/publications.htm [Accessed 20 June 2017].

USAID. (2013). Sustainable Service Delivery in an Increasingly Urbanized World, United States Agency for International Development (USAID) Policy.. Available at: https://www.usaid.gov/sites/default/files/documents/1870/USAIDSustainableUrbanServicesPolicy.pdf [Accessed 4 July 2017].

Wargaming. (2017). In Wikipedia, The Free Encyclopedia.. Retrieved July 4, 2017, from https://en.wikipedia.org/w/index.php?title=Wargaming&oldid=788691680.

What is Intelligence? (2017). In Office of the Director of National Intelligence, Retrieved 13 Nov 2017, from https://www.dni.gov/index.php/what-we-do/what-is-intelligence.

Wilbanks, T. (2007). The research component of the community and regional resilience initiative (CARRI). Presentation at the Natural Hazards Center, University of Colorado-Boulder.

Wolfel, R., Richmond, A., Read, M., & Tansey, C. (2016). It's in There: Rethinking Intelligence Preparation of the Battlefield in Megacities/Dense Urban Areas. *Small Wars Journal.*. Available at: http://smallwarsjournal.com/jrnl/art/it's-in-there-rethinking-intelligence-preparation-of-the-battlefield-in-megacitiesdense-urb. [Accessed 4 July 2017].

Zolli, A. and Healy, A.M. (2013). Resilience: Why things bounce back. Simon and Schuster.

# Augmenting Intelligence: What Augmented Reality Technology Means for the Intelligence Community

**Meghan A. Check**

**Abstract** In addition to its usefulness in industrial and military environments, augmented reality has plenty of intelligence applications. Mapping vast datasets like social networks over geospatial information in real time to individual users or teams has the potential to make fieldwork more manageable and less risky. This paper details the origins and development of augmented reality systems, their current planned uses, and their potential uses. Its objective is to make informed predictions about how the intelligence community will adopt augmented reality systems, the challenges they will face in implementation as well as address broader ethical concerns and the potential to influence the global balance of power.

## 1 Introduction

One of augmented reality's earliest notable publicly available applications was the much anticipated but ultimately unsuccessful Google Glass. Critics of the product raised concerns about privacy and security, and it was scrapped with no clear replacement (Barr 2015). More recently, apps like Snapchat and Instagram have brought a more simplistic version of augmented reality to the masses. While augmented reality technology may seem cutting edge to the public, the Department of Defense has been funding the development of military-grade augmented reality systems that will likely never be accessible to ordinary consumers.

The state of augmented reality (AR) technology today is the result of years of private and public research and development. In addition to the main private sector tech giants like Microsoft, Apple, and Google, and defense industry institutions like Boeing, Lockheed Martin, and BAE Systems, start-ups have recently entered the scene, sometimes working in collaboration or partnership with the public sector. The Defense Advanced Research Project Agency (DARPA), the Office of Naval Research for Science and Technology (ONR), Army Research Laboratory (ARL), and the Central Intelligence Agency are all players in AR development. The United

M. A. Check (✉)
Georgia Institute of Technology, Atlanta, Georgia, USA



M. E. Kosal (ed.), *Technology and the Intelligence Community*, Advanced Sciences and Technologies for Security Applications,
https://doi.org/10.1007/978-3-319-75232-7_13

**Fig. 1** Simplification of Milgram's reality-virtuality continuum (Milgram et al. 1995)

States intelligence community's venture capital arm, In-Q-Tel, has commissioned startups to help the government come out ahead of the curve in technological innovation (Marshall 2013).

## 2 Technology Overview

Augmented reality (AR) is a variation of virtual environments (EV) "that enhances a user's perception of and interaction with the real world" (Azuma 1997). "AR can be thought of as the "middle ground" between VE (completely synthetic) and telepresence (completely real) (Azuma 1997; Milgram and Kishino 1994; Milgram et al. 1995)." One source defines AR as any system that combines the following three characteristics, it: combines real and virtual, is interactive in real-time, and is registered in three dimensions (Azuma 1997). Although, augmented reality can apply to senses other than sight (such as touch or smell); for the purpose of this paper, the augmented reality technologies discussed here will be primarily visual (Carmigniani et al. 2010).

A helpful way to conceptualize augmented reality is through Milgram's Reality-Virtuality Continuum, illustrated in Fig. 1.

### *2.1 History*

Although largely regarded as a new technology, the concept of augmented reality is a relative of virtual reality, which dates back over fifty years. Cinematographer Morton Heilig envisioned an interactive cinematic experience; in 1962 he built a prototype, The Sensorama, which was an enhanced simulator machine designed to engage all of the viewer's senses (Heilig 1962). In 1968, computer scientist and MIT Ph.D., Ivan Sutherland made the leap from virtual reality to augmented reality with a see-through, head-mounted display known as the "The Sword of Damocles," thus combining the virtual world with the real world (Sutherland 1968). Sutherland did much of his early research at Harvard University with funding from the US Air Force, the Central Intelligence Agency, and Bell Laboratories. Fred Brooks and Henry Finch at the University of North Carolina also conducted early research (1971) in the field on grants from the US Atomic Energy Commission and the National Science Foundation (Myers 1998). David Mizell and Tom Caudell of Boeing coined the term "augmented reality" sometime after visiting a Boeing engineering and manufacturing facility in Everett, Washington in 1990. While observing the assembly of electrical wire bundles, Mizell wondered aloud if a virtual reality-like headset would alleviate the inefficiency of using non-uniform foam-board diagrams. Mizell and Caudell envisioned a system with three main components: a see-through head-mounted display, a head position/orientation tracker, and a wearable computer; they initially referred to the new project as "see-through virtual reality" (Barfield and Caudell 2001). Louis Rosenberg's "Virtual Fixtures" was an early augmented reality system designed to assist operators with tasks in remote (but not virtual) environments, such as manufacturing setups involving human-operated robotics (Rosenberg 1993).

In the 1990s, DARPA funded a project, GRIDS: Geospatial Registration of Information for Dismounted Soldiers, which was the precursor to the systems of today although, it encountered problems when operating outdoors (Azuma 1999). Similarly, Steven Feiner, Blair MacIntyre, and Doree Seligmann's augmented reality system prototype, KARMA (Knowledge-based Augmented Reality for Maintenance Assistance) debuted in 1993 (Feiner et al. 1993). Augmented reality remained within the industrial and defense research spheres for several years before breaking into the gaming world in the early 2000s with ARQuake, a first-person shooter game that functioned on a wearable computer platform (Thomas et al. 2002; Carmigniani et al. 2010). Although augmented reality for medical purposes was first considered in the early 1990s, it did not make its debut in operating rooms until 2012, in the form of Freehand SPECT, an intraoperative nuclear imaging modality (Navab et al. 2012).

## 2.2 *How AR Works*

Augmented reality (AR) is not solely a device, but rather a *system* made up of multiple components that achieve a functional purpose by working in concert with one another. AR systems typically include the following components: GPS (Global Positioning System); inertial sensors such as gyroscopes or accelerometers; active sources (active transmitters and receivers); passive optical recorders such as a camera with video capability; electronic compass and tilt sensors; terrain information, and a network through which all of the AR components communicate.

The system works by registering the longitude, latitude, and elevation of objects within the user's field of vision, transmitting that the geo-registered information to the associated network, and storing the related information either for display or use later (Madrigal 2014) (Fig. 2).

Despite the progress that augmented reality has enjoyed over the past few decades, the technology still faces a number of challenges, particularly in outdoor environments. For instance, the display must be focused at infinity essentially, which presents a challenge for reaching an adequately sized field of view as well as sufficient brightness to operate outdoors. Additional challenges include "pose estimation," or the ability of the AR device to accurate determine the user's location so that the projected images and information are accurate, as well as maintaining an

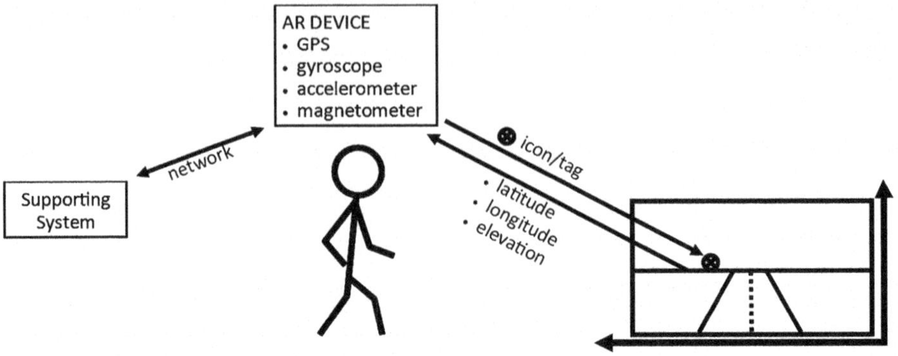


**Fig. 2** A rendering of how augmented reality technology functions in the field

accurate azimuth estimate near magnetic disturbances stronger than the earth's magnetic field (Madrigal 2014).

### *2.3 Related Technologies*

Another area of technological research and development that falls under the umbrella of augmented reality in the broadest sense of the term is "human performance modification," (HPM) which "can encompass actions ranging from the use of 'natural' materials, such as caffeine or khat as a stimulant, to the application of nanotechnology as a drug delivery mechanism or in an invasive brain implant" (National Research Council 2012).

As far as Human Performance Modification (HPM) goes, the Department of Defense currently employs HPM to return soldiers to their baseline via regular courses of action like medical care for wounded soldiers or the use of caffeine to help war-fighters stay awake but as of yet has not seriously explored augmented humans to preform beyond their natural capacities (Brimley et al. 2013).

## 3 Methodology

Since augmented reality is a developing field and therefore provides too limited data for quantitative analysis, this chapter is qualitative in nature and exploratory in design. It will familiarize the reader with history and functionality of augmented reality technology and with the degree to which the technology is expected to be developed beyond the prototype stage, generate predictions about the role augmented reality technology can play in intelligence operations based on the nature of the technology's existing and planned applications. Through consideration of the ethics of dual use technologies as well as emerging global trends in innovation, this chapter establishes a jumping-off point for more in-depth research (Bhattacherjee 2012).

## 4 Current Applications

### *4.1 Industrial*

As described earlier, modern augmented reality technology's first practical application was manufacturing assistance. In a survey conducted by PriceWaterhouse Coopers, over one third of manufacturers indicated that they use virtual and augmented reality technology for product design and development while nearly one third used it for safety and manufacturing skills training. Other uses by manufacturers include: maintenance, repair or equipment operation; data and information access; remote collaboration; virtual assembly/improved process design; customer engagement and communications; and supply chain collaboration/communications (PricewaterhouseCoopers 2016).

### *4.2 Consumer*

One of the most well-known examples of augmented reality technology in recent popular culture was Google Glass. The initial consumer venture failed, but Google is pursuing other augmented reality applications (Google Developers 2016). Despite their early prominence in the field, Google is far from the only private sector company that offers augmented reality technology tools and applications. Many smartphone and tablet apps provide information about local attractions, such as restaurant ratings or crime; other apps, such as PokemonGo, Snapchat, or Instagram, offer games or filters that use the smartphone's camera and superimpose premade images over photos or recordings. It is worth noting that many of these apps are rudimentary at best when compared to the systems available to industry and the military. Despite this, the development of consumer-geared augmented reality applications contributes to the innovation and mobilization of AR technology.

### *4.3 Military*

Training is augmented reality's primary military application thus far. In fall of 2015, Marines at Marine Corps Base Quantico incorporated augmented reality technology into a live-fire training exercise. The Office of Naval Research developed the technology used, the Augmented Immersive Team Trainer (AITT), which includes a laptop, software, battery pack, and helmet-mounted display; although it is not dissimilar from virtual reality headsets, AITT superimposes virtual objects onto the actual environments versus digital ones (Fabey 2015). The value in the AITT program is that it enables soldiers to experience battlefield conditions while still in training, setting realistic expectations about and reactions to what they will encounter when deployed.

New uses for augmented reality technology in the military include Lockheed Martin's F-35 Lightning II helmet that enhances the pilot's situational awareness by showing sophisticated warning signals and even allowing the pilot to essentially "see through" the aircraft, eliminating the disadvantages of a stationary perspective (Wolverton 2013).

In addition to the already-deployed augmented reality tools in the military, many more are in the works. DARPA's Urban Leader Tactical Response, Awareness, and Visualization (ULTRA-Vis) program is the result of roughly five years of research and development. DARPA worked with two contractors for the project, Applied Research Associates for the software, and BAE Systems for the hardware. It enables soldiers to "visualize the location of other forces, vehicles, hazards, and aircraft in the local environment even when these are not visible to the soldier. In addition, the system can be used to communicate to the solider a variety of tactically significant (local) information including imagery, navigation routes, and alerts" (Madrigal 2014).

# 5 Emerging Applications

## *5.1 SPACE*

In addition to traditional military and defense applications, augmented reality technologies have been slated for deployment in space. Boeing Phantom Works teamed up with NASA to create an augmented reality device for use aboard the International Space Station. Augmented reality technology would be especially useful in space when there is often a time delay in communication, such as between the International Space Station and Earth or in the future, Mars and Earth; AR would enable astronauts to access technical instructions quickly when time is of the essence (Memi 2006). Since NASA's expertise with augmented reality is mainly related to the system software, they've enlisted Osterhout Design Group (ODG) to develop the hardware; the technology is scheduled for terrestrial testing before it makes its way into space (Tilley 2015).

### 5.2 Emergency Response

Augmented reality has great potential for emergency and disaster management services. In a sector that already relies heavily on geospatial information systems to complete its objectives, much of the groundwork for augmented reality has already been laid. Identifying areas whose topography was altered in the wake of a disaster is crucial for providing timely and safe emergency responses. Furthermore, search and rescue as well as less immediate rehabilitation and reconstruction operations rely on access to the most current maps and situational awareness information. The ability for emergency workers to update digital maps and tag objects or locations of concern would increase efficiency while minimizing the operational impact of obstacles (Doolin et al. 2013).

### 5.3 Intelligence

Although there are many military applications for augmented reality that have equal usefulness within the intelligence community, the full potential of AR technology in intelligence work has not yet been explored. Relatively unbound by the rigid hierarchy and regimented procedures of the military, the intelligence community has the latitude to use augmented reality in a host of unique situations and environments.

By combining the technological capabilities of augmented reality with geospatial information and intelligence as well as existing intelligence on social networks on a platform operated by a human, the intelligence community has the potential to not only greatly increase the efficiency of their intelligence gathering operations but to prioritize targets in real-time and analyze on-the-go.

It is important to consider that since intelligence operations rely in some part on relevant already gathered intelligence (such as GEOINT or IMINT), the accuracy of the input information as well as the precision of the information gathered by the AR device is crucial to producing a quality intelligence product. Furthermore, AR devices require an individual or 'human in the loop' to make sense of the input information as well as to operate the device; augmented reality is not a substitute for human involvement but rather a way to enhance an individual's ordinary capability and make them a more effect operative.

Figure 3 shows the potential interaction between the intelligence disciplines and augmented reality. The figure is based on the current visual functionality of augmented reality and has the potential to include additional disciplines as technology advances. Geospatial intelligence (GEOINT) is the primary intelligence requirement for augmented reality to function in an intelligence related capacity; imagery intelligence (IMINT) could be useful alone or combined with GEOINT to collect updated GEOINT and possibly new IMINT. Since AR requires a 'human in the loop,' the contextualization that the aforementioned GEOINT and IMINT could

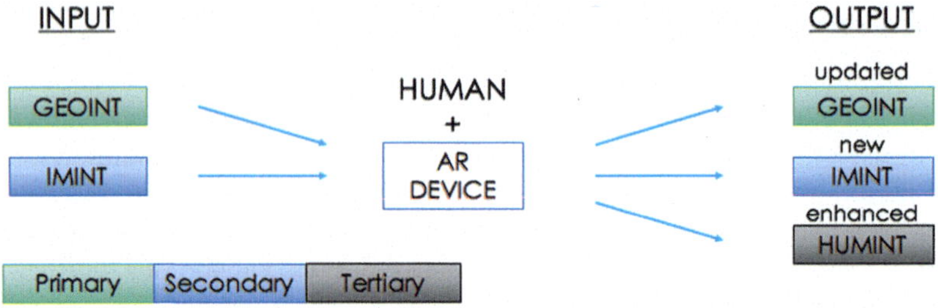


**Fig. 3** The interaction of intelligence disciplines with AR technology

bring to human intelligence (HUMINT) means that HUMINT collected by the AR device and operator would be of a higher grade than information collected manually.

Many of the potential opportunities for augmented reality in the intelligence field are underpinned with potential pitfalls. Web GIS and freedom of data could seriously advance the potential for use of AR in denied areas, but the 'crowd-sourcing' of data for intelligence purposes is inherently dangerous; much like OSINT (open source intelligence), this information would rely heavily upon human analysis to ascertain its validity, possibly sacrificing efficiency and hampering the objective. Furthermore, while cloud computing could make AR much more mobile, operating on a more vulnerable network poses a cyber security risk (Sensor and Systems 2013).

The recent advances in augmented reality technology make it an attractive investment for the intelligence community. As is increasingly the case in a world growing steadily more dependent on technology, it is necessary to carefully weigh both the benefits and risks of new technologies prior to their deployment.

## 6 National Security Implications

### *6.1 Ethics*

The introduction of advanced technologies to new applications is often fraught with ethical dilemmas. "Disruptive" technologies are the science fiction of the past; those that have only recently or perhaps not yet come into existence pose difficult questions about their potential to which we do not yet have the answer (Brimley et al. 2013). For instance, the fact that many disruptive technologies are "dual use," and have both civilian and intelligence applications, poses an ethical dilemma.

Augmented reality occupies an area on the periphery of dual-use technology, technology that "can be used for both civilian and military applications and/or can contribute to the proliferation of Weapons of Mass Destruction (WMD)" (European Commission 2016). While the technology in its current form does not yet have exclusively military functionality, as a military and intelligence gathering tool it has the potential to interact with non-combatants for intelligence or military purposes.

In a world where open source intelligence (OSINT) and big data is routinely mined to develop actionable intelligence, augmented reality may seem inconsequential. However, society's increasing reliance on technology, especially online technology, for everyday tasks casts an ever-wider net around the general public in terms of digital data collection. Much of the data stored by tools like GPS or devices within the internet of things is geospatially registered. When incorporated into intelligence gathering practices, this provides potentially unprecedented access for foreign intelligence organizations into a country's everyday goings-on.

As stated above, while the US currently does not plan to use HPM to alter soldiers for the purpose of increasing their abilities past that of which they are naturally capable (e.g., IQ enhancement), the US anticipates that other countries, possibly adversarial ones, will pursue such programs (Brimley et al. 2013). Since what we may consider to be science fiction today may one day become reality, it would be prudent for the US to develop policy positions about HPM technologies and techniques, either adopting or deterring, so as not to be caught off guard when they are confronted with in on the battlefield (Brimley et al. 2013).

## *6.2 Globalization and Balance of Power*

Historically, wealth, geopolitics, and military strength have determined the global balance of power; however, the end of the Cold War ushered in a multipolar world less bound by the constructs of old. It is not only the fall of the Soviet Union and the rise of non-state actors that have brought about this paradigm shift but also technological innovation and advancement, spurred on by globalization. Will the twenty-first century be the stage for a technological arms race?

The World Economic Forum's Global Competitiveness Report assesses countries on a variety of variables to determine their competitiveness in relation to each other. Table 1 shows the top ten countries ranked by "government procurement of advanced technology products," which represents the degree to which "government purchasing decisions foster innovation." "Capacity for innovation" refers to the ability of companies to innovate. "Quality of scientific research institutions" places a country's research institutions within a spectrum from "extremely poor" to "extremely good." "R&D Innovation" is a composite category from the Robinson Country Intelligence Index that incorporates all of the Global Competitiveness Report's Innovation (Pillar 12) variables; in addition to this listed in the chart, it includes: "company spending on scientific research institutions," "university-industry research collaboration," "availability of scientists and engineers," and "patent application (hard data)" (World Economic Forum 2015). As seen in Table 1, the Robinson Country Intelligence Index, the product of a collaboration between the World Affairs Council of Atlanta and Georgia State University's Robinson College of Business, weights all variables across all countries for whom the data is available so that a score of 1000 represents the most positive possible score and a score of zero is the lowest possible score (Brown et al. 2015; World Affairs Council of Atlanta 2017).

**Table 1** World Economic Forum "Government procurement of advanced technology products" and related "global competitiveness indicators" top ranking countries in 2016 as scored by the Robinson country intelligence index (Brown et al. 2015)

| | Government procurement of advanced technology products | | Capacity for innovation | | Quality of scientific research institutions | | R&D innovation | |
|---|---|---|---|---|---|---|---|---|
| (of 199 countries) | Rank | Score | Rank | Score | Rank | Score | Rank | Score |
| United Arab Emirates | 1 | 1000 | 15 | 864 | 27 | 765 | 27 | 765 |
| Qatar | 2 | 949 | 19 | 837 | 14 | 882 | 14 | 882 |
| Malaysia | 3 | 900 | 13 | 875 | 23 | 823 | 23 | 823 |
| Singapore | 4 | 881 | 20 | 835 | 10 | 901 | 10 | 901 |
| Luxembourg | 5 | 822 | 10 | 877 | 24 | 806 | 24 | 806 |
| Germany | 6 | 802 | 5 | 934 | 11 | 900 | 11 | 900 |
| India | 7 | 793 | 39 | 700 | 36 | 711 | 36 | 711 |
| Israel | 8 | 789 | 4 | 957 | 3 | 966 | 3 | 966 |
| Rwanda | 8 | 789 | 54 | 644 | 79 | 541 | 79 | 541 |
| China | 10 | 787 | 44 | 662 | 40 | 691 | 40 | 691 |
| United States | 11 | 774 | 2 | 967 | 5 | 935 | 5 | 935 |
| Indonesia | 12 | 766 | 32 | 738 | 41 | 674 | 41 | 674 |

**Table 2** The top 5 countries from the bloomberg innovation index in the five categories most closely related to AR technology, as well as the overall category (Bloomberg 2015)

| Rank | Overall | R&D | Manufacturing | High-Tech Companies | Research Personnel | Patents |
|---|---|---|---|---|---|---|
| 1 | South Korea | South Korea | Switzerland | United States | Finland | South Korea |
| 2 | Japan | Israel | Ireland | China | Iceland | Japan |
| 3 | Germany | Finland | Singapore | Japan | Denmark | China |
| 4 | Finland | Sweden | Germany | South Korea | Israel | United States |
| 5 | Israel | Japan | Austria | Canada | Singapore | Germany |

Table 1 clearly demonstrates that countries who previously have not enjoyed much of a determining factor in the global balance of power show great technological potential. That is not to say that current world powers like the United States and China are lagging behind in technological innovation but rather that their contemporaries are changing.

The Bloomberg Innovation Index ranks the "50 most innovative countries" in the world based on six main categories: research and development, manufacturing, high-tech companies, postsecondary education, research personnel, and patents. Table 2 shows the top five performers in each category related to augmented reality.

South Korea, the country that Bloomberg ranked number one overall, is noteworthy because much of their research and development is company-driven rather than strictly within the public sector. That said, government and university funded research in the augmented reality field is not a contraindication of innovation as the efficiency of commercially motivated innovation may result in faster innovation. In Bloomberg's high-tech category, the United States finished leaps and bounds ahead of its competitors due, in large part, to the size of its technology sector. Although quantity is not a substitute for quality, quality in large quantities, as found in the US, is a sign that the country is not relinquishing its status as a technological leader any time soon. The research personnel category is made up entirely of relatively small countries and proves the strength of the trend of globalization; each of these countries rely on free trade to diversify and expand their specific industry specializations. Regarding patents, it is no surprise that industrialized countries with large economies dominate the top of the list. Patents are a key part of the technological 'arms race' in a world where innovation via collaboration can be hampered by the need for countries to protect their intellectual property (Bloomberg 2015).

Despite the fact that the bulk of the early research in augmented reality was completed in the United States, globalization and technological proliferation have expanded the geographic boundaries of the field. Recent conferences such as the Institute of Electrical and Electronics Engineers (IEEE) International Symposium on Mixed and Augmented Reality and the International Conference on Virtual and Augmented Reality in Education were recently held in Mexico; previous conferences have mostly been held in the United States, Germany, Japan, and South Korea. Both events have included participants from the United States, Mexico, Germany, Austria, Japan, and others (VARE 2015; ISMAR 2016).

## 7 Challenges and Recommendations

Undoubtedly, augmented reality technology presents a unique opportunity for the intelligence community to enhance its field operations, but the impact of this evolving technology extends well beyond the field. While incorporating augmented reality into the realm of intelligence operations is inevitable and necessary, it presents several challenges.

Firstly, what happens when disruptive technology is itself disrupted? The increasing dependence on online digital technology for intelligence applications as well as the cloud computing trend presents both an advantage and potential pitfall to the intelligence community. While cloud computing and increasing connectivity makes technology and intelligence tools more portable, and increases the efficiency of the collection and analysis process, it also introduces an inherent vulnerability. The intelligence community must be at the forefront of technological innovation and cyber security in order use devices like augmented reality, whose existence makes them a prime target for foreign intelligence. From a policy perspective, this will mean targeted funding for programs like DARPA as well as sufficient funding and focus on cyber security across the seventeen intelligence agencies.

Secondly, globalization has spurred the proliferation of new technologies, marking a golden age of innovation. As traditional military force becomes less relevant to the global balance of power and technology becomes an increasingly essential part of each country's social, economic, and governmental infrastructures, smaller countries who excel at technological innovation are poised to take up new roles as global influencers, setting the stage for an imminent paradigm shift in our understanding of the balance of power. To this end, government agencies should seek to tap the innovation of the private sector through public-private partnerships. The intelligence community has already demonstrated its willingness to partner with Amazon Web Services to supplement its own technological systems. Such collaboration will help the US maintain its competitive edge in the augmented reality space; something that is especially important with regard to the rise of China.

Finally, the most significant challenge for the deployment of augmented reality systems may be ethics-related. In the modern day, it has proven difficult to strike the appropriate balance between privacy and security. The availability of big data has opened a proverbial Pandora's box of private advancements and public set-backs for the intelligence community. It should be noted that the exploitation of big data presents as much, if not more, of a risk when employed by foreign actors. For this reason, it is necessary that the US increase its capacity for innovation, fund research and development initiatives, and strengthen public-private partnership opportunities in order to maintain its competitive advantage in this field.

## References


Azuma, R. (1997). A Survey of Augmented Reality. *Presence: Teleoperators & Virtual Environments, 6*(4), 355.

Azuma, R. (1999). The Challenge of Making Augmented Reality Work Outdoors. In Y. Ohta & H. Tamura (Eds.), *Mixed Reality: Merging Real and Virtual Worlds* (pp. 379–390). Berlin: Springer-Verlag.

Barfield, W., & Caudell, T. (Eds.). (2001). *Fundamentals of wearable computers and augmented reality* (p. 449). Lawrence Erlbaum Associates: Mahwah.

Barr, A. (2015). Google glass gets a new name and hires from Amazon. *The Wall Street Journal.* [online] sec. DIGITS. Available at: http://blogs.wsj.com/digits/2015/09/16/google-glass-gets-a-new-name-and-hires-from-amazon/.

Bhattacherjee, A. (2012). *Social science research: Principles, methods, and practices* (Vol. Book 3. Textbooks collection). University of South Florida. Available at: http://scholarcommons.usf.edu/oa_textbooks/3.

Bloomberg.com, (2015). *The Bloomberg Innovation Index.*. Available at: http://www.bloomberg.com/graphics/2015-innovative-countries/.

Brimley, S., FitzGerald, B. and Sayler, K. (2013). Game changes: Disruptive technology and U.S. Defense Strategy. In: *Disruptive defense papers*. Center for a New American Security. Available at : http://www.cnas.org/sites/default/files/publicationspdf/CNAS_Gamechangers_BrimleyFitzGeraldSayler.pdf , 17-18.

Brown, C., Cavusgil, S., & Lord, A. (2015). Country-Risk Measurement and Analysis: A New Conceptualization and Managerial Tool. *International Business Review, 24*, 246–265.

Carmigniani, J., Furht, B., Anisetti, M., Ceravolo, P., Damiani, E., & Ivkovic, A. (2010). Augmented Reality Technologies, Systems and Applications. *Multimedia Tools and Applications., 51*(1), 341–377.

Doolin, C., Holden, A., and Zinsou, V. (2013). Augmented government: Transforming government services through augmented reality. GovLab. Deloitte. Available at: http://www2.deloitte.com/content/dam/Deloitte/us/Documents/public-sector/us-fed-augmented-government.pdf.

European Commission. (2016). *Dual-Use Export Controls*. Available at: http://ec.europa.eu/trade/import-and-export-rules/export-from-eu/dual-use-controls/.

Fabey, M. (2015). Marines Test ONR Augmented Reality Tech. In *Aerospace Daily & Defense Report*. 253(47).

Feiner, S., MacIntyre, B., & Seligmann, D. (1993). Knowledge-Based Augmented Reality. *Communications of the ACM., 36*, 53.

Google Developers. (2016). *Glass*. Available at: https://developers.google.com/glass/.

Heilig, M. (1962). *Sensorama Simulator*. 3,050,870. 10. Available at: http://www.mortonheilig.com/SensoramaPatent.pdf.

ISMAR. (2016). *Conference committee*. Available at: http://www.ismar2016.org/.

Madrigal, A. (2014). *How DARPA's augmented reality software works*. The Atlantic. Available at: http://www.theatlantic.com/technology/archive/2014/05/hows-darpas-augmented-reality-software-works/371652/.

Marshall, M. (2013). CIA venture arm backs spike, which lets you make precise models of anything. VentureBeat.. Available at: http://venturebeat.com/2013/09/29/cia-venture-arm-backs-spike-which-lets-you-make-precise-models-of-anything/.

Memi, E. (2006). Now see this. *Boeing* Frontiers.

Milgram, P., & Kishino, F. (1994). A taxonomy of mixed reality visual displays. *IEICE Transactions on Information and Systems, E77-D*, 1321–1329.

Milgram, P., Takemura, H., Utsumi, A., & Kishino, F. (1995). Augmented Reality: A Class of Displays on the Reality-Virtuality Continuum. *SPIE, 2351*, 282–292.

Myers, B. (1998). A Brief History of Human-Computer Interaction Technology.. *Interactions*.

National Research Council (U.S.), ed. (2012). Human performance modification: Review of worldwide research with a view to the future. City: Washington, D.C. National Academies Press.

Navab, N., Blum, T., Wang, L., Okur, A., & Wendler, T. (2012). First Deployments of Augmented Reality in Operating Rooms. *Computer, 45*, 48.

PricewaterhouseCoopers. (2016). Product Design Most Common Application for VR/AR. *PwC*.. Available at: http://www.pwc.com/us/en/industrial-products/next-manufacturing/augmented-virtual-reality-uses-manufacturing.html.

Rosenberg, L. (1993). Virtual fixtures: Perceptual tools for telerobotic manipulation. In *VRAIS*. pp. 76–82.

Sensors and Systems. (2013). *Will the move toward Web GIS provision an operational augmented reality?* Available at: http://sensorsandsystems.com/will-the-move-toward-web-gis-provision-an-operational-augmented-reality/.

Sutherland, I. (1968). "A Head-Mounted Three Dimensional Display." City. Salt Lake City, Utah. The University of Utah. pp. 757–764. Available at: http://design.osu.edu/carlson/history/PDFs/p757-sutherland.pdf.

Thomas, B., Close, B., Donoghue, J., Squires, J., De Bondi, P., & Piekarski, W. (2002). First Person Indoor/Outdoor Augmented Reality Application: ARQuake. *Personal and Ubiquitous Computing., 6*, 75–86.

Tilley, A. (2015). NASA will be taking these augmented reality glasses into space. *Forbes*. Available at: http://www.forbes.com/sites/aarontilley/2015/03/11/nasa-odg-augmented-reality-in-space/.

VARE. (2015). 2015 International Conference Virtual and Augmented Reality in Education. Universidad de La Laguna. Available at: http://eventos.ull.es/event_detail/2027/speakers/2015-international-conference-virtual-and-augmented-reality-in-education-varea15.html.

Wolverton, M. (2013). Augmenting reality. Trajectory magazine.. Available at: http://trajectory-magazine.com/component/k2/item/1621-augmenting-reality.html.

World Affairs Council of Atlanta. (2017). Robinson country intelligence index. Georgia State University.. Available at: http://rcii.gsu.edu/.

World Economic Forum. (2015). Technical notes and sources. *Competitiveness Report*, 2014–2015. http://wef.ch/1uzFHXn. Accessed 4 May 2016.

# Analyzing Public and State Reactions to Global Surveillance Disclosures: Using Ethical Frameworks to Gain Understanding

**Janille Smith-Colin and Nabil Kleinhenz**

**Abstract** Secretive surveillance activities are carried out by government agencies in an effort to protect national security. Upon discovery, however, there can sometimes be significant ramifications on diplomatic relations, potentially impacting international cooperation and security. In the wake of the Snowden leaks, various levels of reaction have been observed from nation states and the public. This work attempts to understand the differences observed in reactions to recent intelligence leaks and explore theories around intelligence ethics. Ethical standards used by intelligence officials differ from those applied by the general public. Following the surveillance disclosures by Edward Snowden, the rhetoric among public officials was heated. Global opinions strongly view NSA surveillance activities of foreign citizens or heads of state as "unacceptable" (unless specifically targeting a terrorist). However, the magnitude of action taken by a nation-state in response to the leaks correlated with the level of public opposition to surveillance activity in that country.

**Keywords** Ethics · Ethical frameworks · Intelligence gathering · Intelligence community · Surveillance · Edward Snowden · Intelligence leaks

## 1 Introduction

Secretive surveillance activities are carried out by government agencies in an effort to protect national security. When some of these activities are discovered, however, there can be significant ramifications on diplomatic relations as trust is broken, potentially impacting international cooperation and security. In the wake of the

J. Smith-Colin (✉)
Southern Methodist University, Dallas, Texas, USA
e-mail: jsmithcolin@smu.edu

N. Kleinhenz
Georgia Institute of Technology, Atlanta, Georgia, USA
e-mail: nkleinhenz@gatech.edu



M. E. Kosal (ed.), *Technology and the Intelligence Community*, Advanced Sciences and Technologies for Security Applications,
https://doi.org/10.1007/978-3-319-75232-7_14

Snowden leaks various levels of reaction have been observed from nation states and the public. This work is motivated by a desire to understand the differences observed in reactions to the Edward Snowden leaks, and by a broader interest in theories around intelligence ethics. What factors influence reaction in the public sphere and what methods are available for systematically conducting an analysis of cases such as the Snowden leaks? What role does ethics play in the decision-making processes of the intelligence community, and how can a knowledge of ethical frameworks shed light on the actions taken within the intelligence community?

The goals and objectives for this work were four-fold. First, to identify the dimensions of intelligence ethics, apply them to contemporary cases of intelligence gathering, and use these ethical frameworks to better understand intelligence activities. Second, to elucidate the differences between intelligence ethics from the standpoint of the individual and the nation. Third to investigate public perception around the acceptability of intelligence activities and to investigate the corresponding reaction from nations impacted by the surveillance disclosures. Last, to review the range of technological methods used by the National Security Agency (NSA) and the corresponding international response.

Theories in intelligence ethics include realism (destroy our enemy), idealism (do no evil), utilitarianism (maximize the public good), and the just war theory which emphasizes just cause, just means, and proportionality. The ladder of escalation, developed by Johnson, builds on the theories of intelligence ethics, and the factors impacting views on intelligence ethics to rank intelligence activities according to their level of harm (Johnson 2006). Levels of harm or thresholds range from Threshold 1 to Threshold 4. The guiding analytical principles of the ladder of escalation are that it ranks intelligence activities according to increasingly serious levels of international law violations, infringements on national sovereignty, and breaches of ethical and moral code. The ladder of escalation reveals that many of the intelligence activities known to the public including wiretapping, photo-reconnaissance, and unmanned aircraft operations are in fact considered to be routine intelligence gathering activities within the intelligence community.

A democratic free society requires open and ongoing debate around government undertakings including the actions of the intelligence community (IC). Intelligence decisions are informed by theories of ethics including public attitude towards the intelligence community itself. Other factors contributing to the ethical view of intelligence activities include the following: view of the enemy, political system of the target nation, sense and severity of the threat, intended use for intelligence information, and the prevailing climate and conditions with respect to war. Intelligence activities, however, because of their inherently secret nature, are often spared intense public scrutiny. In the absence of the public system of checks and balances, philosophies held by intelligence officers and the prevailing views on intelligence ethics and morality, often guide national intelligence decision-making.

In 2013, National Security Agency (NSA) contractor Edward Snowden leaked thousands of documents revealing the global surveillance activities of NSA which included harvesting millions of emails, instant messaging contact lists, tracking and mapping cell phone locations, accessing phone conversations including those of 35 world leaders, and PRISM, a program that allows court-approved access to personal communications through online accounts including Google and Yahoo accounts. In the information age of the twenty-first century however, where leaks such as those perpetrated by Edward Snowden have become more common, the public now has greater opportunity to weigh in on actions taken by the intelligence community in the name of national security.

The Snowden leaks represent a broad range of surveillance activities recently disclosed to the public. Based on response to the Snowden leaks, it is clear that there is a relationship between public perceptions around the acceptability of surveillance activities and the corresponding response taken by the target nation being surveilled. The public opinion regarding the execution of surveillance activities by the NSA, as measured by Pew Research Polls is correlated to the severity in the response displayed by the nation being targeted. For example, when it was revealed that the text messages of Mexican presidential candidate Enrique Peña were being monitored, the Mexican Foreign Ministry merely asked the US in a diplomatic note for an "exhaustive investigation" into the matter (Archibold 2014). This muted response may be connected to the fact that only 66% of Mexicans polled believed that NSA surveillance of their country's leader was "unacceptable." However, when Brazil's president Dilma Rousseff discovered her personal communications were being monitored, her response included cancelling her visit to the White House, making an impassioned speech to the UN General Assembly to introduce a resolution on the right to privacy in the information age, and planning to lay an undersea cable to avoid monitoring of internet traffic by the US. In Brazil, 83% said surveilling their leader was unacceptable.

The ladder of escalation provides a systematic means by which to analyze intelligence activities using an ethical lens. An application of the ladder of escalation to the surveillance activities revealed in the leaks by NSA contractor Edward Snowden in 2013 allows for the ranking of cases according to their level of harm. Cases ranked at Threshold 1 involved routine intelligence activities, Threshold 2 involved modest intrusions, such as distance surveillance of foreign nation civilians or corporations, Threshold 3 involved more intrusive methods such as wiretaps of personal communications belonging to prominent political leaders, and Threshold 4 could hypothetically involve the assassination of a head of state. The Snowden leaks reviewed in this analysis ranked mostly at Threshold 2 and 3.

The review of literature suggests that the ethical standards used by intelligence officials differ from those applied by the general public. In response to the Snowden leaks, the rhetoric among public officials was heated as ascertained through official statements. Global opinions strongly viewed NSA surveillance activities of foreign citizens or heads of state as "unacceptable" (unless specifically targeting a terrorist). The magnitude of action taken by a nation-state in response to the leaks seemed to be correlated with the level of public opposition to surveillance activity in that country.

## 2 Intelligence and Ethics

Controversies involving the detention, mistreatment, and torture of suspected terrorists have challenged the intelligence community, and the nation as a whole to identify ways of reconciling the competing needs to work in secrecy in a manner that often violates individual human rights, all while ensuring the safety and security of a democratic society (Born and Wills 2010). Since 9/11 the decision to carry out intelligence operations has come under increased scrutiny, contributed to in part by the following trends – the shifting nature of the functions of intelligence services, the growing demand for accountability, increased respect for human rights, recent controversies arising from actions taken by Western nations in the war against terrorism, and concerns expressed by national and international parties in light of these controversies (Born and Wills 2010).

As a direct result, some of the most recognized books dealing exclusively with the topic of ethics and intelligence were published in 2006 following the 9/11 attacks (Goldman 2010). Prior to this, literature on the topic of intelligence, ethics, and morality, was dispersed in journals, articles, and conference proceedings (Goldman 2013). The first conference on intelligence and ethics was held in Washington D.C. in 2006, and "Ethics of Spying: a Reader for Intelligence Professionals" was published later that year. Literature examining the interaction between ethics and spying dates back to the 1970s. In 1977, the Central Intelligence Agency (CIA) compiled a bibliography, "Morality and Ethics: Intelligence and Security in Our Democracy," containing 99 entries from academic journals and magazines (Goldman 2013).

Ethics is commonly defined as the discipline dealing with what is good and bad and with moral duty and obligation; a set of moral principles; a theory or system of moral values; or the principles of conduct governing an individual or a group. More specifically, intelligence ethics have been defined as "a set of behavioral guidelines based on certain beliefs……regarding the role of intelligence in society" (Shpiro 2010). There are those who are of the view that the notion of ethics in the intelligence community is a contradictory one. Some view the work of the intelligence community and intelligence officers as inherently based on lying, cheating, stealing, and deceiving (Andregg 2007) in the name of national security. In spite of this, there is an increasing view among current and former intelligence officers that the study of ethics does have a role in the intelligence community.

Ethics is one of the means by which intelligence agencies seek accountability for their actions. In reviewing ethics in an intelligence context, three points are made: (1) ethics are not the same as the law; (2) the ethics of intelligence work are not necessarily synonymous with a person's personal ethics – that is institutional ethics are different from individual ethics ("Intelligence and National Security" 2007), and (3) given its fundamental mission, working in the intelligence community should be considered ethical (Goldman 2013). The study of ethics generally focuses on rules, regulations, policy, and law. However, there are many intelligence actions that are not legislated. In these cases, ethical frameworks often guide decision-making activities.

Ethical issues emerge at each stage of the intelligence cycle. The intelligence cycle was formulated by the Central Intelligence Agency, and it outlines six stages for the intelligence process: planning and direction, collection, processing, analysis and production, dissemination, and new requirements (Central Intelligence Agency 1983). Planning and direction includes decisions made by senior level intelligence officers with respect to strategic planning and prioritizing, operational planning and mission approvals, targeting policies, and recruitment policies. Planning and direction requires ethical considerations as decisions are taken to protect national security and ensure the safety of citizens. The moral dilemma here is whether to protect national security or to focus resources on actions that may result in violations of individual rights. The vast majority of ethical issues with respect to intelligence focus on collection (Born and Wills 2010). Often debated is the use of spies and informers, intrusions of privacy, and of course torture. Ethical considerations with respect to analysis and production focus on the role and responsibility of the analyst not to overstate, underreport, or exaggerate findings for political gain. Intelligence officers have a responsibility to report the facts unembellished facts, as the stakes are often quite high. Dissemination of intelligence involves the communication of findings, sharing of intelligence with domestic and foreign governments (Born and Wills 2010). A primary concern at the dissemination point of the intelligence cycle is the unauthorized sharing of intelligence information. Other considerations include the protocols for international and domestic sharing.

There are two ethical paradigms at play in the intelligence community: personal or individual ethics and government or nation state ethics. In the intelligence community, one's history of moral and ethical conduct can determine whether a security clearance can be issued (Goldman 2013). The intelligence community upholds individuals to high moral and ethical standards. The ethical code held by individuals often weighs the needs of the person against the needs of the community. Individual ethics is also often concerned with how friends and family view one's actions and whether they would approve of the specific behavior or not. A different set of factors is taken into consideration when nations are faced with the decision of whether to collect intelligence information, engage in covert action, or respond to wartime insights. The section that follows provides a brief overview of these ethical frameworks.

## 2.1 *A Look at Frameworks for Intelligence Ethics*

Ethical frameworks provide guidelines for democratic societies that engage in intelligence gathering activities during peacetime and covert operations during wartime. Realism, just war theory, consequentialism, and idealism are all approaches used to evaluate the ethical merits of intelligence gathering activities. Each of these

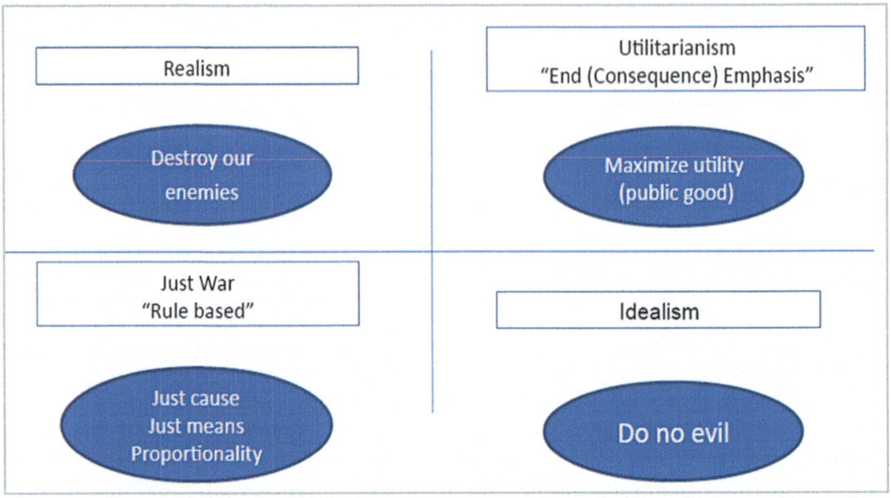


**Fig. 1** Ethical frameworks: a view for intelligence (Charters 2006)

frameworks is discussed in more detail below. Figure 1 identifies the defining characteristic of each approach.

### 2.1.1 Realism

The realist perspective on ethics informs much of the current day discourse around intelligence gathering. The realist views the activities of the intelligence community as morally necessary. The national security of the nation depends on these activities, and governments are required to do whatever is possible to subvert the possible threat. Realists believe that any action is acceptable while in pursuit of broader national security goals.

### 2.1.2 Utilitarianism

Utilitarianism assesses the ethics of action solely based on its outcomes (Charters 2006). The goal is to maximize the public good or utility. Utilitarians are end or consequence-focused. The determination as to whether an activity is ethical or not is based on a cost-benefit analysis of the actions versus the results. The goal under utilitarianism is to have the overall benefit outweigh the overall harm. If the overall harm exceeds the overall benefit, the action is considered unethical (Charters 2006). Individual motivations are not taken into consideration under a utilitarian ethical framework. The danger here of course is that actions that are intrinsically immoral can be taken if it is perceived that the total benefits outweigh the total harm. The needs of the community are therefore considered above/paramount to the perceived or real needs of any one individual. Individual rights are also of less concern under this framework. The benefit of a utilitarian framework is that it focuses on harm (Charters 2006). Any action resulting in harm is subject to analysis under a utilitarian framework. It is not the motivation for action that triggers the consideration but the end result. An action resulting from damaging behavior that only impacts an individual or organization can therefore be deemed not unethical. In contrast, acquiring information through means that are not questionable but that as an end result present a threat to the greater public good would under utilitarian thinking be considered unethical.

### 2.1.3 Just War

Just war theory is applied primarily under conditions of war. Under theories of just war the prevailing belief is that there are times and conditions under which immoral/unethical behavior is justified. Just war theory makes an attempt to outline the conditions under which this unethical behavior should be considered. Though focused mainly on warfare and international conflict, just war theory can also be applied under peacetime conditions to evaluate covert action – action often deemed unethical (Hulnick and Mattsauch 2006). Individuals who believe in just war theory have allowed for measured resolute acts in the face of morality. Just war theory allows nations to consider immoral activities if and when the conditions are right. Just war theory sees war and covert actions under peacetime as a necessary evil, and it is therefore not intended to judge war in and of itself. Just war theory looks at individual actions of conflict within a wartime or peacetime situation (in the case of conflict) and judges these. Actions of war and covert action in peacetime must meet the following seven criteria under the just war paradigm to be considered moral (Hulnick and Mattsauch 2006):

1. Just cause – only actions taken from a defensive posture are acceptable
2. Just intention – actions must be taken for the purpose of achieving peace
3. Last resort – all alternatives other than war must be considered to justify an act of war
4. Formal declaration – Nations have the authority to order force, not individuals
5. Limited objectives – The objectives for war are peace. Not destruction of a nation.
6. Proportionate means – The level of response should be proportionate to the threat. Only what is required
7. Noncombatant Immunity – The wellbeing of civilians should be protected. Force should only be directed to representatives of government.

Just war theorists do not think of war as a necessary evil. Instead they think of war as the better of two worse options: war and leaving a worse aggressor unchecked.

### 2.1.4 Kantian Ethics and Idealism

Under Kantian ethics, individuals are not to be treated as a means to an end or as a source of information that can be disposed of or mistreated. Kantian ethics requires one to consider how other individuals are being treated (Charters 2006). Kantian ethics does not give consideration to outcomes or the consequences of action. Kantian ethics are concerned with the motive behind action. Rational behavior reigns supreme under Kantian ethics. As long as an action has been freely and rationally debated, this ethical framework considers the action moral regardless of the harm it may cause (Jones 2010).

Idealism is of the Kant doctrine. Idealists believe that acting morally requires action guided by principles that are irrespective of consequences. "Do no evil" and "Do unto others as you would have them do onto you" are common refrains of the idealist school of thought. These beliefs are held regardless of the circumstance, and without exception. They are universal and absolute (Jones 2010). General practices of coercion and spying are therefore deemed immoral under the idealistic ethical framework.

## *2.2 Factors Influencing Intelligence Decisions*

Individual views on the legitimacy and moral appropriateness of intelligence operations are dependent on one's socialization, education, evolving political and international perspectives, and peer group influences (Johnson 2006). When taken by nations or representatives of nations the decision to engage in intelligence operations is influenced by a different set of factors. The ethical perspectives held by the decision-maker can impact the decision to engage in intelligence activities. The idealist would reject all options that could result in harm to the target nation, as the idealist believes in the mantra "do no evil." On the contrary, the realist would take any action necessary to secure the safety and security of the nation. Democratic nations are generally spared more adverse actions unless there is a known threat to the national security of the enacting nation. Totalitarian nations are viewed more harshly.

The view of the target nation therefore has influence on the decisions taken. An adversarial regime or totalitarian regime known for civil right abuses is more likely to be subject to intelligence activities that result in harm. The political agenda of leadership can also impact the national approach towards intelligence activities and gathering. For example, under the Regan Administration there was a hard line approach towards communism and therefore an increased willingness to escalate intelligence activities; in contrast, and almost ironically so, under the Nixon administration there was a more restrained view towards intelligence activities and therefore less of a likelihood that intelligence activities would be escalated (Johnson 2006). Table 1 includes a list of factors known to influence intelligence decision-making. Each of these factors can be grouped based on the type of information

**Table 1** Influences on intelligence decision-making

| | |
|---|---|
| Who | View of the enemy – {Adversarial review} |
| | Target regime – {free/open or closed/totalitarian; ally or enemy} |
| | Leadership personality – {views on intelligence ethics} |
| What | Sense of threat to national security |
| | Severity of threat – {ISIS vs. Greenpeace} |
| How | Data collection methods |
| Why | Intended use |
| When | Conditions of war – {wartime vs. peacetime} |
| Other | Attitudes of the public – {education, socialization, global perspective} |

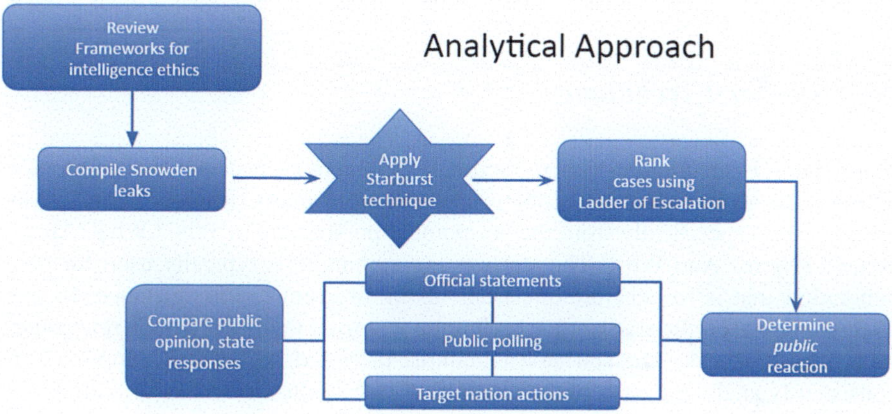


**Fig. 2** Analytical approach (*Adapted from* Beebe 2015)

providing influence; the information considered includes the Who? What? How? Why? and When?

## 3 Analytical Approach: Methodology

This paper uses a mixed-method approach involving a case study analysis to study the relationship between the level/degree of intelligence activity and nation state response (including the public and decision-makers). The main components of the analytical approach include (1) a review of ethical frameworks used to understand and influence intelligence activity, (2) compilation, review and analysis of the Snowden leaks, (3) application of the starburst technique, (4) ranking of cases on the ladder of escalation, and (5) determination of public and official response to the Snowden leaks. Figure 2 provides some additional insight into the steps taken to complete this analysis.

Findings from the literature have been presented in earlier sections and included a review of ethical frameworks relevant to the intelligence community, a discussion of ethical considerations and the intelligence cycle, and a look at factors influencing intelligence decision-making. The Snowden leaks were chosen as a case study for analysis because of the wide range of intelligence activities captured under the Snowden leaks, the ready availability of information on the Snowden leaks, and because of the disparate levels of national response to the Snowden leaks. Six countries with readily available media coverage on the leaks were chosen for analysis. The key analytical methods used in this analysis are the starburst technique and the ladder of escalation combined with obtaining relevant survey data and information contained in media. A description of each of these methods is provided below. Results from the in-depth analysis of the Snowden leaks are presented in the sections that follow.

## 3.1 Starburst Technique

Starbursting is a technique used to facilitate structured brainstorming. The process allows analysts consider the problem at hand from many different perspectives (Beebe 2015) specifically from the six-pointed perspective of Who? What? How? When? Where? And Why? The starbursting technique is typically used for idea generation that is to generate questions about an event or circumstance. In this analysis, however, the starbursting technique was used to dissect details relevant to each Snowden leak. The starburst technique provided a systematic approach to consistently analyzing each of the qualitative case studies, where information was available.

## 3.2 Ladder of Escalation

The nuclear strategist Herman Kahn introduced his ideas around the "essential-ladder metaphor" in 1965. The metaphor was presented as a means for understanding the overt hostile acts carried out by one nation against another (Johnson 2006). The ideas put forth by Kahn have since been adopted by other strategists and presented in various forms of the ladder of escalation. The ladder of escalation is a framework that allows intelligence collection activities to be analyzed based on the harm that they cause and the justification for taking the action. One of the benefits of the ladder of escalation is that it allows very different actives to be grouped together based on their level of harm therefore allowing for a more direct comparison of these activities (Bellaby 2014). Within the ladder of escalation, increasing levels of harm require increasing levels of justification for a specific action. The ladder of escalation, builds on the theories of intelligence ethics, and the factors impacting views on intelligence ethics to rank intelligence activities from Threshold 1 through Threshold 4.

The guiding analytical framework for the ladder of escalation is that it ranks intelligence activities according to increasingly serious levels of international law violations, infringements on national sovereignty, and breaches of ethical and moral codes (Johnson 2006). Also incorporated into the ladder of escalation are the influences on the use of covert operation as defined by Johnson (2006) and described above – ethical perspectives, view of the enemy, target regime, leadership personality, imminent threat, severity of the threat, and short-term and long-term effects. The ranking offered by the ladder of escalation includes the following: Threshold 1 involves routine intelligence activities, Threshold 2 involves modest intrusions, such as distance surveillance of foreign nation civilians, Threshold 3 involves more intrusive methods such as wiretaps of personal communications of prominent political leaders, and Threshold 4 could hypothetically involve the assassination of a head of state (Johnson 2006).

Decision-making is a key component of the intelligence process. "The process should involve elected officials with national security experience, assisted by well-trained intelligence and foreign policy specialists who understand the possibilities – as well as the dangers and limitations – of using clandestine agencies in support of democratic values" (Johnson 2006).

Johnson proposes 11 guidelines to assist with decision making relative to intelligence gathering activities and more specifically covert action.

1. Diplomatic solutions should be considered first
2. Operations should align with publically stated policy objectives
3. Covert operations should not embarrass if released
4. Extensive consideration should be given to the nation being targeted, not just the action to be taken
5. Decision-making processes should be followed and the correct persons should be at the table
6. The laws of the United States should never be broken
7. Only the most low level/low threshold operations should be pursed against other democracies – unless they are engaged in activities that are known to threaten
8. Against non-democratic regimes policymakers should also aspire to low-level/ low-level threshold operations
9. Foreign officers and intelligence officers should be held to the highest ethical and moral standards
10. Policymakers should avoid high-threshold actions such as wars, coups, and forceful interventions
11. America's standing as a shining star of freedom and democracy should not be placed in jeopardy as a result of covert operation

Table 1 provides a description of the thresholds found in the ladder of escalation and corresponding examples of intelligence activities and covert operations for each threshold level (Table 2).

**Table 2** The ladder of escalation (A sample)

| Threshold Level | Description | Example |
|---|---|---|
| Threshold 1 | Routine intelligence operations | Exchange of intelligence information between known intelligence officers |
| Threshold 2 | Modest intrusions | Violations of airspace |
| | | High altitude reconnaissance: By satellite and reconnaissance planes |
| | | Distance surveillance against the target nation |
| Threshold 3 | High risk operations | Intrusive surveillance operations (wire taps for instance) – against prominent political leaders of native country |
| | | Aggressor uses intelligence gathering technology against target nation's highest decision council (a collection operation) potentially resulting in international rift |
| Threshold 4 | Extreme options | Assassination of nation-state president |

# 4 Analysis of Global Responses to Surveillance Disclosures

## *4.1 Background on Edward Snowden Leaks*

Edward Snowden is a former CIA employee and computer professional who began working for the NSA as a contractor in 2013 (Burrough et al. 2014). Over the years he had worked within the intelligence community, he had become increasingly skeptical about the ethical foundations of the mass surveillance work being conducted due to violations of the privacy of American citizens (Glenwald and Poitras 2013). Recalling a moment when he and his colleagues began to have strong ethical doubts, Snowden recounted that they would routinely come across sensitive information about the private life of individuals that had no relevance to security, such as intimate photos which would sometimes be shown to others in the workplace. He also said this behavior was never reported (Rusbridger and MacAskill 2014). Beginning in May 2013, Snowden began leaking thousands of confidential NSA and DoD documents without approval to journalists he trusted – documents that revealed details of a wide-array of surveillance programs used by the NSA to monitor personal communications of American citizens, citizens of other countries, corporations, and high-ranking officials in many countries. Snowden claimed he was motivated by not wanting to "live in a society that does these sort of things [surveillance on its citizens…] live in a world where everything I do and say is recorded… My sole motive is to inform the public as to that which is done in their name and that which is done against them" (Glenwald and Poitras 2013). While Snowden is viewed as both a hero and a traitor, the revelations shed light on the nature of the NSA surveillance programs and brought them into global public scrutiny.

It is unknown exactly how many documents were released, but the Department of Defense estimated that 1.7 million intelligence files were stolen by Snowden (Strohm and Wilber 2014). These documents revealed that the NSA was harvesting millions of emails and instant messaging contact lists, tracking and mapping cell phone locations, and conversations of both American citizens and foreigners abroad including personal communications of 35 world leaders (Gellman and Soltani 2013a, b; Ball 2013). After a 4-month investigation of the documents, the *Washington Post* reported that ordinary Internet users, American and non-American alike, far outnumber legally targeted foreigners in the communications intercepted by the NSA (Gellman et al. 2014). About 90% of account holders found in the large cache of intercepted conversations were found to be not the intended targets but simply caught in the wide net being cast by the NSA, One of the surveillance programs disclosed was a data-mining program code-named PRISM, which reportedly gave the NSA, FBI, and Government Communications Headquarters (Britain's NSA equivalent) "direct access" to the servers of Internet companies including Google, Apple, Microsoft, Facebook, and others (Ray 2016).

## 4.2 Global Response

In 2014 the Pew Research Center polled 44 countries asking the question: "According to news reports, the American government has been monitoring communications, such as emails and phone calls, in the US and many other countries. In your opinion, is it acceptable for the American government to monitor communications from ____________?" The questions filling in the blank included "Your country's leaders, your country's citizens, American citizens, and suspected terrorists" ("Global Opposition" 2014). The wording of the above question is significant, as determining whether the NSA's actions are "acceptable or unacceptable" could be understood as "right or wrong" and the responses may be indicative of the public's view of the ethics of the situation.

Figure 3 shows the global medians excluding the US in response to the questions posed. The overall results indicate widespread global opposition to the surveillance activities of the NSA. The strongest level of opposition was towards monitoring "citizens from your country," (81% said unacceptable) followed by "leaders of your country" (73%), and American citizens (62%). However, only 29% of people viewed monitoring of terrorist suspects as unacceptable, showing that the public is not opposed to surveillance as long as a suspected terrorist is being specifically targeted. In the case of American survey respondents, less than half are opposed to surveilling foreign leaders (43% unacceptable), while closer to half of Americans said spying on other country's citizens was unacceptable (47%). Americans strongly opposed monitoring of American citizens (61% unacceptable) but strongly supported monitoring of suspected terror suspects (24% unacceptable).

Other findings included a decline in the view that the US respects the personal freedoms of its people, but a median of 65% still express a positive opinion about the US and the overall ratings for the US changed little from the previous year's 2013 survey prior to the Snowden leaks ("Global Opposition" 2014). Despite the surveillance disclosures, President Obama maintained international popularity with

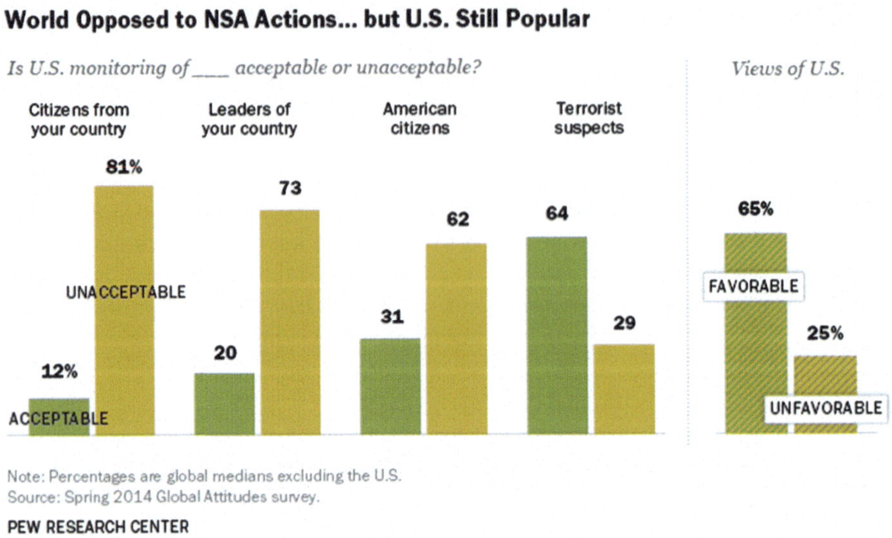


**Fig. 3** Global medians in response to survey questions on NSA actions. Taken from Pew Research Center ("Global Opposition" 2014)

56% of global respondents saying they have confidence in him to do the right thing in world affairs. However, the biggest declines in his ratings between 2013 and 2014 occurred in the two nations where it was widely revealed that the US had listened to private phone conversations of their national leaders: Germany (88% confidence in 2013 to 71% in 2014) and Brazil (69% in 2013 and 52% in 2014) ("Global Opposition" 2014).

Therefore, while the overall image of the US did not change a great deal, it is worth investigating how individual countries responded to the surveillance disclosures to determine their effects on public opinion, the rhetoric of political leaders, the resulting action taken by foreign leaders and the various impacts on US-foreign relations.

## *4.3 Analysis of Reactions Based on Surveillance Subject*

In this study, several surveillance subjects disclosed in the Snowden leaks are compared and contrasted. A variety of subjects from different countries including citizens, political leaders and companies were examined. Findings reveal that the public opinion towards the actions of the NSA correlate well with the actions taken by officials in response to the leaks. The results are compiled into Tables 3, 4, 5, 6, 7 and 8. Cases were chosen for which significant media coverage could be found. There were many more cases of international responses to the Snowden leaks than those listed here. The starburst technique was used for each surveillance subject, listing

**Table 3** Surveillance of United States Citizens (Escalation: 2)

| Surveillance Activity by NSA | Official Statements in Response to Leaks | US Public Perception of NSA Activities (% Unfavorable) | Target Nation Action = Low |
|---|---|---|---|
| Who (subject): US citizens | Obama says he will establish a panel on intelligence and communications technologies that would brief and report to the president. | 61% opposed to monitoring citizens of USA | Panel recommends mass collection of phone records be suspended and advises greater oversight of sensitive programs targeting foreign leaders. But attempt to cut NSA funding for phone metadata program is narrowly defeated in congress. Continued debate between intelligence community and privacy advocates. |
| What: Spying on personal communications | DNI director clapper condemned the leaks as having done "huge, grave damage" to the US intelligence capabilities. | | |
| When: Sept 11, 2001 leading to PATRIOT act, domestic spying begins to shift | | | |
| Why: Used for the purpose of "protecting the country" | | | |
| How: Metadata (records of billions of phone calls each day, content of some are listened to with warrant), emails, IM, Facebook posts, raw internet traffic from internet companies. | | | |

who, what, where, when, why and how for each surveillance activity, if available. The rank on the ladder of **escalation** was chosen based on choice of subject and methods used. Official statements displayed were chosen from media reports. Public perception data were obtained from Pew Research Center, showing both the percentage who said monitoring of their country's citizens was unacceptable and the percentage of who said monitoring their country's leader was unacceptable. Cases are ordered in the sequence of the following tables from lowest to highest in terms of

**Table 4** Surveillance of Mexican President Enrique Peña (Escalation: 3)

| Surveillance Activity by NSA | Official Statements in Response to Leaks | Mexican Perception of NSA Activities (% Unfavorable) | Target Nation Action = Low |
|---|---|---|---|
| Who (subject): Enrique Peña Nieto and Mexican government | Mexico's foreign ministry asked the US in a diplomatic note for an "exhaustive investigation" into the matter | 76% opposed to monitoring citizens of Mexico | Mexican foreign minister met with US ambassador to discuss allegations of hacking emails of then-president Felipe Calderon while in office. |
| What: Text messages intercepted, and government email accounts hacked. | | 66% opposed to monitoring leaders of Mexico | |
| When: 2012, and years before | | | |
| Why: During his presidential campaign, text messages referring to appointment he planned to make to his staff if elected. Government emails also gave insight into Mexican political system. | | | |
| How: Hacking | | | |

public opposition to NSA surveillance. The target nation actions are summarized and classified by their level of severity of "**low**", "**medium**" or "**high**", based on the level of activity taken by the target nation and the potential for diplomatic fallout and deterioration of strategic partnerships in the initial months after the leaks.

### 4.3.1 United States Citizens

Only 61% of Americans say NSA activities surveilling US citizens are unacceptable ("Global Opposition" 2014). Monitoring personal communications of American citizens became more commonplace in the wake of the September 11th terrorist attacks, when the USA PATRIOT Act was signed in 2001, for the purposes of "protecting the country" ("President George Bush" 2001). The Snowden leaks beginning in 2013 revealed that the NSA is now collecting metadata records of billions of phone calls each day, and some of the content is listened to without a warrant. Instant messaging, Facebook posts, as well as raw internet traffic from internet companies are also reportedly harvested. On the ladder of escalation, this type of intelligence activity would rank as Threshold 2 because of its routine nature and broad scope.

**Table 5** Surveillance of British Citizens (Escalation: 2)

| Surveillance Activity by NSA | Official Statements in Response to Leaks | British Perception of NSA Activities (% Unfavorable) | Target Nation Action = Low |
|---|---|---|---|
| Who (subject): British citizens | Foreign minister admits Britain's GCHQ was also spying and collaborating with the NSA, defended the two agencies' actions as "indispensable." | 70% opposed to monitoring citizens of the UK | Foreign secretary Hague reassures public that data-mining efforts are not an issue: If you are a law-abiding citizen... nothing to fear in terms of content of calls being listened to |
| What: Personal communications, telephone metadata, emails | | 65% opposed to monitoring leaders of the UK | |
| When: Since 2007 (PRISM) | | | |
| Why: Wide net surveillance for analysis. | | | |
| How: Likely through PRISM, or other data-mining programs | | | |

In the US, official statements in response to the leaks included those by then-Director of National Intelligence James Clapper who said the leaks have done "huge, grave damage" to the US intelligence capabilities ("Intelligence Chief" 2013), and those of President Obama who was critical of Snowden's methods but announced the creation of an independent panel to examine the US government's surveillance practices. The panel's findings recommended that the mass collection of telephone records be suspended, and they also advised greater oversight of sensitive programs like those targeting foreign leaders who are allies (Savage 2014). While President Obama did act on some of these suggestions and recommended Congressional review of others, e.g., an early attempt to cut funding for NSA's phone metadata program suffered a narrow defeat ("News, Opinion" 2013), the role of the NSA and its data gathering remained in contentious debate between the intelligence community and privacy advocates (Ray 2016). Therefore, the severity of response is assessed as "low."

**Table 6** Surveillance of Chinese Company Huawei & Chinese Officials (Escalation: 2 & 3)

| Surveillance Activity by NSA | Official Statements in Response to Leaks | Chinese Perception of NSA Activities (% Unfavorable) | Target Nation Action = Medium |
|---|---|---|---|
| Who (subject): Political leaders, as well as Huawei (world's second largest network equipment provider, competitor to Cisco) | "China has already lodged many complaints with the United States about this. We demand that the United States makes a clear explanation and stop such acts…"- Chinese Foreign Ministry spokesman Hong Lei | 85% opposed to monitoring citizens of the China | Reverse-engineered a powerful NSA program, denial of service attacks against American websites to prevent Chinese internet users from circumventing blocked websites |
| What: Accessed company emails AND software source code | | 85% opposed to monitoring leaders of China | |
| When: Began 2009 | | | |
| Why: To be able to exploit these products used by many NSA targets | | | |
| How: NSA infiltrated network through Shenzhen central office, where all email comes through | | | |

### 4.3.2 Mexican President Enrique Peña Nieto

The Snowden leaks revealed that during the summer of 2012 Mexican presidential campaign, the text messages of presidential candidate Enrique Peña Nieto were being collected by the NSA, including one message from the candidate indicating his choice for a staff appointment. Sixty-six percent of Mexicans said that the NSA's surveillance of their leader is unacceptable. Because the NSA specifically targeted a head of state, this activity would be ranked at Threshold 3 on the ladder of escalation. Mexico's response after this leak was rather muted. Mexico's Foreign Ministry merely asked the US in a diplomatic note for an "exhaustive investigation" into the matter (Archibold 2014). Later in October, 2013, it would be revealed that the NSA was reportedly systematically eavesdropping on the Mexican government for years, hacking into the president's public email account, allowing deep insight into Mexican policy making and the political system. Even after this revelation, the

**Table 7** Surveillance of German Chancellor (Escalation: 3)

| Surveillance Activity by NSA | Official Statements in Response to Leaks | German Perception of NSA Activities (% Unfavorable) | Target Nation Action = High |
|---|---|---|---|
| Who (subject): German Chancellor (Merkel, Schroeder, Kohl) | Foreign Office's Phillip Mibfelder declared "the current situation in transatlantic relations is worse than it was at the low-point in 2003 during the Iraq war" | 87% opposed to monitoring citizens of the Germany | Germany ejects CIA officer in Berlin; eavesdrops on US secretary of state; cancels symbolic cold-war era agreement to allow surveillance to protect US, French, British troops stationed in W. Germany; low altitude flyover of US consulate in Frankfurt to assess intelligence capabilities; joint proposed resolution with Brazil to UN about international law protecting personal data |
| What: Spying on personal communications | Merkel: "We need trust.... Spying among friends is never acceptable." | 90% opposed to monitoring leaders of Germany | |
| When: Since 2002 | | | |
| Why: Possibly in lead-up to decisions on Iraq war | | | |
| How: Infiltrated German networks, intercepts of phone conversations and fax machine | | | |

response was only for the Mexican Foreign Minister to meet with US Ambassador to discuss the allegations that the NSA hacked the emails of then-president Felipe Calderon while in office. It is not clear what further actions have been taken ("Mexico Foreign" 2013). Compared to the more severe reactions of other nations in terms of action taken, this relatively muted response of Mexico is given a "low" severity classification due to the relatively small discernible change in strategic US-Mexico relations. One reason President Peña Nieto may not have wanted to push the issue was he likely had interest in addressing other pressing policy issues at the time, including US immigration policy and Mexico's security infrastructure along its northern and southern borders (Evans 2013). Thus, other political agendas may also play a role in the response of world leaders to the leaks.

**Table 8** Surveillance of Brazilian President & Petrobas Oil Company (Escalation: 3 and 2)

| Surveillance Activity by NSA | Official Statements in Response to Leaks | Brazilian Perception of NSA Activities (% Unfavorable) | Target Nation Action = High |
|---|---|---|---|
| Who (subject): President Dilma Rousseff and top aides, and Petrobas oil company | "Tampering in such a manner in the affairs of other countries is a breach of international law and is an affront of the principles that must guide the relations among them, especially among friendly nations." –Rousseff's speech to UN | 94% opposed to monitoring citizens of the Brazil | Rousseff cancels visit to white house; proposal to UN - resolution on privacy; plans to reroute internet traffic through underwater cable; plans to create national email system; possible (unconfirmed cause) downgrading of commercial ties including purchase of F-18 s |
| What: Spying on personal communications | | 83% opposed to monitoring leaders of Brazil | |
| When: 2011 or earlier | | | |
| Why: Possibly economy and security concerns | | | |
| How: NSA used different computer programs to filter through communications, access to specific emails, telephone calls, texts of Rousseff's top aides, bugging presidential plane | | | |

### 4.3.3 United Kingdom Citizens

Some 70% of citizens of Britain said that monitoring of British citizens by the US government is unacceptable. In the case of the Snowden leaks, the Foreign Secretary admitted that Britain's Government Communications Headquarters (GCHQ), which is similar to the NSA, was collaborating with the NSA in its surveillance activities. The Deputy Prime Minister defended the two agencies' actions as indispensable. Overall there was a quiet response on the part of British citizens when the leaks were shared. Britain's support for civil liberties has been declining since the 1990s, and overall trust in government remains high. In fact, only 19% of the public think that British security services should cut back their surveillance powers. Members of the public tended to believe the recent leaks about the security services were a bad thing. Actions taken by British officials included reassurance to the public that the data-mining operations are not an issue people should be concerned about. British Foreign Secretary Hague said "The net effect is that if you are a law-abiding citizen of this country going about your business and personal life, you have nothing to fear about the British state or intelligence agencies listening to the content of your phone calls or anything like that" (Paramagur 2013). Hague was reportedly not asked difficult questions by parliament after the leaks came out. The minimal actions taken by the UK are evaluated at a "low" severity rating in our assessment.

### 4.3.4 Huawei (Chinese Networking Technology Company) and Chinese Officials

Surveys found that 85% of Chinese said they viewed NSA monitoring of Chinese citizens as "unacceptable." In addition to targeting Chinese political leaders, the Snowden leaks revealed that the NSA was able to infiltrate Huawei, a Chinese company that is the world's second largest network equipment provider and a competitor to Cisco. As opposed to a head of state, surveilling a company would be classified as Threshold 2 on the ladder of escalation.

Why did the NSA spy on Huawei? In the NSA documents, the reason given was that "[m]any of our targets communicate over Huawei produced products. We want to make sure that we know how to exploit these products" (Sanger and Perlroth 2014). Also, it was mentioned that "Huawei's widespread infrastructure will provide the PRC with SIGINT capabilities." The method of surveillance was that the NSA infiltrated the company network through their Shenzhen central office, which all email passes though. The NSA not only accessed company emails but also software source code. One internal NSA document stated "We currently have good access and so much data that we don't know what to do with it..." It is also interesting to note that this type of spying on a competitive corporation may seem outside the scope of national security and more about economic gain. When challenged, the NSA responded saying "our intelligence activities are focused on the national security needs of our country, we do not give intelligence we collect to US companies to enhance their international competitiveness." However it was also noted in an NSA leaked document regarding infiltrating Huawei, that "the intelligence community structures are not suited for handling issues that combine economic, counterintelligence, military influence and telecommunications infrastructure from one entity" suggesting there was some internal doubt about the scope of the surveillance being done.

The response from Chinese officials to initial leaks revealing spying on China included the statement "China has already lodged many complaints with the United States about this. We demand that the United States makes a clear explanation and stop such acts" ("NSA Spied" 2014). By 2015, the actions taken by the government in response to the Snowden leaks included (according to researchers at the University of California and University of Toronto) making use of details of a powerful NSA program that was contained in the leaks. The Chinese reverse-engineered it so that the Chinese government could use "denial of service attacks" against American websites, and block providers that allowed China's Internet users to circumvent websites blocked by government policies (Street 2015). While this action was not taken against the US government directly, it involves more action taken than simply calling for a meeting or investigation, and therefore is classified as "medium" severity in our assessment.

### 4.3.5 Germany

Approximately 87% of Germans said the NSA's monitoring of German citizens is unacceptable, and 90% said monitoring of their leader is unacceptable. When allegations first appeared that the NSA was spying on German citizens, German Chancellor Angela Merkel defended the surveillance practices to angry citizens, saying the United States was "our truest ally throughout the decades and remains so ("German Chancellor Merkel Defends" 2013). No apparent immediate action was taken in response other than German Interior Minister Hans-Peter Friedrich expressing that he would communicate to US Attorney General Eric Holder on his upcoming visit that the situation is being taken seriously in Germany and that that trust needs to be built again, making clear what should be expected from each other as partners and friends ("German Chancellor Merkel Defends" 2013).

However, when it was discovered that the personal phone conversations of the Chancellor were being monitored since 2002, the rhetoric quickly escalated. Merkel made comments such as "Trust needs to be rebuilt…" and "we need trust… Spying among friends is never acceptable, no matter in what situation." (Smith-Spark 2013). Later, Germany's Foreign Office's Phillip Missfelder said "The current situation in transatlantic relations is worse than it was at the low-point in 2003 during the Iraq War" ("US-Germany" 2014). Actions taken by Germany included ordering the CIA's top officer in Berlin to leave the country. This is significant because it meant the US had to withdraw an officer that was the main point of contact with German intelligence services, exchanging key information ranging from terrorist plots to Iranian nuclear ambitions (Miller and Kirchner 2014). German media also reported in August 2014 that German intelligence eavesdropped on phone calls by US Secretary of State John Kerry and his predecessor Hillary Clinton, saying the calls had been picked up "accidentally" ("Snowden NSA" 2015). German government action also included cancelling a Cold-War era agreement to allow surveillance to protect US, French, and British troops stationed in West Germany (although this agreement was referred to as largely symbolic). Other action included a low altitude flyover of the US consulate in Frankfurt to determine intelligence capabilities. Perhaps the most significant action taken was Germany's joint proposal with Brazil to the United Nations concerning establishing international law protecting personal data in the Internet age. Taking into account all of these immediate actions taken by Germany, we would classify their severity level as "high" relative to the responses by other nations, judging by their potential strategic implications and potential negative impact on a strong alliance between two powerful countries.

However, relations with Germany seem to have warmed significantly in the few years following the leaks. Following the ejection of the CIA officer from Berlin, Obama's Chief of Staff was dispatched to Germany to resolve the tensions (Liptak 2016). It also came out that the Germany's BND foreign intelligence service may have assisted in the surveillance of thousands in Germany (Wagstyl 2015). Despite the US providing no outright denial of the monitoring, saying that the Chancellor's phone was not currently being monitored, nor would it be in the future, Germany dropped the investigation into the alleged tapping of Chancellor Angela Merkel's phone by the NSA citing insufficient evidence in 2015 ("Snowden NSA" 2015). As of April 2016, relations with Germany seem to have returned to a very friendly state. Obama arrived in Hanover, sharing a warm embrace with Merkel, and giving kind interviews to local news on his fondness of Merkel, before proceeding to lend Merkel's backing as she faced opposition over her stand on refugees fleeing war in Syria (Liptak 2016).

### 4.3.6 Brazil

Surveys indicate that 94% of Brazilians said US government monitoring of Brazilian citizens is unacceptable, while 83% said monitoring of their leader is unacceptable. The leaks revealed that the NSA spied on both Brazil's largest company, Petrobras, and on Brazil's President Dilma Rousseff and her top aides. The NSA used different computer programs to filter through communications, gain access to specific emails, telephone calls, and text messages of Rousseff's top aides, including by bugging the presidential plane. The targeting of Brazil's largest company and those serving as top officials in managing the economy may have caused Wikileaks to believe economic espionage was at play (Alcindor 2015). When news of the surveillance broke, Rousseff was furious, cancelling her important, rare visit to the White House and making a scathing speech to the United Nations General Assembly to propose a resolution on privacy of personal data in the age of the Internet. In her speech she said "[t]ampering in such a manner in the affairs of other countries is a breach of international law and is an affront of the principles that must guide the relations among them, especially among friendly nations. A sovereign nation can never establish itself to the detriment of another sovereign nation. The right to safety of citizens of one country can never be guaranteed by violating fundamental human rights of citizens of another country" (Borger 2013). This resolution proposed by both Germany and Brazil "to respect and protect the right to privacy, including in the context of digital communications, was adopted by all 193 UN members in December, 2013, with later versions also being adopted. United Nations General Assembly resolutions are not legally binding as they are for the Security Council, but they are representative of international opinion on the issue (Kopczynski 2013). In addition to these actions taken by Brazil, a Brazilian general announced made plans to reroute internet traffic through an underwater cable. Brazil also chose a French-Italian venture to build a satellite for military and civilian use to ensure

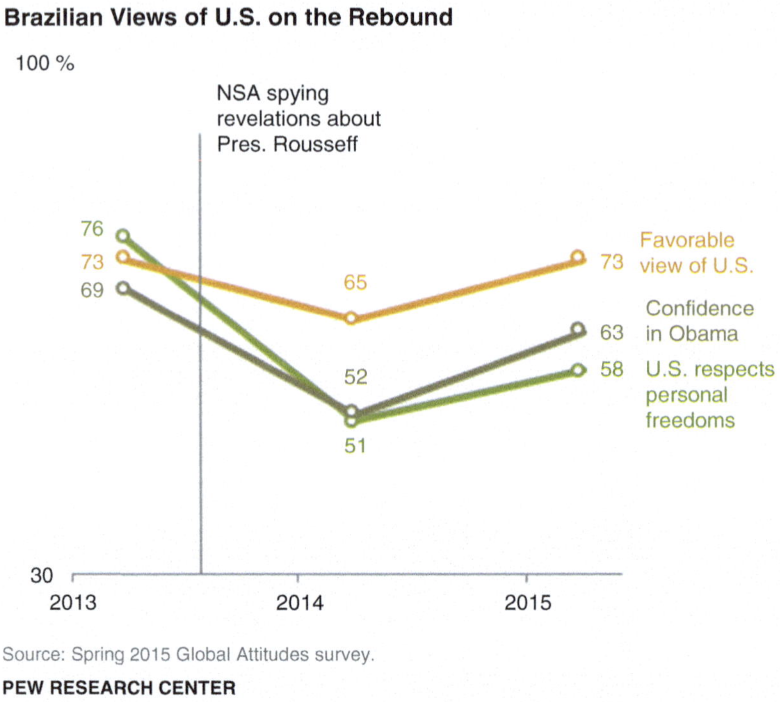


**Fig. 4** Brazilian public sentiment towards US over time. Image from Pew Research Center (Devlin 2015)

sovereignty of important communications. Authorities also ordered Brazil's Postal service to develop a national email system for sharing encrypted messages (Archibold 2014). There was also unconfirmed speculation by analysts that the reason Brazil made the surprising move to purchase Swedish fighter jets over Boeing F-18 aircraft in a $4.5 billion deal in December 2013 was because Boeing's bid was hurt by the recent spying revelations (Brooks 2013).

These actions taken by Brazil are given a "high" level of severity in our assessment, due to their evidence of serious diplomatic fallout (at least initially) and passing of international laws and potential trade reductions. However, by mid-2015, Brazil-US relations had normalized considerably to their positive state. Figure 4 taken from Pew Research Center shows opinion polls of the Brazilian views of the US both before and after the spying revelations, showing that although favorable views of the US dipped the year after the leaks, they were on the rebound the following year (Devlin 2015). Rousseff made an official visit to the US in June 2015. She stated "President Obama and the US government have stated on several occasions that they would no longer engage in intrusive acts of spying on friendly countries. I believe President Obama" (Roberts 2015).

## 5 Conclusions

It was found that most of the NSA surveillance activities revealed in the Snowden leaks and covered in the media would rank at about a 2 or a 3 out of 4 on the ladder of escalation, used to measure their level of breaching of ethical code. Surveilling of citizens both domestic and foreigners at a distance (level 2) has become routine. It was also revealed that monitoring personal communications of foreign heads of state, which ranks high on the ladder at level 3, has also been a commonplace activity, with 35 world leaders reportedly targeted by the NSA. While the overall high level of global opposition to these activities was clear based on public opinion polls, the differences in the nation-state official responses and activity as a result of the surveillance disclosures were found to be relatively aligned with the level of disapproval expressed by citizens of the countries. Countries with lower percentages of people expressing that NSA monitoring of their citizens or leader was unacceptable (such as Mexico and Britain) tended to have muted responses, with only minor actions taken. Countries like Brazil and Germany, whose citizens have higher disapproval for surveillance activities, took more severe actions in response to the surveillance disclosures.

Public opinion regarding NSA activity is correlated to the severity of the nation's response to the leaks - the stronger the opposing public opinion in the state, the more severe were the actions taken by the state. A democracy requires debate over the morality and ethics of intelligence actions, if international security is desired - ethical standards used by intelligence officials are different from ethical standards applied by individuals. A wide range of technological methods were used by the NSA with varied response. Rhetoric among public officials was heated in the months immediately after the surveillance disclosures and global survey respondents strongly viewed NSA surveillance activities as "unacceptable".

Overall, however, the impact on US-foreign relations may have been short-lived, with heated rhetoric and retaliation occurring in the months immediately following the disclosures but with diplomatic reconciliation occurring in the next two years, resulting in maintaining of important friendly alliances in part due to faith in agreements made to no longer engage in "intrusive acts of spying on friendly countries." It is unclear whether the relationship between public opinion and the severity of a nation's response is causal or simply correlated. Perhaps public opinion is swayed by the rhetoric of a leader in response to the leaks, or perhaps leaders take action in response to the level of outrage of their citizens, or combinations of the two or neither is true. The political climate and a country's approval of one's leader may also play an important role. The Snowden leaks raised the level of public scrutiny and looking forward, the preservation of privacy will likely be examined more critically in part due to the adoption of the recent UN resolution on the right to respect and protect one's privacy in the digital age.

Policy makers may wish to consider legislation that makes the scope of surveillance activities more transparent or eliminates them completely. Based on the case studies here in which trust between governments was broken, such as in the case of Germany, in which the Chancellor stated "spying among friends is never acceptable" and in the case of Brazil, whose president said that "Tampering in such a manner in the affairs of other countries is a breach of international law and is an affront of the principles that must guide the relations among them" there is evidence for this kind of deception contributing to increased contention on the world stage, while in these cases it is difficult to find convincing evidence of these surveillance activities promoting well-being or security in global relations. Moreover, abiding by the principles of UN resolution 68/167 on the right to privacy in the digital age, adopted by the General Assembly in December 2013, may help the United States to gain regain the trust of fellow nations which, in adopting this resolution, "expressed deep concern at the negative impact that surveillance and interception of communications may have on human rights."

The negative impacts of surveillance and interception of communication were made all too apparent during the 2016 US National Election. The 2016 US National election has shepherded in a new era for WikiLeaks. It is now widely believed that WikiLeaks, "overwhelmingly focused its surveillance activities on the United States, while seeking support from anti-democratic countries and organizations, and seeking in part to harm the Democratic National Candidate Hillary Clinton while aiding candidate Donald Trump" (Windrem 2017). WikiLeaks, previously characterized as the middle man, or simply the vehicle by which secret information, news leaks, and classified media was published from anonymous sources, is now viewed as an active member of the intelligence surveillance community. On April 13, 2017, CIA Director Mike Pompeo publicly stated: "It is time to call out WikiLeaks for what it really is," he said, "a non-state hostile intelligence service often abetted by state actors like Russia. WikiLeaks walks like a hostile intelligence service and talks like a hostile intelligence service. It has encouraged its followers to find jobs at CIA in order to obtain intelligence (Windrem 2017)."

As in the response to state-on-state or nation-on-nation surveillance, the response to WikiLeaks has differed based on the surveillance target. For example, response to the knowledge that WikiLeaks sought to impact and potentially destabilize the US elections has been less that favorable. While this work examined the ethical implications of state-on-state or nation-on-nation surveillance and the international response to US intelligence surveillance, it is apparent that moving forward, the international surveillance community will need to take into account the role that WikiLeaks and other entities like it play in matters of international security and well-being.

## References

Alcindor, Yamiche. (2015). Wikileaks: US Wiretapped Brazilian Presidents. *USA TODAY.*. Accessed 21 Dec 2017. https://www.usatoday.com/story/news/2015/07/04/wikileaks-us-wiretapped-brazilian-presidents/29694565/.

Andregg, M. (2007). Intelligence Ethics. In L. Johnson (Ed.), *Strategic Intelligence – Volume 2*. Praeger Security International.

Archibold, Simon. (2014). Brazil Angered Over Report N.S.A. Spied On President. *Nytimes.com*.

Ball, James. (2013). NSA Monitored Calls Of 35 World Leaders After US Official Handed Over Contacts. *The Guardian*. https://www.theguardian.com/world/2013/oct/24/nsa-surveillance-world-leaders-calls. Accessed 3 May 2016.

Beebe, S.M. 2015. Cases in intelligence analysis: Structured analytical techniques in action (Second Edition). Press an Imprint of Sage Publication.

Bellaby, R. (2014). *The ethics of intelligence: A new framework*. Routledge.

Borger, Julian. (2013). Brazilian President Rousseff: US Surveillance A 'Breach Of International Law'. *The Guardian*. https://www.theguardian.com/world/2013/sep/24/brazil-president-un-speech-nsa-surveillance. Accessed 4 May 2016.

Born, H., & Wills, A. (2010). Beyond the oxymoron. In J. Goldman (Ed.), *Ethics of spying a reader for intelligence professionals – Volume 2*. United States: Scarecrow Press, Inc..

Brooks, Bradley. (2013). Brazil chooses Swedish fighter jet over Boeing F-18 in $4.5B deal. *The Seattle Times*. https://www.seattletimes.com/business/boeing-aerospace/brazil-chooses-swedish-fighter-jet-over-boeing-f-18-in-45b-deal. Accessed 4 May 2016.

Burrough, B., Ellison, S., Andrews, S. and McCabe, S. (2014). Snowden speaks: A vanity fair special report. *The Hive*. https://www.vanityfair.com/news/politics/2014/05/edward-snowden-politics-interview. Accessed 3 May 2016.

Central Intelligence Agency (CIA). (1983). Factbook on Intelligence, CIA Office of Public Affairs, April 1983.

Charters, D. (2006). Business: The challenge of completely ethical competitive intelligence. In J. Goldman (Ed.), *Ethics of spying a reader for intelligence professionals*. United States: Scarecrow Press, Inc..

Devlin, Kat. (2015). Brazilians' views of US rebound as wounds of NSA scandal heal." *Pew Research Center*. http://www.pewresearch.org/facttank/2015/06/29/brazilians-views-of-u-s-rebound-as-wounds-of-nsa-scandal-heal. Accessed 4 May 2016.

Evans, Michael. (2013). Mexico privately hoped to "put to bed" tensions raised by Snowden leaks. *UNREDACTED*. https://unredacted.com/2013/10/24/mexico-privately-hoped-to-put-to-bed-tensions-raised-by-snowden-leaks. Accessed 4 May 2016.

Gellman, Barton and Soltani, Ashkan. (2013a). NSA Collects Millions Of E-Mail Address Books Globally. *Washington Post*. https://www.washingtonpost.com/world/national-security/nsa-collects-millions-of-e-mail-address-books-globally/2013/10/14/8e58b5be-34f9-11e3-80c6-7e6dd8d22d8f_story.html?utm_term=.490519311600. Accessed 3 May 2016.

Gellman, Barton and Soltani, Ashkan. (2013b). NSA Tracking Cellphone Locations Worldwide. *Washington Post*. https://www.washingtonpost.com/world/national-security/nsa-tracking-cellphone-locations-worldwide-snowden-documentsshow/2013/12/04/5492873a-5cf2-11e3-bc56-c6ca94801fac_story.html?utm_term=.109f77ba75ff. Accessed 3 May 2016.

Gellman, B., Tate, J. and Soltani, A. (2014). In NSA-intercepted data, those not targeted far outnumber the foreigners who are. *Washington Post.*. Accessed 3 May 2016.

German Chancellor Merkel Defends Work of Intelligence Agencies - SPIEGEL ONLINE' *SPIEGEL ONLINE*.( 2013). http://www.spiegel.de/international/germany/german-chancellor-merkel-defends-work-of-intelligence-agencies-a-910491.html Accessed 4 May 2016.

Global Opposition To US Surveillance And Drones, But Limited Harm To America'S Image. (2014). *Pew Research Center's Global Attitudes Project*. Accessed 3 May 2016.

Goldman, J. (Ed.). (2010). *Ethics of Spying: A Reader for the Intelligence Professional – Volume 2*. Lanham: Scarecrow Press, Inc..

Goldman, J. (2013). Teaching About Intelligence Ethics. *Journal of US Intelligence Studies, 20*(2, Fall/Winter).

Greenwald, Glenn, and Laura Poitras. (2013). NSA Whistleblower Edward Snowden: I Don't Want To Live In A Society That Does These Sort Of Things – Video. the Guardian. https://www.theguardian.com/world/video/2013/jun/09/nsa-whistleblower-edward-snowden-interview-video. Accessed 3 May 2016.

Hulnick, S. A., & Mattsauch, D. W. (2006). Ethics and Morality in US Secret Intelligence. In J. Goldman (Ed.), *Ethics of Spying a Reader for Intelligence Professionals*. United States: Scarecrow Press, Inc..

Intelligence and National Security. (2007). 22, No. 1 (February 2007).

Johnson, L. K. (2006). Ethics of covert operations. In J. Goldman (Ed.), *Ethics of Spying a Reader for Intelligence Professionals*. United States: Scarecrow Press, Inc..

Jones, J. M. (2010). Is ethical intelligence a contradiction in terms. In J. Goldman (Ed.), *Ethics of spying a reader for intelligence professionals*. United States: Scarecrow Press, Inc..

Kopczynski, Pawel. 2013. 'To respect and protect right to Privacy': UN votes for end to excessive electronic spying — RT news' RT International. <https://www.rt.com/news/un-resolution-worldwide-surveillance-476/> Accessed 4 May 2016.

Liptak, Kevin. (2016). White House Producer. How Obama, Merkel Learned To Love One Another. *CNN*. https://www.cnn.com/2016/04/24/politics/barack-obama-angela-merkel-germany-europe/index.htm. Accessed 4 May 2016.

Mexico Foreign Minister Meets US Envoy Over Spy Claims. *Mobile.Nation.Co.Ke*. (2013). Accessed 4 May 2016. http://mobile.nation.co.ke/news/Mexico-foreign-minister-meets-US-envoy-over-spy-claims/-/1950946/2047620/-/format/xhtml/-/g9qrh7/-/index.htm.

Miller, Greg and Kirchner, Stephanie. (2014). Germany Orders CIA Station Chief To Leave Over Spying Allegations. *Washington Post*. https://www.washingtonpost.com/world/europe/germany-expels-us-intelligence-station-chief-over-spying-allegations/2014/07/10/dc60b1f0-083c-11e4-8a6a-19355c7e870a_story.html?utm_term=.45aa03ac24a0. Accessed 4 May 2016.

Mitchell, A., Bratu, B., and Deluca, M. (2013) Intelligence Chief Declassifies PRISM Details, Slams "Reckless Disclosures". US News. (June 8, 2013). http://usnews.nbcnews.com/_news/2013/06/08/18850035-intelligence-chief-declassifies-prism-details-slams-reckless-disclosures. Accessed 3 May 2016.

'News, Opinion and Aggregation on Business, Politics, Entertainment, Technology, Global and National – The Wire' (n.d.). https://www.thewire.com/politics/2013/07/amash-amendment-fails-despite-democratic-support/67584. [Accessed 4 May 2016].

NSA Spied On Chinese Government And Networking Firm - SPIEGEL ONLINE - International. *SPIEGEL ONLINE*. (2014). Accessed 4 May 2016. http://www.spiegel.de/international/world/nsa-spied-on-chinese-government-and-networking-firm-huawei-a-960199.html.

Paramagur, Kharunya. (2013). Three Months After Snowden'S NSA Revelations, Europe Has Moved On TIME.Com. *TIME.com*. http://world.time.com/2013/09/27/three-months-after-snowdens-nsa-revelations-europe-has-moved-on. Accessed 4 May 2016.

George W. Bush. (2001). Signs the Patriot Act, Anti-Terrorism Legislation, in the East Room Oct. 26. "With My Signature, This Law Will Give Intelligence and Law Enforcement Officials Important New Tools to Fight a Present Danger," said the President in H' http://georgewbush-whitehouse.archives.gov/news/releases/2001/10/images/20011026-5.html Accessed 3 May 2016.

Ray, M. (2016). *Edward Snowden | Biography & Facts*. [online] Encyclopedia Britannica.. Available at: https://www.britannica.com/biography/Edward-Snowden. Accessed 3 May 2016.

Roberts, Dan. (2015). Brazilian President's Visit To US Will Not Include Apology From Obama For Spying. *The Guardian*. https://www.theguardian.com/world/2015/jun/30/brazil-dilma-rousseff-obama-nsa-spying-apology. Accessed 4 May 2016.

Rusbridger, Alan, and Ewen MacAskill. (2014). I, Spy: Edward Snowden In Exile. *The Guardian*.. Accessed 3 May 2016.

Sanger, David E., and David Perlroth. (2014). N.S.A. breached Chinese servers seen as security threat. *The New York Times Company*.. Accessed 21 Dec 2017. https://www.nytimes.com/2014/03/23/world/asia/nsa-breached-chinese-servers-seen-as-spy-peril.html?_r=0.

Savage, David. (2014). Obama Is Urged To Sharply Curb N.S.A. Data Mining. *Nytimes.com*.. Accessed 30 Aug 2017.

Shpiro, S. (2010). Speak no evil: Intelligence ethics in Israel. In J. Goldman (Ed.), *Ethics of spying a reader for intelligence professionals – Volume 2*. United States: Scarecrow Press, Inc.

Smith-Spark, Laura. (2013). Merkel: Relations with US 'Severely Shaken' over spying claims – CNN. *CNN*. Accessed 4 May 2016.

'Snowden NSA: Germany Drops Merkel Phone-Tapping Probe - BBC News' *BBC News*. (2015). <http://www.bbc.com/news/world-europe-33106044> Accessed 4 May 2016.

Street, Chriss W. China Launches Internet Retaliation Tool Leaked By Edward Snowden. (2015). *Breitbart*. Accessed 4 May 2016.

Strohm, C. and Wilber, D. (2014). *Pentagon Says Snowden Took Most US Secrets Ever: Rogers*. [online] Bloomberg.com. Available at: https://www.bloomberg.com/news/articles/2014-01-09/pentagon-finds-snowden-took-1-7-million-files-rogers-says. Accessed 3 May 2016.

'US-Germany Relations Hit New Low amid NSA Spying Scandal, Official Says - World News' *NBC News*.( 2014). http://www.nbcnews.com/storyline/nsa-snooping/u-s--german-relations-hit-new-low-amid-nsa-n11496. Accessed 4 May 2016.

Wagstyl, Stefan. (2015) Revelations of role in spying on allies turn tables on berlin. *Financial Times*. https://www.ft.com/content/3bf16734-fd5d-11e4-b072-00144feabdc0. Accessed 4 May 2016.

Windrem, Robert. (2017). CIA director Pompeo calls WikiLeaks a 'hostile intelligence service. *NBC News*. 2017. https://www.nbcnews.com/news/us-news/cia-director-pompeo-calls-wikileaks-hostile-intelligence-service-n746311. Accessed 4 May 2016.

# Conclusions

**Anmol Soni and Margaret E. Kosal**

The chapters in this volume provide different cases and examples of the role of technology in the intelligence community and the evolution of organizational and institutional infrastructure to use and adopt technology over time in the community. While technology changes and effects the way intelligence is collected, there is an element of continuity in the underlying organizational and social infrastructure.

The cases presented throughout the chapters of this book touch upon these central concepts in organizational structure, people, leadership, and institutional culture of the intelligence community and broader US government that have interacted with in varying ways with the changing technology and how those themes continue with new and emerging technologies. Technology is not a panacea without the people and organizational structure in place to take advantage of it. This chapter summarizes these, recounting the examples from the different cases to provide a concise conclusion on the technology-intelligence relationship.

A. Soni (✉)
Georgia Institute of Technology, Atlanta, Georgia, USA
e-mail: anmolsoni@gatech.edu

M. E. Kosal
Sam Nunn School of International Affairs, Georgia Institute of Technology, Atlanta, Georgia, USA
e-mail: margaret.kosal@inta.gatech.edu



M. E. Kosal (ed.), *Technology and the Intelligence Community*, Advanced Sciences and Technologies for Security Applications,
https://doi.org/10.1007/978-3-319-75232-7_15

## 1 The Pace of Adoption and Effectiveness of Technology Use for National Security by the Intelligence Community Varies. Critical Factors in Increasing Effectiveness Include Organizational Structure and Organizational Culture, Which Can Play an Important Role in the Process of Technology Adoption

In other words, structure and processes matter. Evidence to this fact is borne out in most cases described in the previous chapters. In the early years of intelligence, following intelligence failures during WWII, the importance of institutional issues was recognized. This led to changes in the organizational structure which became more centralized, and the CIA was created. Efforts were also made to address issues related to inter-agency coordination. Improvements followed quickly, but institutional issues were never completely eliminated. As a result, these appeared again during the Cold War, and later during counter-proliferation efforts when the intelligence community's assessments were frequently distorted due to a range of factors, such as biases, preconceived cultural notions and politicization of issues. For example, during the Cold War, the belief held by the then administrators regarding the primary role of the CIA and the focus on collecting Human Intelligence (HUMINT) led to the slow pace of adoption of new and technologically advanced methods of intelligence gathering and processing.

Schmid's analysis of the role that organizational innovation played in furthering the US response to the launch of Soviet Union's Sputnik I also corroborates this. Bernstein examines the role of technology and how it can be an input to developing the organizational structure of intelligence agencies. Citing the examples of COMINT during the WWII, he makes the point that technology would have potentially exacerbated underlying structural issues. In particular, he argues that structures should be independent of the available technology since the core aspects of intelligence do not vary. This is a theme that re-emerges throughout thethe different cases explored.

## 2 Examining the Role of the "Human in the Loop" Is Essential and Arguably More Important than Any Specific Technology

Individuals play an important role not only in championing initial research and developing the appropriate technology but also in promoting and advancing the use of new technologies in intelligence gathering. The human aspects are critical when decisions need to be made on how to analyze and use the intelligence gathered. And finally, at times, humans can also be the weak link in the process of gathering intelligence.

## 2.1 *Humans Are Key to Conducting Analyses of the Information After the Technology Has Been Used to Collect and Process It*

Evidence for the role of humans in analyzing the information and being able to make decisions based on it are prevalent throughout the cases presented in the preceding chapters.

The comparison of the U2 Incident in 1960 and the RQ-170 incident in Kandahar in 2011 by Ruckle is an example of the role of humans in the technology-intelligence linkage. The use of a risk-reward analysis justification process was in place for aerial surveillance overflights. Although it was not formalized for the early U-2 program, President Eisenhower held tight control over Soviet overflights. Overflights were only performed when all other intelligence options were exhausted and the intelligence needed was vital to national security.

Going forward, more recent advances in technology such as the use of big data, cyber operations, and augmented reality are all poised to deepen the role of the human in the loop. In fact, as articulated by Check, the role of humans – the analysts and officers of the intelligence community and the policy-makers driving choices about how technology is developed, implemented, and used for intelligence and making decisions aided by new technologies – will be paramount as operators in emerging technologies such as augmented reality. Jani and Soni also describe the critical role that humans will play in analyzing the data and learning from field experience that can be used in Big Data analytics. In particular, the role of humans in analyzing the data collected and finding meaning in that analysis using big data techniques will continue to be critical to its role in intelligence. In cyber operations as well, Mahvi points to the role of humans in identifying the targets and combining intelligence from different sources. This helps in putting a larger picture together and using cyber intelligence to complement rather than substitute traditional intelligence.

Finally, the corollary to the all-pervasive role of humans in intelligence gathering is that extraordinary reliance on technology and a lack of focus on the role of the human in the loop can have adverse implications. This was evident in the experiences of the Cold War where an over-reliance on intelligence gathered though technical means led to intelligence failures.

## 2.2 *People Ensure a Continuity in the Process of Decision Making*

Humans have, and will continue to play, a critical role in ensuring continuity in the intelligence community. Tracing the development of the Corona program indicates the prominent role that leadership played in championing the program, overcoming obstacles, and gaining buy-in from decision makers. Despite thirteen mission failures due to technical glitches initially and exceeding its budget, the support of key leadership from participating agencies ensured eventual success. Several people, champions within and outside the intelligence and defense communities, came out in support of the program at different points, ultimately gaining approval from the Eisenhower administration.

In the analysis of intelligence for nuclear counterproliferation, Abou Jaoude touches on the role that technology can have in slowing the decision-making process across multiple states and even leading to inaccurate views of nuclear programs of states. In such cases, the pace at which decision are made hinges on the role of the humans in the process.

## *2.3 People Are Also the Potential Weak Link as They Can Be the Source of Problems*

Analyses of various technology and implementation efforts through the years show us that a technology can aid in the gathering, analysis, or dissemination of intelligence, but it is still subject to the faults of the people and policies that govern it and its use. The best technology won't overcome a broken process.

This is borne out in the discussion of the Cuban Missile Crisis by McGrath. Despite reports from refugees and defectors from Cuba that the Soviet Union had brought missiles to the nation, US intelligence community did not seek to confirm or deny these reports until nearly a month and a half later when US reconnaissance aircraft confirmed the presence of Soviet missiles deployed in Cuba. This considerably delayed the process of decision-making, thereby limiting the available options for response. Failures on the part of humans in the intelligence cycle are also evident in the Iraqi nuclear program in the 1990s, which consequently led the CIA to overestimate the extent of the Iraqi, North Korean, and Iranian programs and follow an approach that Abou Jaoude characterizes as pessimistic and "better-safe-than-sorry."

As McGrath concludes, biases and ongoing pressures to deliver in accordance with the expectations of the administrations have been the primary causes of intelligence failures.

## *2.4 Humans Are the Decision-Makers in the End*

The value of intelligence depends on how the decision makers choose to use the information provided. Even though the gathering of intelligence and its analysis can be completely accurate, and the findings effectively communicated, the final decision still rests on those with authority.

As Abou Jaoude points out, in the case of nuclear nonproliferation, after the intelligence has been collected, recommendations have to be made by leadership in the Executive Branch and, in some cases, in Congress. A considerable level of input is needed in the form of human intelligence. Sudharsan also explores the role of the leadership in determining the appropriate organizational responses to the continually evolving technology and in deciding when and how to restructure.

## 3 Policy Plays an Important Role and Efforts Need to Be Made in Order to Keep up with the Pace of Development in Technology, to Both Take Advantage of Them Effectively and Anticipate & Prepare for Use by Adversaries

The policy-technology relationship has its roots in the very creation of intelligence agencies in the United States. However, change in technology typically precedes policy in all fields, not in the least, its application to intelligence. This, however, has ramifications in adoption, application, and the use of technology in the intelligence gathering and analysis processes.

The development and adoption of wire-tapping technology led to one of the initial discussions in the policy-ethics and technology space. The 1928 decision by the Supreme Court ruling wiretapping legal without a court issued warrant followed by the 1934 decision of the Congress to ban all wiretapping started a policy battle moving back and forth between executive freedom and privacy of the people. Major policies and technologies that affected the balance of power between executive freedom and privacy rights emerged as a result of this debate. The first was the Communications Act of 1934. Over time, pro-wiretapping or pro-surveillance policies became significantly more powerful. Consequently, the regulating policies became stricter as observed in the Foreign Intelligence Surveillance Act (FISA).

Despite the stricter policies, new technological developments allowed the government to be much more efficient in surveillance operations. Developments in computing technology and data analytics no longer meant the government needed personnel to manually listen in on phone calls, but rather, computers could be used to sift through large amounts of data in the form of phone calls, text messages, and other types of electronic data with the goal of identifying patterns related to terrorism. The development of new forms of intelligence gathering and methods of analysis such as open source intelligence, proliferation of big data, and augmented reality are all at different stages of policy design, deliberation, and adoption of legislation.

The debate is further accentuated by the need to maintain a balance between the legal and executive bodies. Changes in policy through the years have intended to both tighten and loosen the authority of the executive branch for surveillance at different points in history. While the balance between executive authority and privacy continue to be debated, one item which is not debatable is the impact technology has had on these policies. The advent of new technologies and surveillance capabilities has significantly expanded the importance and scope of these regulating policies by vastly expanding the government's ability to conduct surveillance. These enhanced capabilities also expand the responsibility and scope of lawmakers to understand and regulate their use.

## 4 Ethics and Public Perception, Both Domestically and Internationally, Can Impact the Use of Technology by the IC. This Is Especially True When Introducing or Using Technology That Is Perceived as New

Increasingly, the activities of operations intelligence organizations are receiving popular attention. This especially came to the fore following the leaks made by Edward Snowden regarding the information collected as part of counterterrorism efforts. This leak has multiple ramifications including, but not limited to, on domestic intelligence-collection policies and on the international perception, and ultimately, foreign relations of the United States. On the domestic front, the conflation of the Snowden leaks with the role of WikiLeaks, is likely to have been one more factors affecting public perception during the 2016 US Presidential Election. On the international front, as Smith-Colin and Kleinhenz explore, there is a relation between the action taken by different nation-states in response to the leaks. The authors find that state response was correlated with the opposition by citizens in the country. Countries where citizens disapproved of US surveillance took more stringent steps in their response to the US.

While changing technologies have often outpaced policy around their appropriate use, there has been a long-standing debate surrounding the balance of power between executive freedom and privacy rights. Public perception of intelligence activities is very different from the institutional view and often from the domestic or international legal implications and limitations.

However, this role of perception and scrutiny of technology is not entirely new. Several points in history have acted as focusing events in different ways in the past. One such case is the launch of Sputnik I by the Soviet Union. In Schmid's application of Posen's model to the US response to the launch of Sputnik I, it is evident that the first level of scrutiny of the space program originated in the political leadership.

The central themes analyzed through the various cases in this book present certain key facts that play a role in determining the trends of technology adoption in the intelligence community. Interestingly, most of these have little to do with the specific technology and more to do with the underlying structural and personal aspects. To borrow from nineteenth century Prussian General and military theorist, Carl von Clausewitz, with the adoption of technological advancements, it is the character of intelligence that might change, not its inherent nature. The inherent nature is determined by the organizational structure, the role of humans involved, and the legal and political institutions.

This also points to some key lessons. In order to facilitate changes and adoption of emerging technologies, due attention needs to be paid to the underlying systemic characteristics. Especially in the case of intelligence, the organizational structure of the different agencies and their horizontal interactions play a role in effective adoption and use of technology. This also needs to be considered when designing new regulations. Typically, there is a lag between the development and adoption of technology, and roll-out of adequate legislation. While there is a need to maintain independence in the development of new technology, a balance could perhaps be maintained in the executive and legislative arms. Further, the role of the 'human in the loop' has continued to remain important and is only likely to get more prominent in the face of technologies such as big data and augmented reality. To facilitate this, the intelligence community needs to communicate findings more effectively with decision-makers and ensure it is impartial in drawing its conclusions and making decisions.

The manufacturer's authorised representative in the EU is Springer Nature Customer Service Centre GmbH, Europaplatz 3, 69115 Heidelberg, Germany. If you have any concerns regarding our products, please contact ProductSafety@springernature.com

Printed and bound by CPI Group (UK) Ltd, Croydon, CR0 4YY
05/10/2026
02228132-0002